Thank you for
your very interesti
and hope to see

S. CORDIER

GW01425213

Understanding Interactions in Complex Systems

Understanding Interactions in Complex Systems:

Toward a Science of Interaction

Edited by

Nicolas Debarsy,
Stéphane Cordier,
Cem Ertur,
François Nemo,
Déborah Nourrit-Lucas,
Gérard Poisson
and Christel Vrain

Cambridge
Scholars
Publishing

Understanding Interactions in Complex Systems:
Toward a Science of Interaction

Edited by Nicolas Debarsy, Stéphane Cordier, Cem Ertur,
François Nemo, Déborah Nourrit-Lucas, Gérard Poisson
and Christel Vrain

This book first published 2017

Cambridge Scholars Publishing

Lady Stephenson Library, Newcastle upon Tyne, NE6 2PA, UK

British Library Cataloguing in Publication Data
A catalogue record for this book is available from the British Library

ISBN (10): 1-4438-9496-6
ISBN (13): 978-1-4438-9496-8

In memory of Cem Ertur

Cem Ertur, full Professor of Economics at University of Orléans, France, sadly passed away in October 2016. As scientific director of the project "Analysis and modeling of interactions in complex systems", he took part in the preparation of the book since its earliest step and impulsed a lot of energy to promote multidisciplinary work. Cem Ertur was a brilliant researcher, always willing to extend the frontiers of knowledge and to better understand the complex world in which we live. His research, recognized by the international scientific community, focused on two fields: structural spatial econometrics with contributions in macroeconomics and international trade on the first hand and methodological contributions in spatial econometrics on the other hand. He will be deeply missed by all those who once have had the privilege to meet him.

Contents

List of Figures

List of Tables

Preface

In common language, an interaction is "the mutual action or influence which may exist between two or more objects, two or more organs, and even two or more phenomena", and it is always followed by one or several effects. In the experimental sciences, physics, chemistry, biology and geology, an interaction "aims to produce a change in the state of the interacting objects, such as particles, atoms or molecules". In the social sciences, Edgard Morin has defined interactions as "mutual actions affecting the behaviour or the nature of the objects, bodies, phenomena on influence"[1]

Complex systems and the inherent interactions between their components are present in almost all scientific fields. For instance, in the social and human sciences, interactions are at the heart of human behaviour, both in their individual and collective dimensions. No matter whether we consider cognitive, economic, linguistic or social dimensions, human behaviour may be conceived as resulting from a self-organizing process, which itself results from the interaction between the constitutive elements of a complex system.[2] As this complexity characterizes all organizational levels, a non-complex, one-dimensional and mono-disciplinary approach may seem quite limited for understanding human behaviour.[3] The study of complex systems in which individuals or groups of individuals interact becomes a major challenge in the social and human sciences and cannot be performed without a true interdisciplinary work, where the interactions between the researchers in the social sciences and in the fundamental sciences appear as the guarantee of an in-depth work with both methodological and conceptual cross-fertilizations.

Another example concerns the life sciences: from the individual perspec-

---

[1] Translated from Morin (1977), *La Nature de la nature* (vol. 1), *La Méthode*, Seuil, Paris.

[2] H. ATLAN, *Le vivant post-génomique ou qu'est-ce l'auto-organisation ?* Odile Jacob, 2011.

[3] E. MORIN, *Introduction à la pensée complexe*, Seuil, 2005.

tive, after decades of development of biological tools for approaches at the
cellular and molecular scales, the advent of "omics" technology (genomics,
proteomics …) allows measuring a great number of individual parameters
in many different situations (pathology, environmental …). These "omics"
approaches generate a large quantity of data, the analysis and interpretation
of which constitute new challenges for biologists. If the functioning of cells
becomes well understood, the interactions between intracellular regulation
pathways or between cells within tissue or organs is addressed in only a limited way. Each level constitutes a complex system and a multi-scale integration represents a considerable challenge. When studying populations, the
understanding of interactions taking into account the temporal and spatial
dimensions integrates mathematical approaches and particularly in ecology
and population dynamics. These approaches represent a major challenge for
understanding and predicting global environmental changes.

The scientific project called "Analysis and modeling of interactions in complex systems" (AMICS), which started at the University of Orléans in 2012,
constitutes the roots of this book. The AMICS project was supported by
the National Center for Scientific Research (CNRS) through its interdisciplinary mission in the framework of a University-wide exploratory project.
Around 20 researchers were directly involved in this project.

The first step of AMICS consisted in a monthly seminar about different
approaches and methods to investigate interactions in various scientific domains: mathematics, physics, computer science, robotics, the life sciences,
geoscience, economics, linguistics, the humanities, and the science of education. Around 20 talks were organized. The objective of these seminars
was to compare the various ways interactions are addressed in each discipline using their own theoretical background and methodology. A crossed
fertilization was expected, and the impulse of interdisciplinary collaborations, by promoting contacts and scientific collaborations.

The second step in AMICS was the organization of an international conference dedicated to interactions in complex systems. It took place from June
17 to June 19, 2013 and this conference involved about 130 participants
from 14 different countries.

For this conference, 6 keynote speakers were invited, each one a specialist
in a different scientific field:

- Henri Beresticky (EHESS- Ecole des hautes études en sciences sociales, CAMS - Centre d'analyse et de mathématique sociales, UMR
  8557 CNRS), "Propagation in inhomogeneous media: From epidemics
  to contagion of ideas"

- Karin Fischer (University of Southern Denmark, Institute for Design and Communication), "Complex processes in human–robot interaction"

- Serge Galam (SciencesPo, CEVIPOF - Centre d'Etudes de la Vie Politique Française, CNRS), "Interacting with a few random liars can jeopardize the democratic balance of a public debate"

- Philippe Gaussier (Université de Cergy-Pontoise, Image and Signal processing Lab – ETIS, UMR 8051 CNRS), "Emergence of capacities of social interactions in autonomous robots"

- Eric Goles (Université Adolfo Ibanez, Santiago, Chili), "Regulatory and segregation networks"

- Alan Kirman (Université d'Aix-Marseille III, GREQAM, and EHESS-Ecole des hautes études en sciences sociales), "Ants and Non-Optimal Self Organization: Lessons for Macroeconomics".

Thanks to the success of the conference, and as the founding members of AMICS, we decided to produce a collective book made of a selection of papers that were presented either in one of the organized seminars or during the conference, and which aims at presenting different disciplinary approaches to deal with interactions in complex systems. In addition to a double blind referee process, as editors of this book we set up an innovative and more interactive process, using a specific platform on which each contributing author could comment and review any of all the other contributions (once the selection of the papers based on the more traditional referee process had been performed). It was difficult to motivate the contributors, probably surprised and certainly not accustomed to such a procedure, but to make the task easier, the editors proposed to each author to review two other papers rather distinct from his/her discipline.

This book is split into 4 parts, each one approaching the interactions in complex systems from a particular perspective. We have gathered contributions by approach rather than by the topic studied; the latter way being, from our viewpoint, less relevant to this type of book.

The first part of the book contains papers dealing with interactions at the **system level**. The included contributions address territory planning, the traveling of populations in the Neolithic era, and interactions in neuron populations.The second part collects articles studying **networks** and proposing different methods for their analysis. This part contains contributions on link prediction, on interaction analysis in learning communities, on the influence of the type of update strategies (parallel or sequential) on the evolution of cooperation among humans, and opinion dynamics or the social function of gossip. Articles included in the third part focus on the analysis of interactions in **social communications**. As such, this part gathers papers studying teacher–student relations, the modeling of a teacher's evaluative speech to study students' interactions and their effects on learning, and the modeling of human communications. Finally, the last part of this book is devoted to the analysis of interactions between **economic agents** in different fields. One contribution develops a method to forecast the employability of students in Earth Sciences; another article studies the importance of language and the interactions between individuals as determinants of market equilibria; a mathematical model is derived to model money asset exchanges in the framework of a complex socioeconomic model in the third article, and the last article in this part examines the choice of study in an evolutionary game and shows under which conditions the population of students splits into two classes of strategies in equilibrium.

Thanks to this book, we realized all the difficulties but also the benefits gained from the confrontation of different methodological approaches in the general analysis of complex systems. Difficulties, because our knowledge is challenged: we are obliged to break free of our disciplinary perspectives and accept different views of the same objects and phenomena, other than those conveyed in each separate field. This experience was also enriching, because it is the disciplinary confrontation that allows the emergence of a thorough understanding of a question, but also the generation of new ideas, methodologies, or even scientific approaches.

We expect that the cross views of interactions in complex systems presented in this book will themselves generate scientific interactions by arousing curiosity and interest in working with colleagues in different disciplines.

We finally would like to thank the "Maison Interdiscplinaire des Systèmes Complexes" (Structure hosted by the University of Orléans (France) and François-Rabelais (Tours-France) dedicated to the promotion of research in complex systems), the Research team DYNACSE , laboratory EPSYLON, (University of Montpellier-France), the Department of Mathematics of the University of Orléans (MAPMO - UMR CNRS 7349) and the "Mission pour l'Interdisciplinarité" of the CNRS for their financial support.

Introduction

« Most likely, if our investigation methods
become more and more efficient,
We will discover the simplicity underneath the complex,
Then the complexity underneath the simplicity, thereafter
The simplicity underneath the complex, and so on,
Without being able to predict which term would be the last.
It has to land somewhere,
And for science to be possible,
We must stop when we find simplicity.
It is the only ground on which we will be able to raise
the structure of our generalisations.
This simplicity being only apparent,
Can we consider this ground will be strong enough?
This is what needs to be searched. »
Henri Poincaré (1902)[1]

Bearing in mind the complexity of biological, physical, chemical, sociological, economics, psychological, … phenomena implies considering that they are made of heterogeneous elements, yet inseparably tied (Morin 2005), in other words, in close interaction. This also means accepting that these phenomena with apparent disorders may be affected by random events, and structure themselves on various time scales (short-term, long-term, fast, slow answer), spatial scales (local, global) as well as different organizational levels (micro, meso, macro) (Dahan & Aubin, 2007). This also implies not being able to consider any study without blending distinctive scientific

---

[1] Translated from Poincaré, H. (1920). La science et l'hypothèse. Ernest Flammarion, Ed., Paris, 2ème édition, (1ère édition 1902), p. 176.

fields, by attempting to identify the most relevant scientific collaborations (multi-, inter-, trans-disciplinary). Thus, the study of complex systems results from a strong paradigmatic commitment which is part of a true revolution in the meaning of Kuhn (1962) and, as we know, did not occur at the same time and under the same conditions for all scientific disciplines. Nevertheless, at this time, it is difficult to disagree with the fact that the scientific community cannot grasp the "Why" and "How" of the phenomena without integrating them into a broader context of action, feedback and interaction.

This explains why it is important for us to introduce this work that gathers a collection of multidisciplinary or interdisciplinary studies dealing with complex systems and their interactions. Throughout this study, numerous definitions of complex systems and their interactions will be brought through diverse perspectives. All of them will nevertheless agree on the fact that a complex system is composed of a large number of components (from *cumplexus*, tied with, tangled up, Morin, 2005) interacting in a non-linear manner and feedback loop, in a way that interactions between local components generate a behavior at a global scale.

The interactions between the different components present properties of self-organization, which appear essential to understand the behavior of the system. The self-organization that can just as well be seen in environmental processes, in embryogenesis and epigenesis, but also in the cognition, the formation of speech, in ethology, in economics ... (Bourgine & Lesne, 2006; Favereau 1989; Odeyeur, 2013, Theraulaz, 1997), may appear structural but equally functional[2] (Atlan, 2011) and responds to two modes of adaptation; on the one hand a spontaneous adaptation of the system facing internal and external constraints which weigh it down and on the other hand an adaptation of a long course which is realized by the phenomena of evolution of Darwinism (Zwirn 2006; Atlan, 2011). But how can we analyze the interactions and the emergence of behavior as well as their adaptive evolution? In these interactions what is the part of the random (Varela 1989) knowing that the random integrates in complex systems such as the necessary element of self-organization, as a source of novelty by the game of disorganization/reorganization it generates (Atlan 2011); it is far from the balance that the interactions of the self-organization can really be put in place (Prigogine, 1994, Bourgine and Lesne, 2006).

To understand the interactions in a complex system it is therefore a question of considering these few properties that we have just skimmed through and of which will find all their magnitudes and depths in the different studies presented. However beforehand, it appears necessary to introduce an epis-

---

[2] The first being of course easier to study and model than the second.

temological and ontological point of view (Morin, 2005; Dahan and Aubin, 2007; Ricard, 2008) in order to draw the contours and the disciplinary issues of what the study of complex systems and their interactions really is. We could legitimately ask ourselves why these philosophical considerations introduce this publication?

The first lies in the fact that all scientific work is part of a certain epistemological positioning of the researcher, who does not refrain from asking himself philosophical questions and participates therefore in his own construction as a researcher, but also as an individual. The construction and the transmission of knowledge involve registers of ontology and ethics, of which it is difficult to ignore the supporters, the limits and the aspirations. Scientific research is not and must not be a decontextualized and disembodied exercise, in which case it would no longer be the work of a researcher, but of a technician only. Research not only demands particularly rigorous scientific and theoretical positioning, but also a commitment to reality from the being involved himself.

This given ontology conditions our account of the world, and therefore the methods that we will invest in to report on the complex reality of this world. This is not without link with what Heisenberg (1984; 2010) considers the regions of reality. When trying to understand a problem, indeed, one is confronted with different regions of reality. Each region of reality is determined by a specific behavior and closed by the explanatory limit of concepts that can apprehend this reality. The considerations of science must therefore aim "to see the regions of reality but also to give exact formulation and without omission of the entirety of the connections which signifies the region" (ibid., p. 35). But it must also be considered that the different regions meet and overlap, and that is the important thing, the discernment of different regions and their occurring laws , according to Heisenberg (1984; 2010), goes through the progression of the objective toward the subjective[3], because "Ultimately, one must still and always realize that the reality that we can talk about is never the reality "in itself", but only a reality of which we can have knowledge, or even in many cases a reality to which we ourselves have given form" (ibid., p. 39). More broadly, our report on knowledge is

---

[3] "So shall he/she (the researcher) start with a part of reality that we can fully detach from and where we can entirely disregard the methods by which we come to a knowledge of its content. But at the top of the arrangement are held, as in the sketch of Goethe, the creative faculties with which we transform ourselves and give form to the world [...]. The "subjective" word one wishes to specify indicates that it may not be possible in a complete description of the connections of a region, to ignore the fact that we ourselves are involved connections. (...) We cannot ignore the fact that our body and the devices we use for observation are subject to the laws of atomic physics. " (Heisenberg, 1984; 2010 p.37)

not independent of the way in which the man is in the world (Bachelard, 1931). This consideration is essential, because it puts the researcher in subjectivity and in all his/her complexity that are both a necessary condition to the creativity and source of obfuscation. It is a warning which is to remind us that the researcher must be alert and conscious and discern in relation to the measures, the scales of analysis and purpose and their level of integration and dependence to understand reality and its laws. This is not without remembering the caution of Bacon (1620; trad. 1857) when he evokes the idols or "false concepts that have already invaded the human spirit and who threw away deep roots, they not only occupy such intelligence that the truth may be accessed with difficulty" (ibid., p. 12). Although warned of all these idols or epistemological obstacles which are confined to the rigorous conduct of the research, the researcher is not protected, especially during the emergence of new paradigm or of the decline of the one that has participated in the construction of any of his scientific career. In connection to this Bacon tells us how "the human spirit as soon as we are seduced by some ideas, either by their charm, or under the influence of the existing traditions and faith, forces all the rest to return to these ideas and to agree with them" (ibid., p. 14). As researchers, we are led to live from the emergence of the interior or the inscription of normal science (Kuhn, 1962) of a new paradigm, that of complex systems (or of complexity) and the decline of others. This experience is sufficiently rich and noteworthy so that we are relatively protected from infatuated personnel and of the entry into "religion" that it provokes. It is therefore a question of asking and understanding the scientific "topic" (eg. the researcher with its purpose of study) as being psychological, social, and even biological. On the basis of this hypothesis, epistemology and ontology constitute from the imprescriptible point of departure a job of disciplinary research for complexity.

From a disciplinary point of view, although the sciences of complexity appear as transdisciplinary (Morin, 2005), the issues relating to the quantification of the level of complexity of a system of any kind and to the understanding of emerging phenomena cannot divest work that is methodological and conceptually interdisciplinary (Resweber, 1991, Dahan & Aubin, 2007). These different tools and conceptual frameworks are derived from scientific fields as varied as cybernetics, dynamic systems, systemic, Artificial Intelligence... and support their interest on any different topics (self-organization, multi-hierarchical levels, interactions, transitions, emergent properties, variability...) (Lucas, 2000). Also it may appear to be more relevant to talk about complexity in the plural, and thereby place the focus methodologically and conceptually.

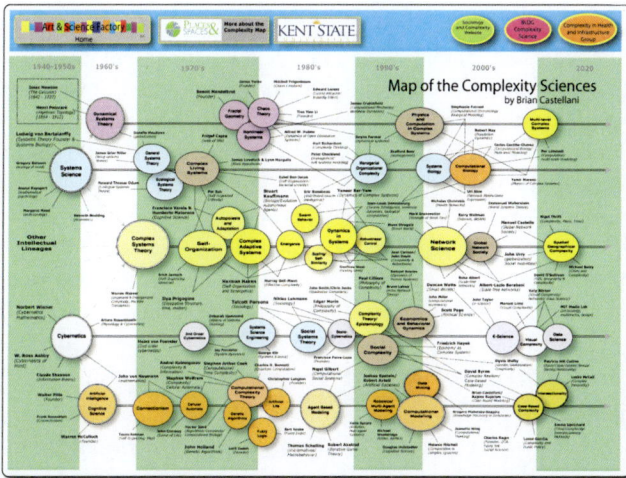

Figure 1: Representation of disciplines with the purpose of the study being the complexity of systems from the Castellani Map of complexity science. Source: http://www.artsciencefactory.com/complexity-map_feb09.html.

Figure 1 presents rather accurately not only all the disciplinary fields rendering account of the complexity, but also the interactions that allow them to inform and to feed each other.[4] When he evokes the sciences of complexity and its paradigmatic status, Edgar Morin invites us in some way not to be tied with any discipline (Morin, 2005). This is in two ways: On the one hand in the direction where it is a question of understanding the complexity looking beyond all disciplines, incorporating all disciplines by exceeding them. It would then be a question of opting for a transdisciplinary glance (Morin, 2005). On the other hand in the sense that if one accepts to investigate the complexity, it is question of accepting the same discomfort, the contradictions, the disorder, and thus take a number of steps from the interdisciplinary side, a form of indiscipline which by a game of harmony and disharmony would account for what is the "true" reality of phenomena.

The study of complex systems has in singularity a strong dependence on mathematical tools. Scientific advances in the field of human, economic and social sciences deal with complexity that does not seem possible to us only in close collaboration between the different disciplines, who are interested

---

[4] Science in general and the sciences of complexity in particular do not escape from the diversity of disciplinary fields and interactions that they can maintain.

from near or from far in the modeling of complex dynamic systems. Applied mathematics are in this sense a privileged partner and this publication presents a large number of studies that show the purpose or method of analysis in the mathematical modeling of complex phenomena. This new look of the borderless science, undisciplined (Morin, 2005), is the promise of real progress at the scientific level. While, all seminars and calls for funding for research on national or international projects on interdisciplinary or thematic transdisciplinary are additional signs. However, the qualifiers multi- inter- or transdisciplinary are often used indiscriminately to qualify the research collaborations (Resweber, 1991). Yet there are very marked differences, which are not without direct methods and the quality of the interactions (Mondada, 2005) and the results which necessarily flow.

While much of the research is part of an explicit interdisciplinary rationale (one may also consider the ever growing number of research groups, groupings of laboratory, name of labs ... Using the term interdisciplinary), researchers despite this are working all the more often on a multidisciplinary level. The multidisciplinary work consists of encountering different disciplines around the purpose of a common study, but this remains within the limits of their methodology and their conceptual framework. The disciplines are then in a strategy of inquiry which has the function of relativizing the part of truth, the portion of truth which each bears witness. It also focuses on an awareness of the limitations of each discipline. This awareness is favourable in the transition towards interdisciplinary work, where the knowledge will therefore enter more directly in contradiction.

Interdisciplinary work is about understanding the «language » of the other discipline, its methodology and conceptual framework, as well as thinking beyond the alternative dialogue of perspectives (multidisciplinary work) in order to truly look at the object of study with a common questioning (Resweber, 1991). Once this confrontation strategy has been applied, a transdisciplinary collective work may effectively gather experts around a joint project, with the aim of « creating new frames of knowledge, broad enough to integrate the interpretations previously discussed » (Resweber, 1991, p. 47). Accessing a transdisciplinary work is a major scientific challenge. New frameworks and new methodologies likely to assimilate and exceed (to transcend in some ways) those already approached appear as the path to breakthroughs. It may sound presumptuous or premature to mention transdisciplinarity yet while it remains a goal to achieve. However in many ways, interdisciplinarity already constitutes a remarkable stage by itself, fraught with challenges. Researchers who have tried will confirm it is time consuming and requests much attention and effort at both levels, epistemological and psychological.

From an epistemological point of view, interdisciplinarity involves thinking beyond simple (sometimes simplistic) analogies. Social Science may be blamed for this when it tends to deal with Physics or Mathematics (Sokal & Bricmont, 1997, Atlan, 2011). Interdisciplinarity also requests a specific positioning among the diversity of scientific disciplines. Indeed, according to Cartwright (1999), the various fields of study have long been ranked and prioritized, and the laws of physics being considered as universal, traditionally used to receive a greater prestige than others disciplines.

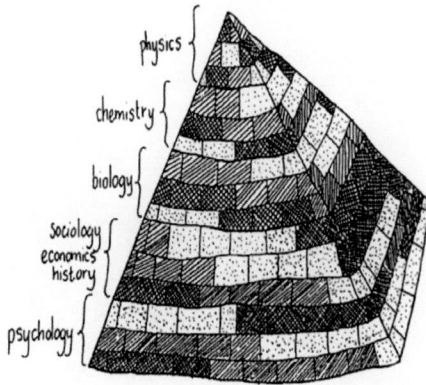

Figure 2: Pyramid view of Science

This concept involves to take into consideration that the laws and concepts of each scientific field can be reduced to those of a more fundamental field, according to a hierarchical classification (Cartwright, 1999). Such a perception allows us to easily understand the difficulties the dialogue between disciplines will inevitably need to face: each discipline demonstrating its authority or locking itself into an identity defense mechanism. This is why it is preferable to choose a non-hierarchical positioning of the "dappled world" (Cartwright, 1999), in reference to whom « The science disciplines are each tied both in application and confirmation, to the same material world (...) » (Cartwright, 1999 ; p.6). The aim is to cooperate between disciplines and eliminate a conceptual hierarchy likely to compromise chances of meeting and dialogue. These two positioning are shown in Figures 2 and 3.

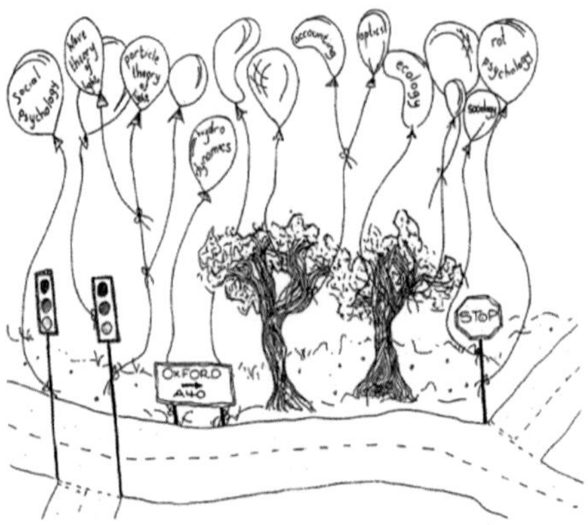

Figure 3. The dapple world

The difficulties experienced may also be psychological or psycho–socio-logical. Indeed, the scientists in collaboration are also human beings in interaction, based on their own « personal equation », and following specific rituals (Goffman, 1967). Whoever has tried interdisciplinarity cannot ignore the essential role played by interpersonal relationships, which may lead to misunderstanding, slowdowns, loss of impetus, or even a premature end of a research program. All searchers' ability to listen and empathy are crucial in order to reconcile the various registers involved in the interdisciplinary collaboration, (Berthoz & Jorland, 2004).

Once most of these difficulties are overcame, which could be related to epistemological obstacles in reference to Bachelard (1938) and to the various idols developed by Bacon (1620, trad. 1857), is it still possible to focus on « How to model? », « How to model an organism, a complex system with a complex behavior? »

Before answering the technical issue raised by this question, based on a range of results exposed further in this research, it is appropriate to establish whether we are interested in the explanation of the phenomena and their causes of appearance, or in the understanding of the process and its purposes. The searcher's position impacts his choice of modeling tools. In this case a fundamental epistemological question needs to be asked, the question of the position chosen regarding the Galilean and Aristotelian science. Indeed, for Aristotle, what matters are the understanding of the phenomena, their purposes and the communication around this understanding. While for

Galilee and the advent of Western science, what matters is the operational and nomic explanation, thus rejecting teleological (or finalistic) explanations. Indeed Contemporary Science gives priority to the « How » compared to the « Why », to the *explanation* of How compared to the *understanding* of Why (Varela, 1989).

Fitting in a scientific process with the Complex systems interactions as subject does not allow us to avoid this reflection. Is it necessary to make a decision between both approaches? Such as suggested by Varela (1989), teleological and nomic explications are not in conflict. It is at once possible to study the laws of a phenomenon (nomic) and the purposes of this phenomenon (teleological) by choosing a *teleonomical* explanation. In this respect « How » and « Why » must work in harmony. In the manner of the *teleonomical* proposition of Varela (1989), it may not appear very relevant to make a choice between both conceptions and to confront them. Understanding the complex systems surely implies this double positioning, incidentally familiar with emerging paradigms.

For whoever is interested in understanding these complex phenomena, regarding the various levels of analysis or organizational levels, we always notice a dual tension between two terms of contradictory appearance. These contradictions are only apparent and quite often come from the level of analysis from where we stand (macro-meso or micro) or from the temporal scale considered (long or short term). They are regularly mentioned in the scientific approach, or more broadly speaking, in its transmission (*Chanced and necessity* from Monod, *Chaos and harmony* from Trinh Xuan Thuan). They even sometimes lead scientists to use neologisms that gather both contradictory terms. To name only the most popular in the field of complexity, *Simplexity* (Berthoz, 2009) exposes the simplifying principles of complex systems; The teleonomical approach (Varela, 1989), to question both, Causes and Purposes; *Glocal Memory* (Goertze, Pitt, Ikle, Pennachin & Rui, 2010), to consider local organizational conditions at a global level.

Are these neologisms and conflicts just some figures of speech, or on the contrary do they tend to show exactly what are the complex systems? It appears that the scientific perspective, especially the one focused on complex systems, does not get any benefit from extreme positions, theoretical withdrawals and methodological crystallizations, that too often tend to lock searchers into restricted debates (such as for instance between the partisans of an inductive or deductive reasoning, or even between the partisans of qualitative or quantitative research methods). Morin (2005) invites us to be undisciplined, in order to endorse a broader perception of the complex thought process. To go even further, this lack of discipline is a major condition for creativity, of a physical intuitionism (Poincaré, 1919). The optimized conditions to generate knowledge are neither gathered in an exclusively deductive[5], nor inductive[6], or even transductive[7] reasoning, but rather in between, with an abductive[8] reasoning which to Pierce (1931-1958, quoted by Cattelin, 2004) seamed to instigate scientific creativity. Abduction stands halfway between deduction and induction, and is open to the reality of facts and experiments through the rigorous frame of deduction. This "halfway", this "in-between" and this compromise is source of an intense creativity, and for some thinkers such as Umberto Eco, the ideal condition of serendipity (Eco, 1999; Catellin, 2004), this aptitude to accidently discover and the sagacity of things we do not look for.

This way, this work constitutes a great example of interactions processing in various systems and levels of collaboration between disciplines, of the observation of phenomena according to various modeling tools. It also reflects the willingness of thinking beyond the boundaries of each discipline to better approach the complexity of this world.

---

[5] Scientific reasoning usually designed to progress from general to specific. Inference drawn from laws and hypothesis.

[6] Scientific reasoning usually designed to progress from singular to specific, including Bacon, considered it was the only valuable scientific approach.

[7] Reasoning not considered as scientific, but rather as a mistake based on analogy. Piaget introduces this reasoning as an irrational approach that allows children to find a logic of things with an assumed false occurrence (Piaget, 1927 ; 1937). This approach is currently considered in Learning sciences as exploratory.

[8] Scientific reasoning that consists of formulating an hypothesis to experiment, without paying attention to the various hypothesis analysis, but only by focusing on the actual facts.

Bibliography

Atlan, H. (2011). Le vivant post-génomique ou qu'est ce que l'auto-organisation? Paris: Editions Odile Jacob (Coll. Sciences).

Bachelard, G. (1931). *L'intuition de l'instant* (2me édition, 1992). Paris : Editions Stock.

Bachelard, G. (1938). *La formation de l'esprit scientifique*. (Edition poche, 2011), Paris: EditionsVrin.

Bacon, F. (1620). *Novum Organum.* Traduit par Lorquet (1857). Paris: Editions Hachette,

Berthoz, A. (2009). La simplexité. Paris: Editions Odile Jacob (Coll. Science)

Bourgine, P. & Lesne, A. (2006). Morphogénèse. *L'origine des formes*. Paris: Editions Belin (Coll. Echerelles).

Cartwright, N. (1999). *The dappled world: a study of the boundaries of sciences*. Cambridge: Cambridge University Press.

Cattelin, S. (2004). L'abduction : une pratique de la découverte scientifique et littéraire. *Hermès, 39*, 179-185.

Dahan, A., & Aubin, D. (2007). Systèmes dynamiques et Chaos : convergences et recompositions, un aperçu historique. In S. Franceschelli, T. Roque & M. Paty (Eds.), *Chaos - Systèmes dynamiques* (pp. 327-356). Paris : Editions Hermann (Coll. Visions des Sciences).

Eco, U. (1999). *Serendipities: Language and Lunacy*, Mariner Books.

Favereau, O. (1989). Organisation et marchés. *Revue Française d'Economie, vol 4* ,1, 65-96.

Goertzel, B., Pitt, J., Ikle, M., Pennachin, C., & Rui, L. (2010). Glocal memory: A critical design principle for artificial brains and minds. *Neurocomputing, 74*, 84-94.

Goffman, E. (1967). *Interaction rituals: Essays on Face-to-Face behavior*. Anchor book

Heisenberg, W. (1984 ; 2010). Le manuscrit de 1942. 1re Trad. 1984 par R. Piper. Dernière Trad. 1998 par P. Lenard. Paris : Editions Allia.

Kuhn, T. (1962). *La structure des révolutions scientifiques*. Paris : Editions Flammarion (Coll. Champs).

Lucas, C. (2000). Quantifying complexity theory. CALResCo Group. http//www.calresco.org/ lucas/quantify.html.

Mondada, L. (2005). Chercheurs en interaction. Comment émergent les savoirs. Coll. Le Savoir suisse, Lausanne : Editions Presses polytechniques et universitaires romandes.

Monod, J. (1970). *Le hasard et la nécessité*. Essai sur la philosophie naturelle de la biologie moderne. Paris; Editions Seuil.

Morin, E. (2005). *Introduction à la pensée complexe*. Paris : Editions Seuil - Points.

Odeyeur, P-Y. (2013). Aux sources de la parole. Auto organisation et évolution. Paris: Editions Odile Jacob (Coll. Sciences).

Piaget, J. (1927). *La causalité physique chez l'enfant*, Paris : Editions F. Alcan.

Piaget, J. (1937). *La construction du réel chez l'enfant*, Paris : Editions Delachaux et Niestlé.

Poincaré, H. (1919). *Dernières pensées*. Paris : Editions Flammarion.

Poincaré, H. (1920). *La science et l'hypothèse*. Paris : Editions Flammarion, 2me édition, (1re édition 1902).

Prigogine, I. (1994). *Les lois du chaos*. Paris : Editions Flammarion (Collection Champs).

Resweber, J. P. (1991). *Le Pari du Transdisciplinaire*. Paris : L'Harmattan.

Ricard, J. (2008). *Pourquoi le tout est plus que la somme de ses parties. Pour une approche scientifique de l'émergence*. Paris : Editions Hermann (Collection Visions des Sciences).

Sokal, A., & Bricmont, J. (1997). *Impostures intellectuelles*. Paris : Editions Odile Jacob (Coll. Livre de poche).

Théraulaz, G. (1997). Auto-organisation et comportement. Paris : Editions Hermes.

Trink Xuan Than (1998). *Le chaos et l'Harmonie*. Paris: Editions Fayard.

Varela, F.(1989). *Invitation aux sciences cognitives*. Paris: Editions Seuil (Collection Points).

Varela,F. (1989). *Autonomie et connaisssance*. Essai sur le vivant. Paris : Editions Seuil (Coll. La couleur des idées).

Zwirn, H.P. (2006). Les systèmes complexes. Mathématiques et biologie. Paris: Editions Odile Jacob (Coll. Sciences).

Part I

Global interactions

The article by G. Nadin surveys the mathematical tools used in archeology to model for instance the propagation of agriculture during the Neolithic period. It starts with the classical reaction-diffusion equations which give a mechanical interpretation to the emergence of "waves of advance' observed in the neolithic colonization of Europe for instance and which show that the propagation phenomenon is not driven by any external force. It then resumes with the hyperbolic reaction-diffusion equations which incorporate a time-delay in the classical reaction-diffusion equations and finally shows that improved models are particular cases of more general kinetic reaction-diffusion equations which are involved among others in bacteria dynamics models.

Urban project is a complex object resulting from interaction of several operations in progress on the territory. The systemic approach developed in the contribution by S. Chemin Le Piolet investigates the modelling of urban system as a possibility to go over contemporary fragmentation and complexity which assign strongly planning of the territory and urban-planning disciplines. These cope with the necessity for renewal of traditional modes of action in favour of reflexive approaches reflecting the experiment applied here. The modelling of urban system realized by the setting-up of an interactive tool of map making is the indicator of the interactions between the various elements of the urban project. This tool is a part of experiments led on the urban observatories contributing to give a new vision of projects management, more flexible and adjustable to the context of uncertainty, and attentive to the unexpected and interactions.

The article by E. Frenod suggests a systemic approach to model territory functioning, represented by the spatial distribution of persons, firms and energy. His model is not intented to be realistic but it allows to describe and understand several aspects of territory behaviour. A territory is a complex system that involves, among others, geographic constraints, people, local government actions and firms, with interactions between all these components. The performed numerical simulations on this toy model indicate that the systemic approach can deal with nonlinearities in the behaviour of the agents but can also account for interactions between several territories. The contribution of this paper is a first step toward a possible software tools to foresee policy impacts, to help making a decision when facing a change in the cultural behaviour and as such, may constitute a powerful tool to manage territory working.

S. Mancini's contribution focuses on modelling the dynamics behind the problem of selecting one of two possible ways of seeing a picture. The author wants to model two quantities, namely reaction time and performance.

Reaction time is defined as the time needed to make a decision while performance is defined as the number of times the good answer is given, no matter the time needed. The decision making is done at the neuronal level. Since neurons in the visual cortex have different skills and are connected, this selection problem cannot be modelled by the description of a single neuron activity, but by different populations of neurons. Up to now, the techniques used to model these two quantities did not allow to have an explicit mathematical expression and required huge computation power. The author uses a partial derivative equation technique which allows to describe the evolution of the probability distribution of neurons populations. The complexity reduction method derived in the paper is a good way of overcoming the difficulty of not knowing the explicit form of the underlying potential and get approximated solutions on the whole domain of definition and not only locally, as the earlier techniques only allowed to.

Chapter 1

Hyperbolic travelling waves in the modelling of Neolithic populations

Grégoire Nadin[1]

1.1   Reaction–diffusion models in archaeology

1.1.1   The wave of advance model

The first use of reaction–diffusion equations in an archaeological framework is due to Ammerman and Cavalli-Sforza in their seminal work Ammerman and Cavalli-Sforza (1984), which investigated the spread of agriculture in Europe during the Neolithic transition. Using carbon 14 datings of archaeological sites, they observed that the diffusion of agriculture starts from the Middle East in 7000 BC and then spreads in Europe at an approximate speed of 1 km/yr. In order to explain the existence of such a linear propagation speed, they used the classical Fisher-KPP equation

$$\partial_t u - D\Delta u = ru(1 - u/u_{max}). \tag{1.1}$$

[1]Laboratoire Jacques-Louis Lions, UPMC Univ. Paris 6 and CNRS UMR 7598, F-75005, Paris. Email: nadin@ann.jussieu.fr

This equation was first introduced in the 30's by Kolmogorov et al. (1937) and Fisher (1937), both in the framework of biological invasions. It has then been used in lots of models, related to genetics, chemistry, combustion, economics, ecology, etc.

It relies on the hypothesis that the population is large enough to be modelled by a continuous density variable $u = u(t, x)$, and that the following interactions between individuals drive their evolution:

- The population disperses with a flux that is proportional to the gradient of the density (Fick's Law), at a rate D;

- At each time step dt, the population gives birth to a number $ru(t, x)dt$ of new individuals, where r is the reproduction rate;

- The environment is assumed to only contain a finite quantity of resources, which leads to a maximal density of population u_{max} and creates the nonlinear death term $-ru^2/u_{max}$.

Due to this nonlinear term, the equation cannot be explicitly solved. The trajectory followed by a given individual cannot be predicted. However, this model gives rise to deterministic asymptotic patterns at the population scale: the level lines of the population density behave like circles at time large enough, $t \gg 1$, containing a surface proportional to t^2. This is characteristic of a complex system.

A fundamental mathematical property of equation (1.1) is the existence of **travelling wave solutions**, that is, solutions of the form $u(t, x) = U(x - ct)$, where U is positive and smooth, $U(-\infty) = u_{max}$, $U(+\infty) = 0$. Here, U is called the profile of the travelling wave and $c \in \mathbb{R}$ is its speed. The shape of such solutions implies that if one shifts the time from t to $t + T$, then the solution keeps the same profile and is only translated by a length cT. It was proved in Kolmogorov et al. (1937) that there exists a travelling wave with speed c if and only if $c \geq c^* = 2\sqrt{Dr}$. Moreover, the travelling wave with minimal speed c^* attracts, in a sense, the solutions of the initial value problem (1.1) associated with compactly supported initial data. In other words, if one introduces a density of population in some given location of the space and lets it evolve through (1.1), then it will propagate in all directions with speed $c = c^*$ and at large times will "look like" the wave with minimal speed c^*. Note that the minimal speed $c^* = 2\sqrt{Dr}$ only depends on the linearization near the steady state $u = 0$ of the nonlinear equation (1.1), it does not depend on the saturation threshold u_{max}. This means that the waves is "pulled" by the edge of the invasion, where $U \simeq 0$.

Coming back to the Neolithic transition problem, Ammerman and Cavalli-Sforza (1984) identified travelling wave solutions as an appropriate mathematical equivalent of the **waves of advance** empirically observed. Using anthropological observations of preindustrial farmers, they estimated the parameters: $D \simeq 1100 - 2200$ km/gen, a generation time $\tau = 25$ years and $r \simeq 0.029 - 0.035$ per year. This gives a propagation speed $c^* = 2\sqrt{Dr}$ between 2.25 and 3.50 km/yr and a mean value of 2.86 km/yr, which has the same order of magnitude as the empirical speed 1 km/yr.

This validates this "wave of advance model" and gives a mechanism from which the linear spreading speed naturally arises, meaning in particular that this transition is not driven by some external force. This model was then used in many works on the diffusion of technologies or languages in early populations. It was improved, in a way that we will describe in the next section, by Fort et al. (2004) in the Palaeolithic framework, in order to show that the speed of the wave of recolonization might not have been limited by climate change but only by the intrinsic characteristics of the population.

1.1.2 The time-delayed model

The two physicists Fort and Mendez (1999) observed that the Fisher-KPP equation (1.1) relies on the Fick law. That is, one assumes that the migration (flux) rate at time t is proportional to the gradient of the density of the population at time t. This approximation is too rough for human populations, for which one expects the migrations to occur at a generational scale: the gradient of population at time t should give rise to a migration flux at time $t + \tau$, where τ is the time of a generation. This gives a delay term in the Fick law, which leads to the new hyperbolic reaction–diffusion equation

$$\frac{\tau}{2}\partial_{tt}u - D\Delta u + \left(1 - \frac{r\tau}{2} + \frac{2ru}{u_{max}}\right)\partial_t u = ru(1 - u/u_{max}). \qquad (1.2)$$

The mechanism leading to this equation is very simple and indeed it is used in many fields, such as forest fire modelling, epidemics, and chemistry (see Fort and Mendez (1999) and the references therein).

Assuming that the propagation speed should be given by the leading edge of the invasion, where $u \simeq 0$, as with the classical Fisher-KPP equation (1.1), Fort and Mendez heuristically derived the following propagation speed for equation (1.2):

$$c_\tau^* = \frac{2\sqrt{rD}}{1 + \frac{r\tau}{2}}. \qquad (1.3)$$

Figure 1.1: Figure from Fort and Mendez (1999): comparison between the wave-of-advance model and the time-delayed model. Each point was associated in Ammerman and Cavalli-Sforza (1984) with a dated archaeological site where evidence of agricultural activities were found, distances being measured as great circle routes from Jericho (the presumed center of diffusion). Dates are conventional radiocarbon ages in years before present (BP). The solid line is the regression by Ammerman and Cavalli-Sforza (1984). The other three lines are least-squares fits with slopes calculated from the classical wave-of-advance model (dashed-dotted line), and the time-delayed model (dotted line). We do not discuss here the dashed line, the interested readers being referred to Fort and Mendez (1999).

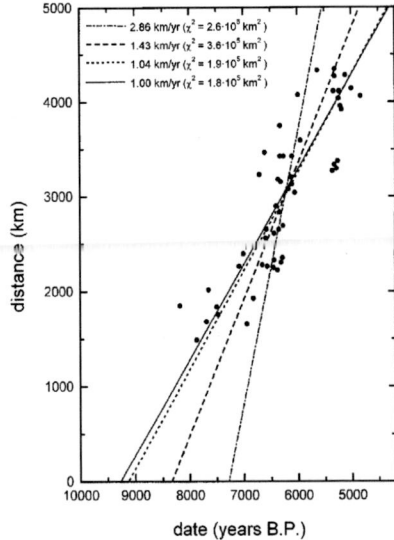

Evaluating r, τ and D as in Ammerman and Cavalli-Sforza (1984), and noticing that the estimate of D should be divided by 4 since the population evolves on a 2d plane and not a 1d line, this gives a propagation speed between 0.84 and 1.24 km/yr, with a mean value of 1.04 km/yr. This is much closer to the empirical speed 1 km/yr, validating this corrected delayed model. These results are displayed in Figure 1.1.

However, this computation of the speed c_τ^* relies on the approximation that only the leading edge of the invasion is important. Indeed, such an approximation is a bit risky from a mathematical point of view: the term $\partial_{tt}u$ changes the nature of the equation. There is a balance between the parabolic and the hyperbolic terms, and it is known that hyperbolic equations might give rise to singular solutions, such as shocks, instead of smooth travelling waves.

1.2 Mathematical investigation

1.2.1 Computation of the propagation speed for hyperbolic reaction–diffusion equations

A mathematical study of equation (1.2) was recently provided by Bouin et al. (2014a), showing that one should distinguish two regimes.

If the delay is sufficiently small ($r\tau/2 < 1$), then the same behaviour as for the non-delayed Fisher-KPP equation occurs. Namely, there exists a smooth travelling wave solution of speed c for all $c \geq c_\tau^*$ and the one with minimal speed $c = c_\tau^*$ is stable, in some sense. We mention here that this framework was investigated mathematically in the older paper Hadeler (1988), using different techniques.

When the delay becomes larger ($r\tau/2 \geq 1$), typical hyperbolic phenomena arise. Namely, there exists a travelling wave of speed c for all $c \geq \sqrt{2D/\tau} =: \widetilde{c}_\tau^*$, and the travelling wave with minimal speed $c = \widetilde{c}_\tau^*$ is discontinuous if $r\tau/2 > 1$, that is, it is a shock. This means that the propagation is not driven by the leading edge of the front. Indeed, as the migration time τ is large, the wave equation operator $\frac{\tau}{2}\partial_{tt}u - D\Delta u$ dominates equation (1.2) and the solution evolves with the associated speed $\widetilde{c}_\tau^* = \sqrt{2D/\tau}$. The shapes of the travelling waves in these various regimes are displayed in Figure 1.2.

As a consequence, the approximation made in Fort and Mendez (1999) is not valid in general. Nevertheless, the parameters involved in Fort and Mendez (1999) satisfy $r\tau/2 < 1$, so that the computation of the minimal speed c_τ^* is correct in this particular setting. But if $r\tau/2 > 1$, then the propagation speed is $\widetilde{c}_\tau^* > c_\tau^*$.

1.2.2 Generalization: A kinetic approach to reaction–diffusion

As already mentioned, equation (1.2) is involved in other fields of applications. Apart from archaeological models, it arises, for example, in bacteria dynamics modelling, where it can be viewed as a simplification of more general kinetic reaction–diffusion equations. Such equations are both interesting from a modelling point of view, since they give a microscopic interpretation to classical reaction–diffusion equations, and from a mathematical point of view. Let us briefly describe these promising prospects in order to conclude this paper.

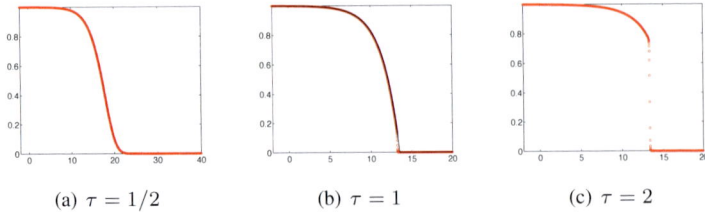

(a) $\tau = 1/2$ (b) $\tau = 1$ (c) $\tau = 2$

Figure 1.2: $r = 2$, $D = 1$, $u_{max} = 1$. When $\tau < 1$, the wave-of-advance (i.e. travelling wave) is smooth and the speed is determined by the leading edge, where $u \simeq 0$. When $\tau \geq 1$, the wave is no longer smooth: the density of the population grows suddenly from $u = 0$ to $u = 3/4$ for $\tau = 2$.

Such a microscopic approach was initiated recently by Cuesta et al. (2012), who introduced Fisher-KPP equations with a kinetic distribution kernel:

$$\partial_t f + v \cdot \nabla_x f - M(v)\rho_f + f = r\rho_f\big(M(v) - f\big), \text{ for all } (t, x, v) \in (0, \infty) \times \mathbb{R}^N \times V, \tag{1.4}$$

where $\rho_f(t, x) := \displaystyle\int_V f(t, x, v)dv$, $V \subset \mathbb{R}^N$ is the space of admissible speeds,

$$M(v) \geq 0, \quad \int_V M(v)dv = 1.$$

Here, f is a density of populations parametrized by the time t, the space x, and their speed v. The left-hand side is associated with the movement of the population with speed v and its random changes of speed with probability $M(v)dv$. The right-hand side models the births and deaths of the population. Such models might be well-fitted to investigating the dynamics of microscopic populations, such as bacteria (see the mathematical biological papers cited in Bouin et al. (2014b)).

If the kernel M is a combination of Dirac masses $M = \frac{1}{2}(\delta_{v_0} + \delta_{-v_0})$, then simple computations show that (1.4) turns into (1.2) under appropriate changes of variables. Equation (1.2) was thus first investigated in Bouin et al. (2014a) as a toy model in order to understand the more general equation (1.4).

In the asymptotic regimes where the change of speeds occurs very frequently, it can be proved that the solutions of equations (1.4) converge to the solutions of the classical Fisher-KPP equation (1.1) under an appropriate rescaling. Cuesta et al. (2012) used this observation in order to construct travelling wave solutions of equation (1.4) in a perturbative regime when the set of speeds V is bounded.

Indeed, Bouin et al. (2014b) showed that such travelling waves exist in general, not only in perturbative regimes, as soon as the set of speeds V is bounded, that is, the speed of each individual cannot be too large. Moreover, one can show that the associated propagation speed is given by the leading edge of the front. The difference from equation (1.2) is that this speed is no longer explicit, but is given by an integral equation involving the kernel $M(v)$. When V is unbounded, that is, when each individual could take arbitrarily large speeds, then travelling waves no longer exist Bouin et al. (2014b). Indeed, in this case the propagation is superlinear. For example, when $V = \mathbb{R}$ and $M(v) = e^{-v^2}/\sqrt{2\pi}$, the leading edge of the front is located approximately at $x \simeq t^{3/2}$. The understanding and rigorous proofs of these phenomena is the subject of works in progress.

Conclusion

In this article, we began by describing mathematical models involved in archaelogy. The classical model introduced by Ammerman and Cavalli-Sforza (1984) gives a mechanistic interpretation of the emergence of "waves of advance", observed for example in the Neolithic colonization of Europe, showing that such a propagation phenomenon is not driven by any external force. These waves are associated with travelling wave solutions of the Fisher-KPP equation. The order of magnitude of the propagation speed derived from this model can be improved by introducing a time delay in the equation, as suggested by Fort and Mendez (1999), giving rise to a hyperbolic reaction–diffusion equation. This equation was rigorously investigated by Bouin et al. (2014a), who identified two different regimes: if the delay becomes too large, then the approximation made by Fort and Mendez (1999) is no longer valid, and shocks, that is, non-smooth travelling waves, arise. Hyperbolic reaction–diffusion equations are indeed simplifications of more general kinetic reaction–diffusion equations, which are involved, for example, in models of the dynamics of bacteria. The investigation of these equations is more involved and might give rise to contrasting phenomena, such as superlinear propagation, which are still not completely understood (see Bouin et al. (2014b)).

Acknowledgements

The author is grateful to professors J. Fort and V. Mendez for having kindly authorized the reproduction of Figure 1.1.

The research leading to these results has received funding from the European Research Council under the European Union's Seventh Framework Programme (FP/2007-2013) / ERC Grant Agreement n. 321186 - "ReaDi - Reaction–Diffusion Equations, Propagation and Modelling" held by Henri Berestycki.

Bibliography

Ammerman, A. J., and L. L. Cavalli-Sforza, 1984, *The Neolithic Transition and the Genetics of Populations in Europe*, Princeton University Press.

Bouin, E., V. Calvez, and G. Nadin, 2014a, "Hyperbolic traveling waves driven by growth", *Math. Models Methods Appl. Sci.* **24**.

Bouin, E., V. Calvez, and G. Nadin, 2014b, "Front propagation in a kinetic reaction–transport equation", Accepted for publication in *Archive for Rational Mechanics and Analysis*.

Cuesta, C., S. Hittmeir, and C. Schmeiser, 2012, "Traveling waves of a kinetic transport model for the KPP-Fisher equation", *SIAM J. Math. Anal.* **44**, 4128–4146.

Fisher, R. A., 1937, "The advance of advantageous genes", *Ann. Eugenics* **7**, 335–369.

Fort, J., and V. Méndez, 1999, "Time-Delayed Theory of the Neolithic Transition in Europe", *Phys. Rev. Lett.* **82**(4).

Fort, J., T. Pujol, and L. L. Cavalli-Sforza, 2004, "Palaeolithic populations and waves of advance", *Cambridge Archaeol. J.* **14**, 53–61.

Hadeler, K. P., 1988, "Hyperbolic travelling fronts", *Proc. Edinburgh Math. Soc.* **31**, 89–97.

Kolmogorov, A. N., I. G. Petrovsky, and N. S. Piskunov, 1937, "Etude de l équation de la diffusion avec croissance de la quantité de matière et son application à un problème biologique", *Bulletin Université d'Etat à Moscou (Bjul. Moskowskogo Gos. Univ.)*, 1–26.

Chapter 2

The urban project: A complex system in search of coherence

Séverine Chemin Le Piolet[1]

2.1 Introduction

This chapter is based on a town-planning investigation related to the coherence of the urban project in a context of uncertainty and complexity, applied to the case of the City of Saint-Etienne in France. Urban planning and territorial planning are guided by the structural principle of territorial coherence (Zepf & Andrés, 2012; Zepf & Novarina, 2009; Jourdan, 2012; Faludi & Peyrony, 2011). However, the complexity and fragmentation inherent in the contemporary conditions of urban development question very strongly this objective (Ascher, 2001; Chalas, 2004; Kokoreff & Rodriguez, 2004). First, we will demonstrate how the urban project can be broached as a complex system resulting from the interaction of several planning operations taking place about the territory. After underlining this systemic character, we will develop the stakes related to the modelling of an urban system. The method developed for this research is based on using an interactive tool realized

[1]PhD in Urban-Planning mention Planning of the Territory, Université Grenoble Alpes, CNRS, PACTE, F-38100. Email: severine.chemin@gmail.com

with the city of Saint-Etienne, and revealing the interactions between different projects of the territory. This research represents an attempt to overcome the uncertainty nowadays involved in urban and territorial planning. It investigates the systemic approach and modeling of an urban project as a tool for decision support and the management of city projects, in the quest for territorial coherence.

2.2   The urban project as a complex system

The urban project is a polysemous and multi-scale concept (Bourdin and Prost 2009, Genestrier 1993, Panerai and Mangin 1999, Rey et al. 1998, Tsiomis and Ziegler 2007) that can be understood at different levels alluding to, at the same time, a project of several architectural buildings, an operation of complex planning at the district level, and a city or urban area project (Merlin & Choay, 2005). An urban project also takes many dimensions: technical, political, economic, and social (Ingallina, 2003). It will be understood here in its spatial dimension as the combination of various planning actions or projects carried out at the city of Saint-Etienne. In this chapter, we will tackle the urban project not as the compilation of the various projects of the territory, but as a combination in which the projects come into interactions that are reinforcing, or weakening, etc. The urban project is consequently seen as a complex system resulting from many interactions between projects (Toussaint & Zimmermann, 1998).

2.2.1   The major spatial components of the city project

The research topic is the project of the city of Saint-Etienne. In Figure 2.1, the municipal boundary of Saint-Etienne is represented by the pink outline, and buildings are in grey. We used these base maps to identify different project areas according to their natures, ladders, etc.

The dominant and general public view of city projects comes mainly through an approach to large projects represented by hatching roses (Figure 2.2). The perimeters of these complex and emblematic urban projects represent exceptional territories in which public actors take part to accelerate the development of economic hubs, the replacement of precarious districts, the design of new districts, etc. This outstanding intervention is motivated by its strategic specificity at the global territorial level. The territories are excluded from the common law, and are subject to operational arrangements (public–private partnerships, global urban planning procedures, etc.) and distinctive

Figure 2.1: The municipal territory of Saint-Etienne. Figure by the author, based on Ville de Saint-Etienne data.

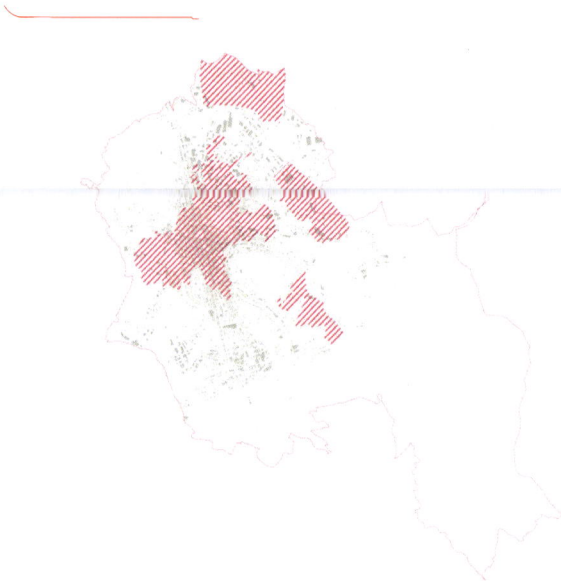

Figure 2.2: The perimeter of major projects.Figure by the author, based on
Ville de Saint-Etienne data.

monitorings (the setting of a specific organization in project mode for example). These are areas in which local actors have great knowledge and about which they often communicate in a territorial marketing logic.

This vision, synthesizing the city project through major projects, seems reductive in a systemic approach. Our approach advocates an expanded vision of the city project that encompasses a multitude of actions, including major projects but also actions that are more common and anonymous, that bring the territory into consideration. The challenge in this systemic approach is to include the diversity of projects that make up the territory, to develop an all-embracing vision, and to surpass the focus usually placed on major urban planning projects. It integrates the common city with the thought of a global urban project. The so-called common city represents 80% of the territory but is not the subject of cultural, residential, or commercial emblematic major public programs, and yet this city is also changing, according to the operations in the more common districts, by private interventions of limited replacement in the plot, etc. These projects are involved in the evolution dynamics of the territory, and therefore of the overall urban project.

In this way, the overall urban project of the city of Saint-Etienne represents a system in which multiple emblematic operations involve such major projects, and more anonymous actions deal with the common city. It therefore appears as a system combining varied elements, such as major projects, cross-functional projects such as transport infrastructures, equipment and utilities projects, and the varied operations encompassed in what we will call *the diffuse*.

- "The Diffuse" (Figure 2.3) provides the most expanded possible vision of the dynamics of the territory. Areas not covered by complex operations still contain varied dynamics regarding private operations and/or public local actions. They are consequently definitely included in the overall dynamics of the urban project, without requiring a strong action of the public actor in terms of "project". These projects are of a smaller scale and of a one-off nature, as shown in Figure 2.3. It may be the renewal of a square in a district, a small public garden, the creation of a block of flats, the setting up of facade restoration, etc.

- Equipment and utilities projects raise the question of territorial balance, through the implementation of various public policies. They embrace projects of equipment and services to the people, such as the kindergardens shown in Figure 2.4, schools, associations, etc., to

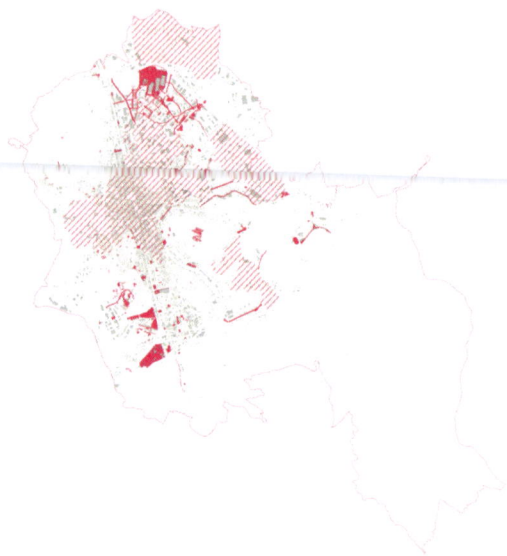

Figure 2.3: "Diffuse projects". Figure by the author, based on Ville de Saint-Etienne data.

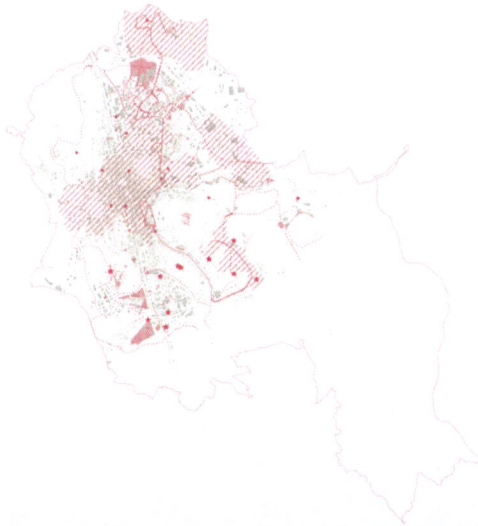

Figure 2.4: List of municipal kindergartens. Figure by the author, based on Ville de Saint-Etienne data.

develop a balanced offer, in terms of equipment and services, to the population in the different districts.

- Transport infrastructure projects strongly impact the territories they cross. Consequently, they represent a high stake for the territories around them and specially for the cross-functional projects. . Examples of public transport (Figure 2.5) and green lane projects (Figure 2.6) in Saint-Etienne are representative of projects of this particular character.

2.2.2   The combination of many projects at the municipal level

Resulting from this treatment of the urban project as a system, we are interested in the interactions between these various elements, in identifying how these elements work in synergy, how the projects aid, or not, each other and the overall project. Thus, an urban project cannot be reduced to the sum of the different projects developed in the area, but rather their combination, in which some of them will weaken, others strengthen, or even annihilate

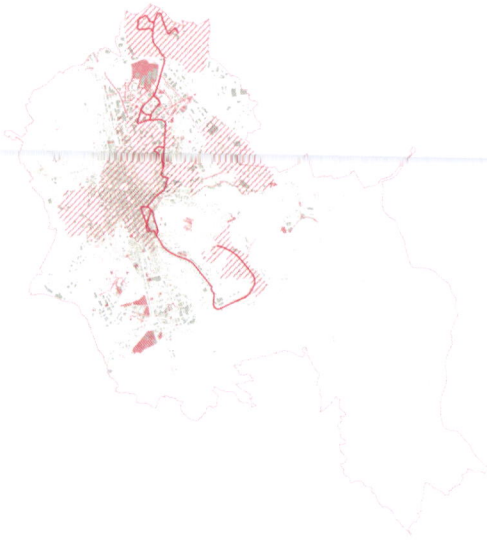

Figure 2.5: Possible drawing for the public transport in Saint Etienne. Figure by the author, based on Ville de Saint-Etienne data.

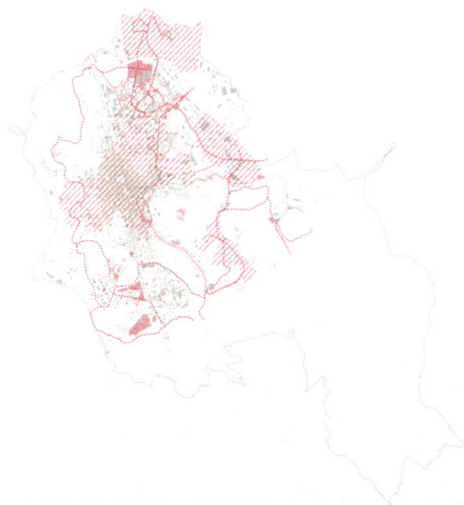

Figure 2.6: Possible drawing for green lane, inter-district link. Figure by the author, based on Ville de Saint-Etienne data.

themselves or each other, etc.

Thus, cross-functional projects can constitute a boosting or a nuisance for other operations in progress, and challenge their progress and future. For example, public transport in a district can be an opening-up factor for a revitalization project of the district; or the setting up of a green lane for a smooth inter-district connection can enhance the operation of the housing nearby. The setting up of a highway can connect spaces, while constituting a nuisance and splitting the areas crossed. Similarly, several "diffuse" actions matched together can impact a large project by producing competing projects, such as a new housing offer with a similar target and quality, office areas, etc.

All these actions or projects, developed and produced by various actors, come into interaction with each other, and impact the overall system. In the field of urban planning, each plan for public areas, any renovation or construction of equipment or various housing projects, impacts the district not only where the operation is taking place but also the city in its entirety. What is actually appropriate is to speak of 'congruence', rather than 'impact' between projects, because in this complex system, it becomes very difficult to assess which element has an impact on any other. Each project coming into resonance with the other actions in the territory can influence the overall strategies of the city project and change the balance of the urban project, notably in terms of the distribution of equipment, access to public utilities, etc.

The challenge, in this systemic approach to the urban project, therefore lies in the consideration of many projects which are its components. This task falls to the public agency who is responsible for the development of the territory, and whose role is to coordinate projects, in order to promote synergy, rather than a weakening of the urban dynamics, and to reduce the inequalities which might occur. These projects, under the aegis of a multitude of actors, with the stakes in the coherence of these actors and their actions, constitute, in our opinion, the current major challenge for city planners.

2.3   The quest for the coherence of the urban project

2.3.1   Coherence and complex systems

The concept of system is explained by Morin as 'different elements' combinatorial association"[2] (Morin, 2005, p. 28). That sociologist and philosopher also points out, 'systemic virtue is to have placed at the centre of the theory, with the notion of system, not a discrete elementary unit, but a complex unit, a 'set' that cannot be reduced to the 'sum' of its constituent parts"[3] (Morin, 2005, p. 29). Several versions of the concept of system allude to the logic and organization that result from its structure. The system appears as "a set of equipment or not, mutually dependent on each other to form an organized whole"[4] (Lalande, 2006, p. 1096). Thus, the concept of 'coherence' is implicitly included in the internal workings of the system.

When it is complex, the system's working doesn't appear very decipherable: it is even intertwined and therefore its logic and coherence escape one. It is the complexity that takes us away from the concept of coherence, as many parameters become intertwined in their interaction. Complexity prevents us from assessing the future state of the system and its various parameters. Complexity is the "characteristic of a system which, due to

[2]Translation by the present author.
[3]Translation by the present author.
[4]Translation by the present author.

the heterogeneity of the processes that take place, has the ability to evolve in different directions, which makes this dynamic difficult to predict from present conditions".[5] (Levy & Lussault, 2003, p. 188).

Thus, taking an interest in the coherence of the urban project as a complex system reflects a certain paradox, which would be similar to breaking through the logic which guides the system and clearing up its complexity. This paradox particularly impacts the disciplines of land settlement and town planning, of which the projection of the "urban system" in time is one of the founding principles. Thus, the urban action collides with the complexity of the system on which it must operate.

2.3.2   Uncertainty and fragmentation of urban systems

It must be noted that land settlement is now occurring in the context of increasing complexity and uncertainty. Urban transformations, changing territories and projects appear extremely unpredictable and question the practices of land settlement. Contemporary dynamics, such as globalization and metropolitanization, have caused a loss of intelligibility of the cities, confusing the keys of reading and action on the territories (Chalas et al. 2009). Multiple uncertainties impact town planning, such as the economic, environmental, and social instability of the territories, the future consequences of innovations in mobility and new communication technologies, etc. Uncertainty and complexity are the paradigms of contemporaneity. In this sense, they radically transform the way we think about territories and the actions to be taken on them.

Moreover, the framework for developing and implementing various land settlement projects is becoming increasingly fragmented. In particular, we are witnessing a fragmentation of the competencies of public actors through the "institutional *millefeuille*" resulting from the decentralization of France. The financial framework constrains the public actors, and the increasingly significant involvement of private actors and people in the production of the city increases tenfold the partnership aspect of these projects.

In this way, an urban project results from the combination of actions led by multiple public and private actors. It appears fragmented as a result of the multitude of projects and actors interacting amongst themselves, and the prediction of future effects becomes extremely complicated. These projects are so numerous that it seems to be impossible to assess how the global urban system will evolve. Especially, within each project, economic, sociological, economical, political and environmental parameters influence their

---

[5]Translation by the present author.

own progress, success, impact, etc.

And yet, the particularity of the discipline is to take place on the territories. Town planning is a "science of action" (Merlin, 2009) based on the principle of beneficial response to the current or future working of the territory. The approach developed here is based on the consideration of the territory as a system. Therefore, the taking into account of the interactions that drive our complex urban systems emerges as one of the major challenges in the discipline, but also becomes the way to expand our thinking of the "coherence" of the urban project.

The analysis of systems traditionally operated through the black box principle (Wiener, 1948), listing the inputs and outputs and deducing the production of the system. This does not seem adapted to the land settlement disciplines. Indeed, the challenge for these disciplines is to influence trends deemed undesirable before they produce too many undesirable effects. The aim is therefore to take into account the dynamics in progress and not wait until they have produced irreversible effects.

The complexity of operating urban systems and the inability of land settlement actors to predict their evolution has been highlighted above. The challenge of the coherence of the urban project is therefore in the broadest possible knowledge of ongoing and future projects in order to "re-act" knowledgeably. This approach to the urban project has distanced itself from the traditional vision of the planner of a land settlement. The integration of the contemporary complexity and uncertainty inherent in urban systems reduces long-term urban anticipation in favour of an informed and reflexive management of the urban project.

This sense of the urban project cannot be limited to only to sloppy and emergency logics. It is vital to manage the urban project, not to undergoe urban developments. This management is based on three elements of the urban project: the expected vision of the future of the territory, strategies deployed to work towards this vision, and the fulfilment of these strategies through projects. In our opinion, the coherence of the urban project lies in the articulation between these elements. This articulation is based on a strong principle of reflexivity in which it is no longer a question of stating an action plan in a linear way, but re-examining the current action based on the first effects produced and new parameters used in the interaction (Ascher, 2004).

In this changing and fragmented context, the job of the public actor appears as the conducting of the general meaning of the action. In this way, he guarantees the coherence of the urban project across the entire city. It is a matter of overcoming the breaking up of the actions and actors of the system by giving a comprehensive and shared vision of the territory. The main challenge for the public actor lies therefore in the broadest possible knowledge of the projects ongoing in the area and in the implementation of a readable global urban project.

2.4   Modelling a complex system: The urban project atlas of Saint-Etienne

2.4.1   Towards a "global" and "informed" management of the urban project

Faced with the challenges of readability and the reflexive management of the urban project, this reflection has led to the modelling of complex urban system of Saint-Etienne. This experiment was managed by the town-planning services of the city of Saint-Etienne and with a tool for the readability and coherence of urban projects, called Atlas, which was created by the urban project in Saint-Etienne.

In 2010, the fragmentation of the urban project in Saint-Etienne was marked daily by the lack of a common representation on the part of the planning actors. Each of them was developing its own communication media on its own projects (large projects, thematic cross-functional projects such as public transport, etc.). The lack of support for representing all projects, including in particular diffuse and private projects, does not enable taking into account the evolution of the common city and the territory as a whole. So, the urban project is the subject of multiple partial representations reflecting the fragmented governance that implements it. The complex urban system is therefore broached in a divided and partial way by each of these actors. From this perception of the system there arises a partial management and vision of the urban project that will have to be overcome by the experiment of the modelling of a complex urban system as a whole.

For the modelling of an urban project as a complex system, the support used is the map. It allows representing each project in its spatial form, so as to reveal possible future interactions through a temporal slicing. It is therefore the development of a partnership tool representing all projects, regardless of the project's ownership. The partnership dimension is here very important because each actor has his own data about his own project, and it is necessary to group them in order for the tool to work.

The modelling of a complex urban system, "The Urban Project Atlas", works as:

- An element-resource: a database, a filing of town-planning studies, a reporting and monitoring of projects;

- A tool for decision support and production analysis that informs us about urban form, the progress of projects, and makes the connection with strategic planning documents and public policies.

2.4.2   Revealing interactions by means of a temporal map

The core of the tool is a temporal map. This map is the indicator of the interactions between the various elements of the system. This is the modelling of different projects in their urban form at the city scale, according to a common programmatic legend and a temporal slicing of the actions, concerning the short term (current year), the medium term (end of the mandate), and the long term (beyond the mandate). The major challenge lies in the completeness and responsiveness of the modelling projects, large and small, private and public. Thus, it is by means of the support of the community database (building permits, housing services, displacements, prospective, etc.) for inputting the private operations, public areas, emblematic town-planning operations, and diffuse actions, with all the projects' ownerships. The base map is the cadastral plan of the territory, with buildings in grey and highways. The outline of the municipality delimits the study area. In a homogenized corporate identity and style guide, all projects of the territory are represented according to the categories of the programs: public areas, housing restoration, new equipment, etc. A particular colour corresponds to each program category.

Thus, knowing and modelling all projects become strategic because together, they will be the "indicator" of the "possible" future interactions. This map reveals the interactions between projects at several levels. At first, the mere fact of representing many projects of the territory on the same map is indicative of the potential interactions between operations previously considered separately. Thus, such a support opens up new perspectives in terms of synergies to strengthen or competitions to regulate between the various ongoing operations. The production of this map allows developing a global vision of the territory and its dynamics, and examining the equilibrium distribution of the various programs in the territory, for example.

Secondly, the most revealing element of these interactions is the temporal slicing done with this map making (Figure 2.7). In fact, each project is assigned a specific temporality, short term (current year), medium term (end of the mandate, 2014) and long term (beyond the mandate). This temporal slicing can render the map making dynamic and therefore refine the interactions between projects. It is this dynamic vision of the fulfillment of "scheduled" projects that will reveal possible interactions and may cause a re-orientation in terms of programming. Seeing that similar projects are planned nearby could, for example, make a project initiator postpone the phasing, or develop another type of product, etc.

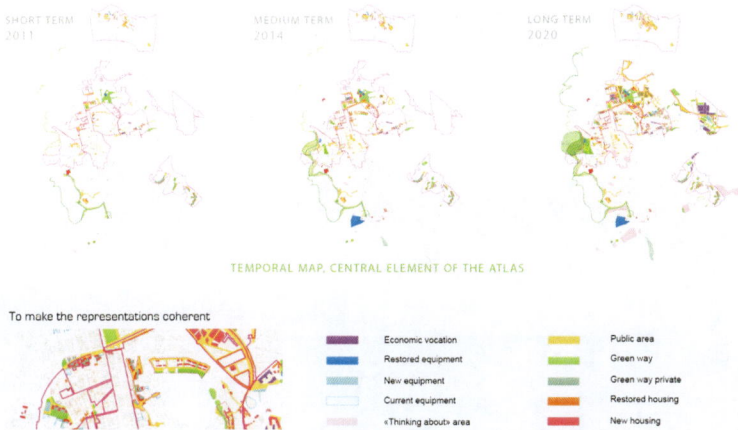

Figure 2.7: Extracts of the temporal map and legend. Source: Ville de Saint-Etienne.

The Urban Project Atlas is an indicator of the ongoing urban dynamics in the territory and a tool for decision support and production analysis. Through its interactivity it connects the "representation of projects" in terms of urban form, content (program, planning, etc.) and progress, with the strategic planning documents and major public policies (Figure 2.8).

This interactive tool, based on the principle of articulation scale, allows including a global vision of the urban project, from the strategy to the details of operations. It is now accessible to various officers of the City of Saint-Etienne and their privileged partners, such as the urban conglomeration Saint-Etienne Metropolis in particular.

2.5   Conclusions

This action-research carried out on the interactions in complex systems allowed us to demonstrate the essential sign of the concept in the field of town planning and land settlement.

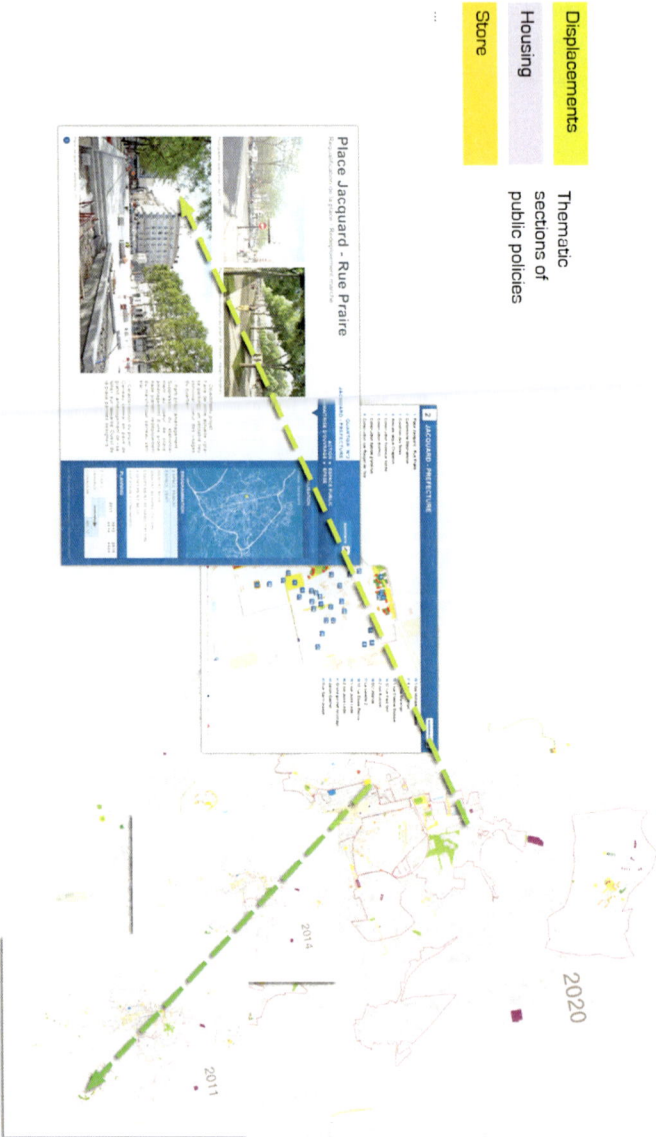

Store

Housing

Displacements

Thematic
sections of
public policies

Place Jacquard - Rue Praire

2011

2014

2020

Figure 2.8: The Urban Project Atlas scales the articulation principle and interactivity. Source: Ville de Saint-Étienne.

A territory is, in essence, an extremely complex object, resulting from the combination of multiple geographic, economic, social parameters, and thus generates a "situation" or context whenever new. André Corboz describes it as a combination of "factors as diverse as the geology, topography, hydrography, climate, forest cover and crops, populations, technical infrastructure, productive capacity, judiciary, administrative zoning, national accounting, service networks, policy stakes, and so on, not only in all of their interferences, but dynamically, in virtue of an intervention project"[6] (Corboz, 2001, p. 10). Town planning and land settlement are complex disciplines by definition, since they involve working on these areas in order to improve their functioning, reduce inequalities, etc. It is in this sense an art of synthesis, the consideration of many factors that have to be reconciled, or at least to reconcile the greatest possible number.

In this way, this chapter dealt with the urban project, the territorial project of the city of Saint-Etienne, as a system confronted with complexity, uncertainty, and the break-up of urban areas. And, whose the quest for coherence, has resulted in a reflection on interactions between multiple projects constituent of the global urban project. The necessity for renewal of traditional modes of action of the land settlement in favour of reflexive approaches has been highlighted by this reflection.

As a consequence, the experiment based on this analysis was led through the modelling of complex urban system of Saint-Etienne represented by The Urban Project Atlas. This interactive tool, whose core is a temporal map making of projects, acts as an indicator of the interactions between the various elements of the urban system.

This experimental and innovative tool is part of multiple investigations with the theme of *watching territories*. These tools, which fall in the category of *urban observatories*, are a way to "enter" the interactions of complex urban systems, in the sense of their understanding as well as their representation. The current rise of these observatory functions reflects the challenge of taking into account the interactions and complexity of urban systems for the territorial sciences.

---

[6]Translation by the present author.

Bibliography

Ascher, F., 2001, *Ces événements nous dépassent feignons d'en être les organisateurs*, l'Aube.

Ascher, F., 2004, *Les nouveaux principes de l'urbanisme*, l'Aube.

Bourdin, A., and R. Prost, 2009, *Projets et stratégies urbaines : regards comparatifs*, Parenthèses, La ville en train de se faire.

Chalas, Y., 2004, L'urbanisme dans la société d'incertitude, in: *L'imaginaire aménageur en mutation*. L'Harmattan, pp. 231–269.

Chalas, Y., C. Gilbert, and D. Vinck, 2009, *Comment les acteurs s'arrangent avec l'incertitude*, Paris: Editions des Archives Contemporaines.

Corboz, A., 2001, *Le territoire comme Palimpseste*, Les Editions de l'imprimeur.

Faludi, A., and J. Peyrony, 2011, *Cohesion Policy Contributing to Territorial Cohesion—Future Scenarios*, in: *European Journal of Spatial Development* **43**, 2–17.

Genestier, P., 1993, Que vaut la notion de projet urbain? *L'architecture d'aujourd'hui* **288**, 40–46.

Ingallina, P., 2003, *Le projet urbain*, Paris: Presses Universitaires de France.

Jourdan, G., 2012, La planification face au défi de la cohérence territoriale, in: *Enjeux de la planification territoriale en Europe*, Presses Polytechniques et Universitaires Romandes, pp. 163–180

Kokoreff, M., and J. Rodriguez, 2004, *La France en mutations, quand l'incertitude fait société*, Paris: Payot.

Lalande, A., 2006, *Vocabulaire technique et critique de la philosophie*, Paris: Presses Universitaires de France.

Levy, J., and M. Lussault, 2003, *Dictionnaire de la géographie et de l'espace des sociétés*, Belin.

Merlin, P., and F. Choay, 2005, *Dictionnaire de l'urbanisme et de l'aménagement*, Paris: Presses Universitaires de France.

Merlin, P., 2009, *L'urbanisme, que sais-je?*, 8th Ed., Paris: Presses Universitaires de France, pp. 3–4.

Morin, E., 2005, *Introduction à la pensée complexe*, Seuil.

Panerai, P., and D. Mangin, 1999, *Projet Urbain*. Parenthèses.

Rey, J., F. Thomas, J.-Y. Toussaint, and M. Zimmermann, 1998, *Projet urbain: ménager les gens, aménager la ville*, Mardaga.

Rochette, D., 1999, *Image et identité urbaines de Saint-Étienne*.

Toussaint, J.-Y., and M. Zimmermann, 1998, Un enseignement de la complexité, in: *Projet urbain : ménager les gens, aménager la ville*, Mardaga, pp. 163–173.

Tsiomis, Y., and V. Ziegler, 2007, *Anatomie de projets urbains : Bordeaux, Lyon, Rennes, Strasbourg,* Editions La Villette.

Ville De Saint-Etienne, 2012, *Atlas du projet urbain: Cahier de méthodologie*.

Ville De Saint-Etienne, 2013, *Saint-Étienne, construire le projet urbain horizon 2020*.

Wiener, N., 1948, *Cybernetics, or, Control and Communication in the Animal and the Machine*, Cambridge, Mass.: Technology Press.

Zepf, M., and G. Novarina, 2009, Territorial Planning in Europe: New concepts, New experiences. *DISP, The Planning Review* **179**, 18–27.

Zepf, M., and L. Andres, 2012, *Enjeux de la planification territoriale en Europe*, Presses Polytechniques et Universitaires Romandes.

Chapter 3

A PDE-like toy model of the functioning of a territory[1]

Emmanuel Frénod[2]

3.1 Introduction

In this paper, by 'territory', ... we mean a countryside or a suburb or a town or a set of towns, suburbs and/or countrysides .... having some coherence and being administered by a common local government.

As explained in Cullingworth and Nadin (2006), town and country planning needs an understanding of the interactions occurring within a territory, at various scales.

With the perspective of providing a solution for this need, our long term goal is to build a software tool that behaves like a territory, and in particular that incorporates its spatio-temporal multi-scale nature, in order to make simulations, explore scenarios, foresee policy impacts, help make decisions when facing changes in the environment or in cultural behaviour, etc..

There are territorial models do not take space into account (see Moriarty

---
[1]This work is supported by MGDIS and AMIES.

[2] Univ. Bretagne - Sud, UMR 6205, LMBA, F-56000 Vannes, France. Email: emmanuel.frenod@univ-ubs.fr, http://web.univ-ubs.fr/lmam/frenod/index.html.

and Adams 1976). Now, many questions in town and country planning are related to the displacementof people: for instance, daily displacements are linked with decision making concerning transport networks, and population migration has consequences for construction policies. Hence, these models are a little limited for helping decision making. On the other hand, spatial models have existed for a long time in Spatial Economics (see for instance Vickerman 1980, Beckmann and Puu 1985, Meardon 2001) but do not seem to be much used for the purpose of town and country planning.

A territory is clearly a complex system that involves, among other things, a geographic area, people, the actions of the local government, and enterprises. It also consumes energy and produces wealth, goods, services, and culture. To comprehend globally the workings of a territory is certainly a very difficult task. Possessing a sufficiently rich mental picture of it without using modeling may be impossible.

Hence, as we want to begin the process of understanding the workings of a territory, we are faced with the following issue: we need to build a model of something that, essentially, we do not understand. To do this, the usual analytic modeling method, that describes precisely the workings of every detail of a system and that assembles these details into a global mathematical model, is ineffective.

We need to turn to the systems approach, which starts from the global and that offers the possibility of considering only a part of a complex system or considering several parts that are not modeled with the same level of detail. To do this, the systems approach considers a complex system as having several compartments that act on each other via action/reaction loops and on themselves via feedback loops. Building a first instance of a systemic model of a given complex system consists in isolating a few parts of the system and assigning them the status of compartment. Then, we have to describe how the parts (or compartments) act on each other and on themselves to define the action/reaction and feedback loops.

If this task is properly achieved, it is possible to associate with each compartment a mathematical object and with every loop a mathematical operator acting on one of these objets. Hence, we finally obtain a mathematical model that can be analysed or simulated using the usual mathematical tools.

Once the first instance of the systemic model of a given complex system is built, it is possible to add a part to it. That is to say, a new compartment and new loops are added to the systemic model and that a mathematical object and mathematical operators are added to the mathematical model. It is also possible to go into the details of one part. This consists in replacing one compartment by several ones and in re-defining the loops.

In this paper we build a mathematical model, which we call a toy model, following the way described above, which embeds several aspects of the working of a territory. In the first step of the systems approach, there arises a systemic model. Then, we translate this systemic model into a mathematical one. For that, we use density functions as the mathematical objects and PDEs (see Farlow 1993) as the mathematical operators. More precisely, the action/reaction and feedback loops are translated into the coefficients and source terms of the PDEs that give the time evolution of the considered densities.

For building the systemic model, we make choices concerning the parts to take into consideration. In these choices, there is some arbitrariness. Yet, we find it advisable to assume that the working of a territory is essentially the consequence of the actions of the people living in it and the economic activity occurring within it. Hence, we choose to use a "People" compartment and an "Enterprises" compartment. Then, we want to put our work into the perspective of the long term goal described in the beginning of this Introduction. Hence we choose to introduce an "Energy" compartment as energy is, on the one hand, used for the movements of people and, on the other hand, is something that consumes the wealth that the economy produces. In addition, we choose to embed our model into a geographical area because of our long term goal and because if space is not modeled, no displacement can be taken into account, and so, no related energy consumption estimated.

Thanks to these choices, the simulations made with the resulting mathematical model allow visualizing the dynamics of people going to work, the consumption of energy and wealth these displacements induce, and the fact that people working in enterprises generate wealth. It also helps to comprehend the result of the way this wealth is split into the part used to improve the enterprises' efficiency, the part allocated to increase their size, and the part that goes to the people.

3.2   Systems approach to the workings of a territory

We assume that the workings of a territory result from the interactions between the people and the enterprises within a geographic area and that this interaction consumes energy. Integrating this within a systems approach re-

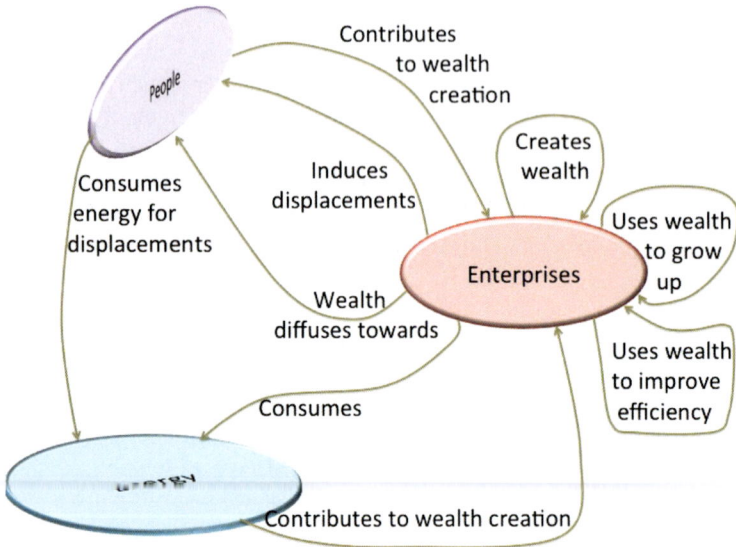

Figure 3.1: Systems approach to the workings of a territory.

sults in a systemic model, represented in Figure 3.1. Three compartments
are considered. The first one ("People") concerns the population, the second one ("Energy") concerns the energy questions, and the last one ("Enterprises"), concerns the world of enterprises. We have, at this level, a non-
restrictive definition of what is an enterprise. It can be any organization
producing goods or services and in which people work.

Then, the compartments influence each other and themselves. This fact
is symbolized by the arrows. For instance, the arrow which is more at the top
of the figure indicates that when people are at work, they contribute to the
production of wealth. In return, the arrow just below means that displacements are induced by enterprise locations and the arrow on the left indicates
that energy consumption is induced by these displacements. The three arrows that point from the "Enterprises" compartment to itself symbolize that
enterprises create wealth and that this created wealth is used by the enterprises for growing and to improve their efficiency. The two last arrows, on
the bottom, express that enterprises consume energy to create wealth.

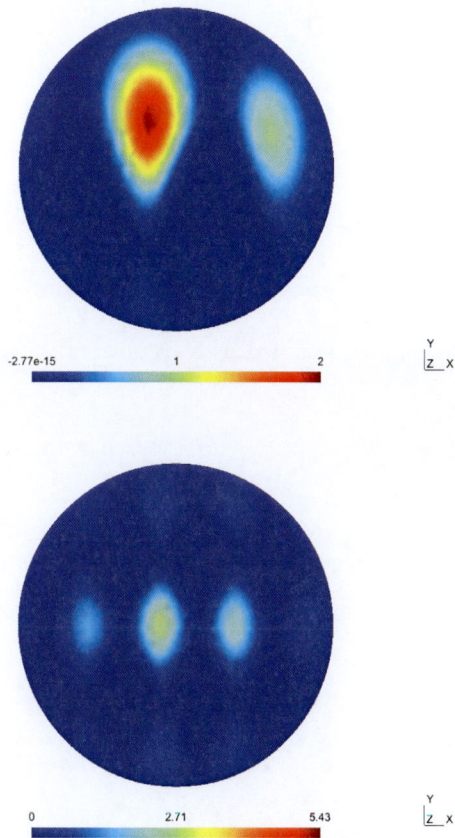

Figure 3.2: Initial population density $\mathcal{P}_0$ (up) and job-station density $\mathcal{E}_0$ (down).

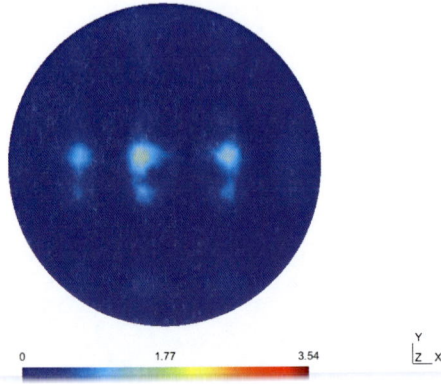

Figure 3.3: Job-station density $\mathcal{E}_0$ times efficiency indicator i_0.

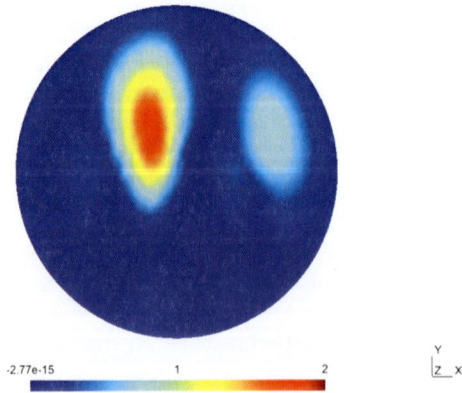

Figure 3.4: Displacements from home to work and back (Animation available on: http://youtu.be/LkPlVT-a4pg).

3.3   The toy model

From the systemic model of the previous section, we now build the toy model. In this toy model, all quantities are dimensionless, meaning that they have no units and that their order of magnitude is 1.

In order to provide the reader, as quickly as possible, an interpretation of the equations that are introduced, we made simulations during the model writing. They were done using Freefem++, which is a Finite Element software tool. All the simulations were made over the same time interval $[0, T)$, defined for a real number $T > 0$, which is driven by the variable t.

The geographic model of the toy model is a disk D, provided with coordinates x and a boundary ∂D. The vector of norm 1, orthogonal to ∂D and pointing outside D, is denoted by μ.

On this disk, several densities, which depend on time, are defined:

- $\mathcal{P} = \mathcal{P}(t, x)$ is the Population Density.

- $\mathcal{E} = \mathcal{E}(t, x)$ is the Job-station Density.

- $\mathcal{W} = \mathcal{W}(t, x)$ is the Wealth-that-goes-to-people Density.

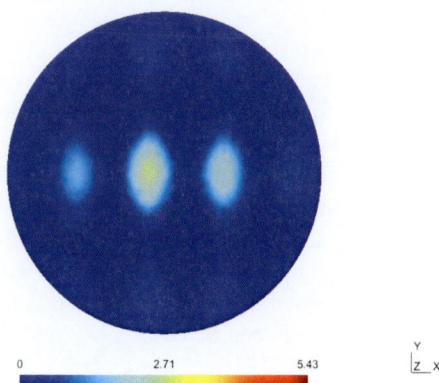

Figure 3.5: Evolution of job-station density $\mathcal{E}$ (Animation available on: http://youtu.be/PorgK4q5Ds0).

For these densities, the initial values are defined as follows: $\mathcal{P}_0 = \mathcal{P}_0(x)$ is the Population Density when everybody is at home at night, $\mathcal{E}_0 = \mathcal{E}_0(x)$ and $\mathcal{W}_0 = \mathcal{W}_0(x) = 1$.

An example of the initial density $\mathcal{P}_0(x)$ is drawn on the left of Figure 3.2 and an example of $\mathcal{E}_0(x)$ is drawn on the right of Figure 3.2. The simulation to come will be done with these initial conditions.

An Efficiency Indicator $i = i(t, x)$ is also defined on the disk. It ranges within $[0, 1]$ and measures the efficiency of the enterprises at transforming people's work into wealth at every point of the disk; it is initialized at i_0. The product of the initial Job-station Density $\mathcal{E}_0$ of Figure 3.2 with the initial Efficiency Indicator i_0 that will be used in the simulation to come is given in Figure 3.3.

From the initial Population Density $\mathcal{P}_0$ and Job-station Density $\mathcal{E}$, two attractors are built. The first one, $\mathcal{A}_{\mathcal{E}} = \mathcal{A}_{\mathcal{E}}(t, x)$, attracts the Population Density towards the Job-stations. It is defined as

$$\mathcal{A}_{\mathcal{E}}(t, x) = \frac{\displaystyle\int_D \mathcal{P}_0(x)\,dx}{\displaystyle\int_D \mathcal{P}_0(x)\,dx + \alpha \int_D \mathcal{E}(t, x)\,dx}\left(\mathcal{P}_0(x) + \alpha\mathcal{E}(t, x)\right), \quad (3.1)$$

where α is a constant larger than 1. The second one, $\mathcal{A}_{\mathcal{P}} = \mathcal{A}_{\mathcal{P}}(x)$, makes the population go back home. It is defined as

$$\mathcal{A}_{\mathcal{P}}(x) = \mathcal{P}_0(x). \quad (3.2)$$

With these two attractors, we can simulate the daily motion of the population. For this, we firstly adopt the convention that the length of a day is 1. Then, we define two functions $t_m = t_m(t)$ and $t_e = t_e(t)$ depending only on time and periodic of period 1. $t_m(t) > 0$ for those instants that corresponding to the morning displacements, when people go to work. It is 0 otherwise. $t_e(t) > 0$ for instants corresponding to the evening displacements, when people go back home, and is 0 otherwise. Secondly we write the equation meaning that in the morning the Population Density is attracted by the Job-station Attractor and in the evening by the Home Attractor:

$$\frac{\partial \mathcal{P}}{\partial t}(t, x) = t_m(t)\left(\mathcal{A}_{\mathcal{E}}(t, x) - \mathcal{P}(t, x)\right)$$

$$+ t_e(t)\left(\mathcal{A}_{\mathcal{P}}(x) - \mathcal{P}(t, x)\right), \ \forall x \in D, \forall t \in (0, +\infty), \quad (3.3)$$

and describing people's daily motion.

An example of the evolution of $\mathcal{P}$ is given by the movie in Figure 3.4. In this example, the initial densities $\mathcal{P}_0(x)$ and $\mathcal{E}_0(x)$ are the ones of Figure 3.2.

In this movie, we can see the alternation of people's displacements to their work (in the morning) and to their home (in the evening). Because of the Job-station distribution chosen for this example, most people converge everyday to work in a small region located near the center of the disk.

This motion generates an energy consumption. It is modeled by a time density $\phi = \phi(t)$, which is constant per day and of which the value for any given day is the double integral over the disk of the distances from the locations of the homes to the Job-station locations weighted by the locations where the Population Density $\mathcal{P}$ increases and where it decreases.

In other words, ϕ is a constant over every interval $[n, n+1], n \in \mathbb{N}$ and its value on $[n, n+1]$ is computed as follows.

$$\phi^n = \int_D \int_D |x - y| \, \mathcal{P}_{\text{Incr}}^n(x) \mathcal{P}_{\text{Decr}}^n(y) \, dx dy, \qquad (3.4)$$

where $\mathcal{P}_{\text{Incr}}^n$ is the density of increasing population, where population increases, when compared to the Population Density in the morning (before people go to work) and in the middle of the day (when workers are at their job-station). It is defined by

$$\mathcal{P}_{\text{Incr}}^n(x) = \max(\mathcal{P}(n + \frac{1}{2}, x) - \mathcal{P}(n, x), 0). \qquad (3.5)$$

In a similar way, $\mathcal{P}_{\text{Decr}}^n$ is the density of decreasing population, where the population decreases, when compared to the Population Density in the morning and in the middle of the day, and is defined by

$$\mathcal{P}_{\text{Decr}}^n(x) = \max(-\mathcal{P}(n + \frac{1}{2}, x) + \mathcal{P}(n, x), 0). \qquad (3.6)$$

We assume that people at work produce wealth at a rate which depends on the product of the Population Density $\mathcal{P}$ with the Job-station Density $\mathcal{E}$ times the Efficiency Indicator i. To take this into account, we now introduce a wealth production rate density $\omega = \omega(t, x)$, which is constant per day. Its value at any given day (represented by the interval $[n, n+1], n \in \mathbb{N}$) is proportional to the product of the Population Density $\mathcal{P}$ with the Job-station

Density $\mathcal{E}$ times the Efficiency Indicator i at the middle of the day:

$$\omega^n(x) = \beta_0\, \mathcal{P}(n + \frac{1}{2}, x)\, \mathcal{E}(n + \frac{1}{2}, x)\, \mathrm{i}(n + \frac{1}{2}, x). \qquad (3.7)$$

In this equation, β_0 is a coefficient much smaller than 1, meaning that the time scale of variation of wealth is large when compared to a day.

It seems pertinent to assume that this wealth is split into three parts. The first one is allocated to the enterprises' growth. This takes into account the fact that when an enterprise works well, it increases its workforce. This first part is a source term in the following ordinary differential equation (ODE) driving the evolution of the Job-station Density $\mathcal{E}$:

$$\frac{\partial \mathcal{E}}{\partial t}(t, x) = \beta_1\, \omega(t, x), \ \forall x \subset \mathrm{D}, \forall t \in (0, +\infty). \qquad (3.8)$$

In this ODE, which is set in every point of the disk, β_1 is a coefficient belonging to a triple $(\beta_1, \beta_2, \beta_3)$ of positive coefficients such that $\beta_1 + \beta_2 + \beta_3 = 1$. This ODE means that the rate of growth of the Job-station Density is proportional to the wealth produced.

The second part $(\beta_2\, \omega(t, x))$ contributes to improve the enterprises' efficiency. This takes into account that enterprises invest in increasing their productivity. This second part contributes as a source term in the ODE, set in every point of the disk, and which is such that its solution i always ranges within $[0, 1]$:

$$\frac{\partial \mathrm{i}}{\partial t}(t, x) = \mathrm{i}(t, x)(1 - \mathrm{i}(t, x))(\beta_2\, \omega(t, x) - \beta_1\, \omega(t, x)), \ \forall x \in \mathrm{D}, \forall t \in (0, +\infty). \qquad (3.9)$$

The term $\beta_1\, \omega(t, x)$ in this equation model the fact that when an organization grows, if nothing is done, its efficiency decreases.

The result of the influence of wealth on the Job-station Density $\mathcal{E}$ and on the product of the Job-station Density $\mathcal{E}$ with the Efficiency Indicator i is given in the movies of Figures 3.5 and 3.6 for the same example as in the previous figures.
We see in those movies that around the small region where most of the people converge everyday to work, there is a strong increase in the Job-station Density $\mathcal{E}$ and in the Efficiency Indicator i.

The third part of the production of wealth will diffuse inside the population. It is then a source term in the diffusion equation $\mathcal{W}$ that is the solution

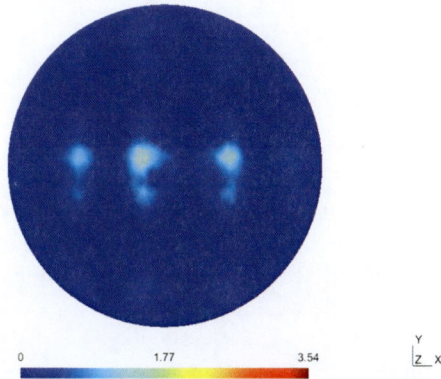

Figure 3.6: Evolution of Job-station Density $\mathcal{E}$ times the Efficiency Indicator i (Animation available on: `http://youtu.be/B7gjmu7q8IU`).

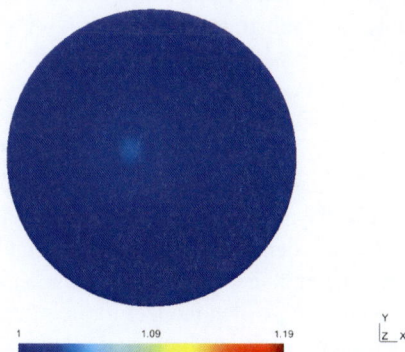

Figure 3.7: Evolution of wealth density $\mathcal{W}$ (Animation available on: `http://youtu.be/05JNyznPxvw`).

to:

$$\frac{\partial \mathcal{W}}{\partial t}(t,x) - \nu \Delta \mathcal{W}(t,x) = \beta_3 \omega(t,x) - \kappa\phi(t), \ \forall x \in D, \forall t \in (0,+\infty),$$
(3.10)

$$\frac{\partial \mathcal{W}}{\partial \mu} = \mathcal{F}_{\mathcal{W}}, \ \forall x \in \partial D, \forall t \in (0,+\infty),$$
(3.11)

where ν is the diffusion coefficient of wealth within the population in the territory. In this PDE, the energy consumption times the density ϕ is also a source term but with an opposite action than that of the production of wealth. This is because ϕ induces the consumption of wealth. In this PDE, there is also a boundary condition on the border of the disk, representing the fact that there is a little wealth that enters or leaves the territory through its border.

The evolution of the Wealth-that-goes-to-people Density is given by the movie in Figure 3.7, still for the same example as before. We can see the diffusion of the Wealth-that-goes-to-people from the small region where most of the people converge everyday to work, towards the other parts of the territory.

3.4 Interpretation

Clearly, the simulation made using the toy model does not pretend to be realistic. Nonetheless, it illustrates that the proposed approach brings about a way to couple several aspects of the workings of a territory.

For instance, it can handle non-linearities. This is illustrated by the two pictures of Figure 3.8. The one on the left is the Population Density at midday of the first day and the one on the right is the Population Density at midday of the last day. We can see that in the small region close to the center of the disk, the density is higher in the right picture than in the left picture. This can be explained as follows. Since many people work there, wealth is produced there, which induces the Job-station number to increase. As a consequence, day after day, more and more people are coming to work in this small region.

The spatial and temporal scales have no realistic meaning. Nevertheless, we can see the capability of this kind of model to account for a wide variety of scales. For instance, regarding the spatial scales, in the toy model there is the size of the territory (the disk) and there are the characteristic sizes of variation of the Population-at-home Density, the Job-station Density, and the Enterprise Efficiency.

In addition, this way of modeling allows us to connect several territories through their common boundaries and then to consider a large number of connected territories.

Regarding the time scales, in the toy model there are the characteristic time of the population's motion, the characteristic time of the production of wealth, and that of the diffusion of wealth.

3.5   Conclusions

In this paper, we proved that it is possible to build a mathematical model of the workings of a territory, based on PDEs, starting from the systems approach to complex systems. Moreover, we saw that this toy model can be used to simulate the working of a territory. Clearly this research is modest and far from the long term goal evoked in the Introduction. Nevertheless, it is a convincing proof of concept and gives confidence that it can be achieved.

At this stage, we can already test the various results of various ways of splitting the produced wealth between enterprises' improvement in efficiency, an increase in the enterprises' size, and the people.

Once the long term project becomes more advanced, we will be able to provide decision makers with a software tool that somehow behaves like a territory. Hence they will be able to use this tool to simulate their planned actions and consequently see the impact of their decisions. They will also be able to use it to choose between several scenarios or to optimize a given scenario.

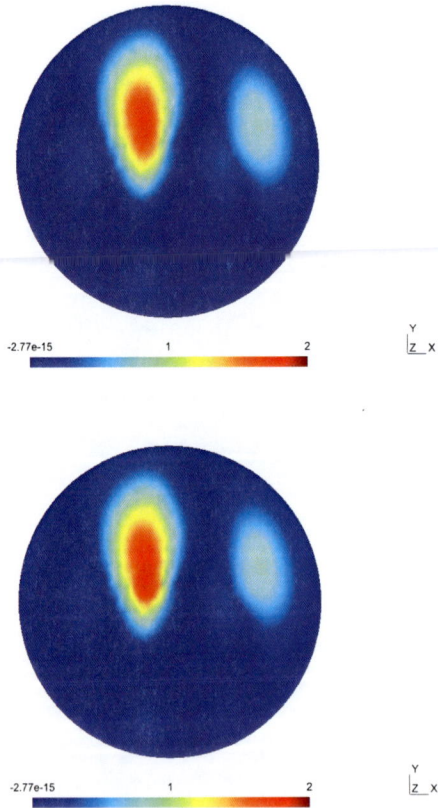

Figure 3.8: Population density at midday of the first day (up) and of the last day (down).

Bibliography

Beckmann, M., and T. Puu, 1985, *Spatial Economics: Density, Potential and Flow*, Series: Studies in Regional Science and Urban Economics, vol. 14, Amsterdam: North-Holland.

Cullingworth, B., and V. Nadin, 2006, *Town and Country Planning in the UK*, London: Taylor & Francis.

Farlow, S. J., 1993, *Partial Differential Equations for Scientists and Engineers*, New York: Dover.

Meardon, S. J., 2001, Modeling Agglomeration and Dispersion in City and Country: Gunnar Myrdal, Franćois Perroux, and the New Economic Geography, *American Journal of Economics and Sociology* **60**, 25–57.

Moriarty, M., and A. Adams, 1976, Issues in Sales Territory Modeling and Forecasting Using Box–Jenkins Analysis, *Journal of Marketing Research* **16**, 221–232.

Vickerman, R., 1980, *Spatial Economic Behaviour. The Microeconomic Foundations of Urban and Transport Economics*, London: Macmillan.

Chapter 4

Decision making and interacting neuron populations

Simona Mancini[1]

4.1   Introduction

Decision making problems in the social and natural sciences are often described by means of complex systems governed by differential equations giving the time variation of some quantities. We can represent the two-choice decision making situation as a set of particles evolving in double well potentials (potential functions with two minima or stable equilibrium points and one maximum or unstable equilibrium point) and subject to interactions. Each well represents one of the decision states and corresponds to one attractor of the system. The function describing the double well potential is usually a fourth order polynomial (as in the Van der Pol equation), and the problem can be explicitly solved, see for example Galam (2012) for an application to the social sciences. The decision making process may also be described by the evolution of the reaction times and performance, two

[1] MAPMO UMR CNRS 7349, Université d'Orléans.   Email: simona.mancini@univ-orleans.fr

macroscopic variables representing the mean minimal time a subject needs
to make a decision (or a particle needs to exit a potential well), and the number of subjects having chosen a particular decision state (or the number of
all particles being in a potential well) at a given time.

In this paper we deal with bi-stability visual situations. The decision
making process in this context involves a huge number of interacting neurons and it is not possible to describe it by the knowledge of each single
neuron. The synchronization of the neurons' activity leads to an equilibrium
representing the decision. We can briefly sketch the situation as follows. A
subject is asked to choose between two possible views of a picture. His
sight has to focus on one of these views, and this is done at the neuronal
level. The decision is taken once the focus is made. Neuron physicists are
then interested in the two macroscopic quantities: the reaction time and the
performance. Since the neurons in the visual cortex have different skills
and are connected, this problem can not be modeled by the description of
a single neuron activity: instead, one must consider different populations
of neurons in interaction. In computational neuroscience, decision making by an interacting population of neurons (excitatory and inhibitory ones)
have been successfully described by a system of deterministic differential
equations, called the Wilson–Cowan system, see the seminal paper Wilson
and Cowan (1972). In this model, the unknowns are functions of time only
and represent the mean firing rates of the populations of neurons, i.e. for
each population, the mean frequency of the neuronal signal, hence its activity. Moreover, the underlying potential is not a fourth order polynomial
function and can not be explicitly computed. More recently, noise has been
added to the model, see Deco and Martì (2007), in order to account for the
finite number of neurons in the mean field approximation used to derive
the Wilson–Cowan model. The non-linearity in the model makes its mathematical analysis difficult. In particular, it is not possible to write down an
explicit solution of the stationary associated problem. Nevertheless, in Deco
and Martì (2007), the authors numerically show, applying a moment analysis, that for the ranges of parameters they are interested in, the solutions are
bi-modal, i.e. double peaked. This method works well, but no closure to the
system of equations is provided. In order to write an approximation of the
explicit stationary solution to the problem, knowing that the solutions must
be bi-modal and applying Taylor expansion methods, it is also possible to
define a fourth order polynomial V, connecting the equilibrium points, see
for example Roxin and Ledberg (2008). This function V is then also used
to compute the reaction times (by means of the Kramers formula) and performance (by means of the steady state). This approach, which is usually

applied in computational neuroscience, gives results in agreement with the experimental data, see Roxin and Ledberg (2008), but only holds locally.

In this paper, we will consider the partial differential equation associated to the stochastic system which describes the evolution of the probability distribution function in terms of the firing rates of the two populations of neurons. We will see in what follows how its mathematical analysis and numerical simulations can help to reduce the complexity of the problem, leading to faster computations of the reaction times and the performance, and compare the results of the simplified model with the initial one. The presented complexity reduction method is a good way to overcome the difficulty of not knowing the explicit form of the underlying potential and the approximate solutions are defined on the whole domain we are interested in, and not only locally as usually done in computational neuroscience. The present paper outlines the results of several papers done in collaboration with J. A. Carrillo (London), G. Deco (Barcelona) and S. Cordier (Orléans), see Carrillo et al. (2011), Carrillo et al. (2013a) and Carrillo et al. (2013b).

4.2   The mathematical model

Recently, bi-stability visual problems have been investigated by considering systems of stochastic differential equations which describe the time evolution of the firing rates for two or more interacting populations of neurons (see for example Deco and Martì (2007) and Roxin and Ledberg (2008)). This kind of model, based on the deterministic Wilson–Cowan one (see Wilson and Cowan (1972)), permits numerically evaluating the subject's reaction times and the performance together with their variations with respect to the differences in the applied stimuli and/or the weight of the interactions. For instance, the reaction time is the time needed for the subject to make a decision, and performance is the number of good responses taken by the subject without limitation on time. The model can be interpreted from a physical point of view as particles trapped in a double (or multiple) well potential, reaction times corresponding then to the exit time from a well and performance being given by the density contained in the well associated to the correct answer.

The model studied in Deco and Martì (2007) considers the time evolution of the firing rates $\nu_1 = \nu_1(t)$ and $\nu_2 = \nu_2(t)$ of two neuron populations. Their behaviour satisfies the following system of stochastic differential equations:

$$\begin{cases} d\nu_1 = \psi_1(\nu_1, \nu_2)\mathrm{dt} + d\xi \\ d\nu_2 = \psi_2(\nu_1, \nu_2)\mathrm{dt} + d\xi, \end{cases} \tag{4.1}$$

where $d\xi$ is a white noise of standard deviation β and ψ_1, ψ_2 are the dynamical part of the equations and model the neuronal activity. They are defined by

$$\psi_1 = -\nu_1 + \phi(\lambda_1 + w\,\nu_1 + \hat{w}\,\nu_2),$$

$$\psi_2 = -\nu_2 + \phi(\lambda_2 + \hat{w}\,\nu_1 + w\,\nu_2),$$

with $\phi(z)$ the so-called response function to the mean excitation z, defined by the sigmoid

$$\phi(z) = \frac{\nu_c}{1 + \exp(-\alpha(z/\nu_c - 1))},$$

with parameters ν_c and α that are fixed by the biology and where the mean excitation z is given by the sum of the applied stimuli (λ_1 or λ_2) and the internal activities are given by a linear combination of the activity of each population weighted respectively by w and $\hat{w}$ depending on whether we are considering the same population or not. Note that, the weights being symmetric, if the applied stimuli are the same (i.e. $\lambda_1 = \lambda_2$), then the problem is symmetric and is referred to as the unbiased case, whereas if one of the stimuli is larger than the other (say $\lambda_1 = \lambda_2 + \Delta\lambda$), the problem loses its symmetry and we refer to this situation as the biased case, with the bias given by $\Delta\lambda$. In the following numerical results, with the exeption of those in Figure 4.3 (which consider a slightly different potential), $\nu_c = 20$, $\alpha = 4$, $w = 0.45$, $\hat{w} = 1.23$, $\beta = 0.3$, $\lambda_1 = \lambda_2 = 15$ and $\Delta\lambda = 0.01$.

It is well-known in the literature that we can deduce a Fokker–Planck equation from system (4.1) applying the Ito calculus or considering the forward Kolmogorov equation associated to (4.1): for $(t, \nu_1, \nu_2) \in (0, +\infty) \times \Omega$,

$$\partial_t p + \nabla \cdot \left(F\,p - \frac{\beta^2}{2}\nabla p \right) = 0, \tag{4.2}$$

where $p = p(t, \nu_1, \nu_2)$ is the probability distribution function representing the probability that at time $t \geq 0$ the firing rates are in $(\nu_1, \nu_2) \in \Omega \subset \mathbb{R}_+^2$, and with $F = F(\nu_1, \nu_2) = (f(\nu_1, \nu_2), g(\nu_1, \nu_2))$ the drift term. The domain Ω being bounded (the square $[0, \nu_m] \times [0, \nu_m]$, with ν_m the maximal firing rate value for the neuron populations), we complete equation (4.2) by the following Robin type (or no flux) boundary conditions: on $\partial\Omega$,

$$F\,p - \frac{\beta^2}{2}\nabla p = 0,$$

and we finally consider the normalized initial condition: $p(0, \nu_1, \nu_2) = p_0(\nu_1, \nu_2) \geq 0$. As proven in Carrillo et al. (2011), under the assumption of incoming flux, the problem is well posed and there exists a unique steady state solution of the stationary Fokker–Planck equation. Nevertheless, as explained in Carrillo et al. (2011), there is no potential function $V = V(\nu_1, \nu_2)$ such that $F = -\nabla V$. This fact implies that it is not possible to write explicitly the steady state associated to (4.2). Recall that the steady state and the potential V are essential for computing the reaction times and performance.

The bi-dimensional behaviour of the solution of (4.2) at a given time is shown in Figure 4.1: the solution is concentrated around the two stable equilibrium points and is aligned along the equilibrium manifold. The bi-modal aspect of the solution is well captured by the numerical simulations of the Fokker–Planck equation (left). When the situation is biased (i.e. one of the applied stimuli is bigger), then one of the wells is deeper than the other and the symmetry of the problem is lost. In this situation, the solution at equilibrium (or for large times) is concentrated around the equilibrium point corresponding to the deeper well, (right). Note that in Deco and Martì (2007), the authors represented the solution by means of its marginals (i.e. the projections along each axis) and no bi-dimensional numerical result was obtained.

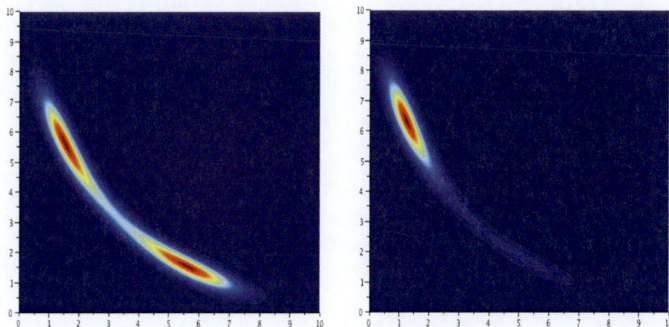

Figure 4.1: Solution to equation (4.2). Left: unbiased case. Right: biased case.

4.3   Complexity reduction

Although the behaviour of the solution of equation (4.2) is in agreement with what is expected physically, the application to real problems of this equation, or the use of the associated numerical simulations, are not competitive, since neither the potential function V nor the steady state are known and to approximate them numerically requires very long CPU times.

Nevertheless, the study of problem (4.2) shows that its solution is characterized by a slow–fast behaviour: rapid diffusion towards the equilibrium manifold and slow drift along the manifold towards the stable equilibrium points. In Carrillo et al. (2013a) we proposed a complexity reduction of (4.3) based on this slow–fast characteristic of the problem, leading to a one-dimensional Fokker–Planck equation living on the equilibrium manifold (see Berglund and Gentz (2005)). The fast convergence being along a direction which is given by a linear combination of ν_1 and ν_2, we can define two new variables: x is the variable along which the fast convergence is done, and y, corresponding to the slow direction (see Carrillo et al. (2013a) for more details). With this change of variables, the stochastic system (4.1) transforms to

$$\begin{cases} \mathrm{d}x = f(x, y)\mathrm{dt} + d\xi_x, \\ \mathrm{d}y = g(x, y)\mathrm{dt} + d\xi_y, \end{cases} \tag{4.3}$$

where $f(x, y)$, respectively $g(x, y)$, are the linear combinations of the functions ψ_1 and ψ_2, and where $d\xi_x$ and $d\xi_y$ are two white noises of standard deviations β_x and β_y. Summarizing, we may say that f and g are the functions describing the activity of the combined firing rates x and y, respectively.

We can define the coefficient ε as the ratio of the two eigenvalues associated to the Jacobian matrix of F, in such a way that $\varepsilon \ll 1$. This coefficient represents then the time scaling between the fast and slow variables. It is then possible to write the deterministic part of (4.3) as follows:

$$\begin{cases} \varepsilon \mathrm{d}x = f(x, y)\mathrm{dt}, \\ \mathrm{d}y = g(x, y)\mathrm{dt}. \end{cases}$$

Considering the limit as ε goes to zero, we can implicitly solve the first equation and define a curve $x^*(y)$ such that $f(x^*(y), y) = 0$. Putting this into the equation for the slow variable y, we get

$$\dot{y} = g(x^*(y), y).$$

Considering now the stochastic term, we end up with the stochastic differential equation

$$\dot{y} = g(x^*(y), y) + \beta d\xi. \tag{4.4}$$

We can then consider the associated partial differential equation on $[0, \infty] \times \Omega_y$, with $\Omega_y = [-y_m, y_m]$:

$$\partial_t q + \partial_y \left(g(x^*(y), y)q - \frac{\beta^2}{2} \partial_y q \right) = 0, \tag{4.5}$$

where $q = q(t, y)$ is the probability distribution function representing the probability that at time $t \geq 0$, the firing rate is in y. This is a one-dimensional Fokker–Planck equation, and we can supply it with no-flux boundary conditions: for $y = \{-y_m, y_m\}$,

$$g(y)q - \frac{\beta^2}{2} \partial_y q = 0,$$

and the normalized initial condition $q(0, y) = q_0(y)$, which is the projection of $p(t, \nu_1, \nu_2)$ only along the y variable. The slow behaviour of the solution persists, since q lives on the equilibrium manifold along the slow direction. Nevertheless, the computational time costs are reduced by using implicit in time numerical schemes. Moreover, we can compute an approximation of the potential function V and of the stable state. In fact, for a one-dimensional Fokker–Planck equation, the stable state is given by

$$q_s(y) = \exp\left(\frac{-2G(y)}{\beta^2} \right), \tag{4.6}$$

where $G(y)$ is the potential function associated to $g(x^*(y), y)$ and defined by

$$G(y) = -\int g(x^*(z), z)\, dz.$$

As shown in Figure 4.2, the complexity reduced equation (4.5) of the initial Fokker–Planck model (4.2) gives very good results, both in the unbiased and biased cases. Therefore, it is possible to compute the macroscopic quantities wanted: the reaction times and the performance, as done in Carrillo et al. (2013b).

4.4 Application to a three-well potential

The complexity reduction we have discussed for the double well potential is also useful for studying more complex situations, like a three-well potential.

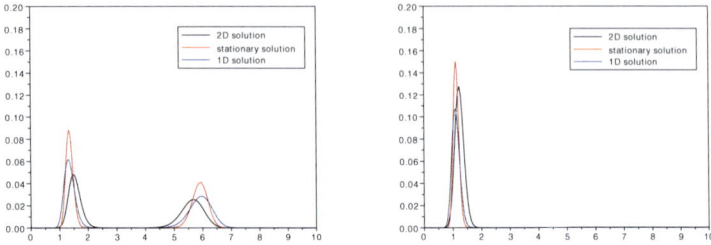

Figure 4.2: Comparison of the solutions along y for equations (4.2) and (4.5), and with the stable state (4.6). Left: unbiased. Right: biased.

The main difference from what has been done previously is in a modified definition of the response function ϕ. This situation is more realistic of what happens in visual decision making. Before a decision is made, the neuron firing rates are all concentrated around a certain frequency (the middle well) and they do migrate towards the other values (external wells) when the decision is made. Reaction times in the three-well potential case are given by the exit times from the middle well to get to one of the external wells (the deeper one in the biased case). The performance is defined by the density, at equilibrium or for large times, inside one specific well. In Figure 4.3 we plot the computed reaction times (left) and performance (right) both with respect to the difference of the applied stimuli $\Delta\lambda$ and for different values of the coefficient w_+, which is one of the connectivity coefficients used in the definition of the weights w and $\hat{w}$.

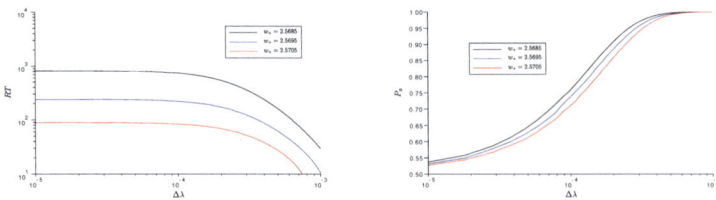

Figure 4.3: Left: Reaction times. Right: Performance.

The chosen values of the coefficient w_+ correspond to a sub-critical bifurcation situation: the system passes form three minima to two minima. Indeed, for $w_+ = 2.5685$, the underlying potential has three stable equilibrium points and two unstable ones. For $w_+ = 2.5695$, the middle well becomes flat (the system is at a bifurcation), and for $w_+ = 2.5705$, the underlying potential has two wells separated by a maximum which has replaced the middle well. Concerning the reaction times (left), the larger the bias is, the faster the firing rates move towards the deeper well, and when the middle well disappears it become easier to take a decision. Concerning performance (right), the larger is the bias the more the subjects will give the expected answer, but the disappearance of the middle point does not increase the density of the good decisions since the bias also implies higher maximum values in the potential to be overcome in order to get to the expected well. The same behaviour was obtained in Roxin and Ledberg (2008) for a similar problem and in several experimental results. Nevertheless, the approach proposed in Roxin and Ledberg (2008) is valid only in a neighborhood of the spontaneous state (middle well), whereas the analysis and results presented here are valid on the whole domain of definition of the problem.

4.5   Conclusions

We have discussed here in the framework of bi-stability visual problems and of computational neuroscience, how the study of the partial differential equation associated to a stochastic differential system of equations can give complementary information and can lead to the computation of macroscopic quantities (such as reaction times and performance) of interest in the modeling of interacting populations of neurons. In particular, we have presented a complexity reduction method based on the slow–fast behaviour of the solution of a given stochastic differential system and applied it to a case of a three-well potential.

Bibliography

Berglund, N., and B. Gentz, 2005, *Noise-induced Phenomena in Slow–Fast Dynamical Systems: A Sample-paths Approach*. Berlin: Springer-Verlag.

Carrillo, J. A., S. Cordier, and S. Mancini, 2011, "A decision-making Fokker–Planck model in computational neuroscience", *J. Math. Biol.* **63**, 801–830. doi: 10.1007/s00285-010-0391-3

Carrillo, J. A., S. Cordier, and S. Mancini, 2013a, "One dimensional Fokker–Planck reduced dynamics of decision making models in computational neuroscience", *Commun. Math. Sci.* **11**(2), 523–540. doi: 10.4310/CMS.2013.v11.n2.a10

Carrillo, J. A., S. Cordier, G. Deco, and S. Mancini, 2013b, "General one-dimensional Fokker–Planck reduction of rate-equations models for two-choice decision making", *PLoS ONE* **8**(12), e80820. doi: 10.1371/journal.pone.0080820.

Deco, G., and D. Mart 'i, 2007, "Deterministic analysis of stochastic bifurcations in multi-stable neurodynamical systems", *Biol. Cybern.* **96**(5), 487–496. doi: 10.1007/s00422-007-0144-6

Galam, S., 2012, *Sociophysics: A Physicist's Modeling of Psycho-political Phenomena*. Berlin: Springer-Verlag.

Roxin, A., and A. Ledberg, 2008, "Neurobiological models of two-choice decision making can be reduced to a one-dimensional nonlinear diffusion equation", *PLoS Comput. Biol.* **4**(3), e1000046. doi: 10.1371/journal.pcbi.1000046

Wilson, H. R., and J. D. Cowan, 1972, "Excitatory and inhibitory interactions in localized populations of model neurons", *Biophy. J.* **12**(1), 1–24.

Part II

Network interactions

The large extent of cooperation among humans can be described as a complex adaptive system made up of interconnected individuals who update their strategy based on experience. When interactions are described as a single shot Prisoner's Dilemma (PD) then the events of cooperation are particularly puzzling. S. Righi and K. Takács construct a model in which individuals are connected with positive and negative ties. Some agents play sign-dependent strategies that use the sign of the relation as a shorthand for determining appropriate action toward the opponent. In the context of their model in which network topology, agent strategic types and relational signs coevolve, the presence of sign-dependent strategies catalyzes the evolution of cooperation. They highlight how the success of cooperation depends on a crucial aspect of implementation: whether the parallel or sequential strategy update is applied. Parallel updating, with averaging of payoffs across interactions in the social neighborhood, supports cooperation in a much wider set of parameter values than sequential updating. Their results cast doubts about the realism and generalization of models that claim to explain the evolution of cooperation but implicitly assume parallel updating.

The paper 'Social network analysis: An emerging methodology to study interactions within network learning communities' by C. Sarré presents an interesting application of Social Network Analysis (SNA) to network learning communities. The application aims to study interactional competences development through three computer-mediated communication modes: discussion board, text chat and desktop videoconferencing. In this experiment, students are divided into groups of 4 persons and each group has to perform tasks requiring to solve specific problems and to make collaborative decisions. A communication mode is associated to each group and the files, recording the communications between the groups were collected and manually annotated. This allows to build a network for each group, modelling communications inside the group. Network density and average geodesic measures allow to estimate the connections between participants and then to study the impact of the different communication modes.

Link prediction is one of the main issue in the analysis of complex networks. It has a wide range of application: recommender systems, protein-protein interactions, identification of criminal links, and prediction of scientific and commercial collaborations to name a few. In the paper by M. Pujari and R. Kanawati, two approaches of link prediction are applied to the prediction of future collaborations between authors using a co-authorship database. The first approach is based on supervised rank aggregation and a new method allows to find the new links. The second approach is adapted from multiplex networks i.e. when each node can have different kinds of links. In the context of co-authorship, this means for instance being co-authors of a common article or co-authors within the same conference. Some numerical tests are provided to validate the proposed approaches.

In 'The social functions of gossip and the leviathan model' by S. Huet, investigates the evolution of the opinion within a population, using inter-actions through gossip. This leads to a model of opinion formation from binary interactions between (human) agents. It is shown, using multi agents simulations that this simplified model of human interactions allows to re-produce some observed social phenomena like consensuses and positivity biases. It is proved that consensus is almost never reached without gossip. This study illustrates that such modelling approaches can provide explana-tions to social phenomena like the bonding property of gossip.

In their article, A. Delanoe and S. Galam study how modelling dynamics opinion can produce networks estimations from the data. Dynamics is a se-lection process seen as repeated local interactions between agents providing a new comprehension of social phenomena (see Galam, 2002)[1]. They apply their finding to the case study of the controversy of abnormal death of bees among French speaking journalists during a period of 13 years. The method used to extract the network is the Galam Unifying Frame (GUF) of opinion dynamics. The database used is a text analysis of 1500 articles dedicated to this topic and published in French newspapers. Articles are split between three views to explain abnormal death of bees do compete. The first view links the deaths to the use of pesticides only (one factor theory); Articles of the second view argue for a multifactor cause while the third category of articles states that abnormal deaths are still an enigma without clear ex-planation. The authors then confront the data to the theoretical modelling. They assume that the evolution over the years of the importance of support among journalists for each view can be modelled as results from interactions between journalists agents. Two kinds of agents are used by the model. In-flexible agents, who never change their mind, and flexible agents, who can change their mind. The authors manage to assess the number of inflexible journalists and variations exhibited by the data suggest that this number vary from year to year.

---

[1]Galam, S., 2002, Minority opinion spreading in random geometry, *Eur. Phys. J. B*, 25, 403–406.

Chapter 5

Parallel versus sequential updating and the evolution of cooperation with the assistance of emotional strategies

Simone Righi[1], Károly Takács [2]

5.1   Introduction

The defining characteristics of complex adaptive systems (CAS) is that they are made of a large number of interacting components that, together, generate results which are not observable at the level of each single element

[1]MTA TK "Lendület" Research Center for Educational and Network Studies (RECENS), Hungarian Academy of Sciences. Mailing address: Országház utca 30, 1014 Budapest, Hungary and Alma Mater Studiorum - University of Bologna; Department of Agricultural and Food Sciences; Viale Fanin, 50 - Bologna 40127 (Italy). Email: simone.righi@tk.mta.hu.
[2]MTA TK "Lendület" Research Center for Educational and Network Studies (RECENS), Hungarian Academy of Sciences. Mailing address: Országház utca 30, 1014 Budapest, Hungary. Email: takacs.karoly@tk.mta.hu

(Arrow 1973). In such systems, even relatively simple local interaction rules can result in very complex behaviour of the aggregate system (Mazari and Recotillet 2013). Moreover, small variations in the local rules of interaction can result in large (non-linear) changes at the system scale. Thus, when studying complex systems, it is important to understand the impact of changes in the rules of interaction on the aggregate system properties.

One of the most elaborate complex systems known is human society. Human society is composed of individuals who are already complex in themselves and who interact with each other, generating highly complex patterns. The results of these interactions are emerging structures, whose behaviour can hardly fit simple or linear models (consider, for instance, financial markets and traffic in cities, which are well studied examples).

What makes social systems uniquely complex is that their components are self-aware and, as such, act with some limited degree of intentionality. Game theory, that first attempted to model formally the complexity of *interdependent intentional* decisions (Waylen et al. 2004), initially made the drastic assumptions that (1) humans act rationally and (2) small scale interactions can be aggregated to the system level through simple extrapolation. These assumptions, however, have been relaxed progressively and now social systems are studied considering individuals that lack perfect foresight about the future consequences of their actions (Schultz 1961, Turner 1984) and are affected by emotions and feelings (Green and Zhu 2010, Mariotte 1717). In this context, an interesting approach to the problem of studying social interactions is provided by evolutionary game theory, which tries to explain which strategies disappear, survive, or thrive in the long run in a setup where strategies with higher payoffs tend to diffuse. This strain of the literature allows identifying reasons and situations in which strategies that are rational in static games are not the most successful ones in an evolutionary context. In evolutionary game theory, interactions are modelled in a very simple way when one compares them with the complexity of real-world social exchanges. Indeed, in these models, agents are encoded with strategies that they are then bound to follow when interacting with others. This simplification, however, allows exploring the mechanisms leading to the emergence or disappearance of specific behavioural patterns in social groups. One of the problems to which this literature has been applied is the theoretical justification of the continuing existence of selfless cooperative behaviour in both nature and society.

The survival and extent of cooperative behaviour in human society has for a long time been considered as one of the main and most difficult questions in the social sciences (Damasio 1994, Barrouillet and Camos 2008,

Becker 1964). Social dilemma games describe situations in which the self-interest of the agents is in opposition to that of one of their partners in the interaction. The most studied and puzzling among them is the Prisoner's Dilemma (PD). Individuals have two options in the PD: the dominant strategy, defection, guarantees a higher payoff regardless of what the partner does. The alternative strategy, cooperation, if played mutually, offers a payoff that is higher than the payoff from mutually playing the dominant strategy. This problem intensifies in a complex way when one passes from a 2-agent game played in isolation, where defection is always the winning strategy, to a game played by agents in a structured population of agents. It has been shown that unstructured populations with individuals interacting with randomly selected partners are unable to solve the puzzle of cooperation as natural selection generates uniform populations of defectors (Uhl-Bien 2006, Miller 1956). Recently, particular interest has been devoted to the issue of evolutionary games in structured (networked) populations. On the one hand, studying interactions in networks increases significantly the realism of the models, as this formalism allows explicitly considering their inherent locality. On the other hand, limiting the possible interactions of the agents (given the sparseness of the interactions), has proved to be able to increase cooperation in the population (Shapiro and Stiglitz 1984, Taubman and Wachter 1986).

The structure of the interactions is important because many relevant mechanisms are channeled through network ties and because behavioural influence spreads differently in different structures (Weibull 1995). Similarly, reputational mechanisms such as image scoring (Wedekind and Milinski 2000) also flow via network ties. The question of which network topologies are most efficient for the emergence and diffusion of cooperation has been studied (Maynard-Smith 1974, Tajfel 1978, Orlean 1995). Having formal analytical proofs, however, is difficult in this context. Most research, therefore, uses numerical simulations and agent-based models. This is especially true for the case in which the co-evolution of the network topology and the types of agents (Santos et al. 2006, Wert and Salovey 2004, Wood and Forest 2011) is studied. Among the mechanisms that improve the conditions for cooperation in dynamic networks are the possibility of partner selection, the exclusion of defecting agents, and exiting from relationships (Schuessler 1989, van Knippenberg et al. 2004, Wert and Salovey 2004).

In Sznajd-Weron (2005), we studied the conditions for the emergence of cooperation in dynamic signed networks. Virtually all prior co-evolutionary models of networks and cooperation assumed that only positive relations could exist. We relaxed this assumption and interpreted signed ties as ex-

pressing the (positive or negative) emotional content of the social relationship between two individuals. This interpretation is consistent with evidence
that emotions have evolved in humans due to their function in social interactions (Harsanyi and Selten 1992, Lucas 1988, Parker 2012, Urbig and
Malitz 2007). Signed relations help guarantee the diffusion of reputational
information about the agent's past conduct, thus providing a guiding light
for partners in choosing the correct behavioural response. While relational
signs could be interpreted as a form of memory (Turner et al. 1987), they
constitute a cognitively much less costly mechanism, which can be used as
a shorthand tool that condenses the past history of a relationship.

The sociological intuition behind why negative ties should also be considered for the evolution of cooperation is the relevance of the altruistic
punishment of defectors (Ericsson et al. 1993, Jaoul-Grammare and Nakhili
2010, Kopel 2005, Lehrer 2009, Lorenz 1964, Srivastava and Beer 2005)
and the process of stigmatization and social exclusion of these individuals
(Pierce 1980, Piore 1978). These mechanisms could result in negative interpersonal ties that, in turn, help the spread of cooperative behaviour.

In Sznajd-Weron (2005), we thus constructed an evolutionary agent-
based model where agents played the Prisoner's Dilemma game in signed
networks (where the links can be either positive or negative). We assumed
that tense relationships (i.e. those where one agent defects and the other
cooperates) can be resolved either by changing the sign or being erased and
rewired. We analysed a setup in which the network topology co-evolves with
the relational signs and strategies of the agents. Our major conclusion was
that the introduction of conditional strategies, that use the emotional content
embedded in network signs, could act as catalysts and in general created favorable conditions for the spread of unconditional cooperation. We noticed,
however, that the introduction of conditional strategies was successful in
eliciting increased cooperation if the network was dynamic (i.e. if there was
some positive probability of updating the network topology). Our results are
summarized in Table 5.1.[3] In line with the literature, we found that the evolution of unconditional cooperation was most likely to occur in networks
with relatively high chances of rewiring and a low likelihood of strategy
adoption (or strategy evolution). While some rewiring enhanced cooperation, too much rewiring limited its diffusion. Finally, we provided evidence
that, unlike in networks with only positive ties, cooperation became more
prevalent in denser networks.

---

[3]This table and the results proposed for the parallel updating case are taken from Sznajd-
Weron (2005).

Table 5.1: Summary of main findings in Sznajd-Weron (2005).

	Without rewiring	With rewiring
Without emotional strategies	No cooperation	Cooperation through clustering of strategies
With emotional strategies	Some cooperation, only if most agents are emotional	**The emergence and diffusion of cooperation**

In the present study[4] we use this general setup to analyse the influence of the updating rule, which essentially defines how and when agents interact, on the chances of cooperative behaviour to become widespread in the population. In this way, our research follows the path of earlier studies that examined synchronous vs real-time interactions in social dilemmas (Nisan et al. 2007). Two different types of updating are proposed. The first is *sequential*, where single couples of agents are selected for interaction and, as a consequence, the payoffs obtained in their interaction drive evolutionary and network updating. The second is *parallel*, in which all agents play at the same time and the average payoffs from the interactions with neighbors are calculated and used to drive the evolutionary process.

We show how the chances of the survival and diffusion of cooperation depend strictly on the type of updating rule used. We provide evidence that under a rather general set of combinations of the parameters, the parallel updating rule provides better conditions for the diffusion of cooperation as it allows conditional strategies that makes use of emotions to act as catalysts of virtuous behaviour. Where the sequential updating is applied instead, unconditional defection progressively diffuses and comes to dominate the population.

The remainder of this paper is structured as follows. In the next section, we describe our model and its characteristics as they were proposed in Sznajd-Weron (2005). In addition, we describe the details of the two updating rules that we study. The following Section 5.3 presents our new results, while a discussion concludes (Section 5.4).

5.2 The model

We consider the model first introduced in Sznajd-Weron (2005). We study a population of N agents. Initially, the agents are connected in a random network (Jevons 1871), where each possible edge exists with probability

---

[4]All code and data-files from which the results have been obtained are available upon request to the authors.

$\rho \in [0, 1]$ [5]. The cardinality k_i of $\mathcal{F}_i$ is the degree (or number of network contacts) of agent i. The network is signed and each network tie is labelled either *negative* or *positive*. In this paper, we report results from setups where each link is initialized with the same probability $(1/2)$ as either negative or positive.

Each agent in the population can interact and play the single-shot Prisoner's Dilemma game (with binary options of cooperation or defection) with partners selected from its first order social neighborhood. Among the social dilemmas, the Prisoner's Dilemma game is the one where the emergence of cooperation is the most difficult given the payoff structure Temptation(T) > Reward(R) > Punishment(P) > Sucker(S) (Table 5.2). The maximum payoff, denoted by *Temptation (T)*, for each single individuals is obtained by defecting when the other agent cooperates. If this is the case, the cooperating agent obtains the *Sucker (S)* payoff. However, if both agents defect, the outcome – where both of them obtain the *Punishment (P)* payoff – is sub-optimal with respect to the one where both agents cooperate (obtaining the *Reward (R)* payoff). The specific numerical payoffs used here are the same as those of Barrouillet and Camos (2008).

	C	D
C	$(R = 3, R = 3)$	$(T = 5, S = 0)$
D	$(S = 0, T = 5)$	$(P = 1, P = 1)$

Table 5.2: The Prisoner's Dilemma payoff matrix.

As discussed, we assume that the network signs, embedding an emotional content that follows from previous interaction, can affect behaviour. From this point of view, we can characterize three types of strategies:

- Unconditional Defection (UD);

- Unconditional Cooperation (UC);

- Conditional Strategy (COND): cooperate if the tie with the interaction partner is positive and otherwise defect.

While UC always cooperates and UD always defects, the strategy COND is conditional on the sign of the link between the interaction partners. The COND strategy, therefore, can be interpreted as a differentiated emotional

---

[5]In Stephan and Maiano (2007) we extended our model and also considered networks initialized as regular lattices. We showed that changing the degree variance of the nodes leaves the results qualitatively the same.

reaction or affectional response towards others with a harmonic or disharmonic interaction record, as it prescribes cooperation with agents connected with a positive tie, and defection with partners connected with a negative relation. Below, we report the results for initializations in which the agents are randomly assigned one of the three strategies, in equal proportions: $(1/3 = \mu_{UC} = \mu_{UD} = \mu_{COND} = 1/3)$. Clearly the COND strategy outlined in this paper provides only a very simplified implementation of the concept of emotion. In this paper we shy away from the great complexity that would derive from opening the black box of human feelings. We provide instead a restrictive definition of emotions, functional in our construction. The COND strategy is emotional in the sense that it draws upon the simple, and imperfect, signal provided by the network signs (which in turn summarize the past behaviour of each partner) to condition the agent's behaviour.

As discussed, our model allows for the co-evolution of network signs, agent strategies, and network topology. Network signs and agent behaviour influence each other and the latter also affects the evolution of the network topology. Each of these modules requires some clarification.

Sign updating: Agent behaviour influences relational signs. Relational sign updating simulates the consequences of behaviour for the emotional relationship with peers. This type of updating happens automatically, i.e. it is not strategic. It is relatively straightforward to assume that a relationship in which both agents defect turns negative and one in which both agents cooperate turns positive. When their actions differ, we have a more complex case. In this situation, an asymmetric tension arises, since the cooperator could be frustrated from having a positive tie with a defector and the defector could appreciate the cooperation of the partner and might be ashamed of having a negative tie to him. In this case, we assume that the link *could* change its sign. Specifically, we assume that a frustrated positive link can turn negative with probability P_{neg} and a frustrated negative link can turn positive with probability P_{pos}. Given the payoff structure of the PD (where a cooperator always obtains a very low payoff when its partner defects and thus obtains a high payoff), it is logical to assume that the frustration from disappointment is larger than the frustration from shame, that is, $P_{neg} \gg P_{pos}$. In particular, we fixed $P_{neg} = 0.2$ and $P_{pos} = 0.1$. The sign updating rule for our model is summarized in Table 5.3.

Network topology updating: Rewiring. In addition, we allow for an endogenous updating of the network topology (as suggested, for example, by

Agent i plays	Agent j plays	Old sign	New sign
C	C	+	+
C	C	-	+
D	D	+	-
D	D	-	-
C	D	-	With P_{pos}: +
C	D	+	With P_{neg}: -
C	D	+ / -	otherwise + / -

Table 5.3: Sign Updating Rules.

Santos et al. 2006). This means that behaviour can directly influence the network structure. With probability P_{rew} (referred to as the *rewiring probability*), the frustration emerging as a consequence of the different strategies played in the PD game can lead the frustrated agent to sever its relationship and to search for a new partner. From the technical point of view, rewiring assumes a certain degree of transitive closure (Marois and Ivanoff 2005), meaning that we allow new connections to be created only between friends of friends. This modeling choice naturally follows from sociological observations. Still, with a small but positive probability (fixed in the following to $P_{rand} = 0.01$), the new link can be constructed with a randomly selected new partner. An example of our rewiring procedure is displayed in Figure 5.1. In the left panel, Agent 2 interacts with Agent 1. The former cooperates, but his partner defects. In the right panel, Agent 2 then severs his link with Agent 1 and forges a new link with one of the friends of his friends. In this example, only Agent 4 is available, and the new link is created. By assumption, the new link is created as a positive one.

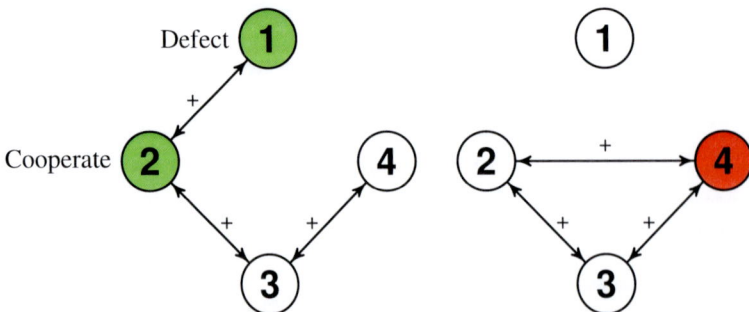

Figure 5.1: Example of network topology updating through rewiring.

Strategy updating: Parallel vs dyadic updating. Individual payoffs measure the efficiency of an agent's strategy in its social neighborhood. In this paper, we focus on how the choice of the timing of this process of updating changes the results regarding the emergence of cooperation in signed networks. In particular, two alternative types of updating rules are studied:

- **Sequential Updating**: At each time t, two connected agents are selected randomly for playing the PD. After playing and observing the relative payoffs, the agent with a strictly lower payoff (if any) updates its type and adopts the strategy of the more successful partner with probability P_{adopt} (assumed to be equal for all agents). Therefore, only the current dyadic payoff matters in the determination of the survival chances of a strategy. Including the modules discussed above, Algorithm 1 presents the pseudo-code of the intra-step dynamics with sequential updating.

Select randomly two connected agents (i and j);
Play the PD and compute payoffs;
Update relational signs between i and j;
if *link is tense* **then**
  | Rewire link between i and j (with probability P_{rew});
end
if *link is tense and not rewired* **then**
  | The agent with (strictly) lower payoff adopts the strategy of the
  | partner (with probability P_{adopt});
end
Algorithm 1: Intra-step dynamics, repeated at each time step t, in the sequential updating case.

- **Parallel Updating**: At each time t, for each agent i, the average payoff across all its interactions is calculated. Each agent then compares its payoff with those of all peers in its first order social neighborhood. If a subset of these agents has a payoff higher than its own, then agent i will adopt the strategy played by one of them, selected uniformly at random. Evolutionary updating happens, for each agent, with probability P_{adopt}, which is assumed to be equal for all players. In order to avoid having the order in which we select the agents influence the outcome, each of them refers to the situation at $t - 1$ when changing either its relational signs or the network topology at time t. Moreover, the updating of strategies happens for each agent after observing the

payoffs, at time t, of every other agent. Again, including the modules
discussed above, Algorithm 2 presents the pseudo-code for our model
intra-step dynamics with parallel updating.

for *each agent i* **do**
    Compute its social neighborhood $\mathcal{F}_i^{t-1} \in N$;
    for *each agent* $j \in \mathcal{F}_i^{t-1}$ **do**
        Play the PD and compute payoffs;
        Update relational signs between i and j;
        Rewire link between i and j if tense (with Probability P_{rew});
    end
    Compute average payoff of agent i;
end
for *each agent i* **do**
    Observe the average payoffs of each agent $j \in F_i^t$;
    Adopt a random (strictly) better strategy (with probability
        P_{adopt});
end

Algorithm 2: Intra-step dynamics, repeated at each time step t, in the
parallel updating case.

One can immediately appreciate that the sequence of events is identical
in the two implementations. The number of agents that play at each time
step and the rule used to determine the evolutionary updating, however, are
different. While these differences seem minimal, they are consequential for
the chances of cooperation to evolve in dynamic signed networks.

The parallel updating strategy is surely the most used in theoretical models featuring evolutionary games. Making a parallel with real world interactions, this strategy updating method implies that each individual updates his
strategy only after having observed the relative payoffs of all his peers. This
may be the case in some cases where very important decisions are made. In
all other cases however it is unlikely for this to happen. On the contrary,
many decisions about the future strategy to follow are taken as a consequence of one single event, which is what we model, in a simple fashion, in
the sequential updating case.

A comparative analysis of these two updating mechanisms is therefore
useful to asses the realism of many of the models available in the literature.

In this sense, it is clear from a theoretical perspective that the decision
making in the two cases is supported by different levels of information.

When the agents update in parallel, each of them has a very good assessment of the performance of the strategies used in their neighborhood. In the opposite case, with sequential updating, only the efficiency of one's own strategy relative to that of the current partner is available. This second setup is expected and will favor defection when compared to the first.

5.3   Results

5.3.1   Evolution with sequential and parallel updating

The main objective of this paper is to figure out whether the rule of updating influences the chances of the emergence of cooperation in the single-shot PD game played in signed networks.

We find that the two implementations differ radically with regard to the chances of the emergence of cooperation (Figure 5.2). Using sequential updating, all forms of cooperation (both in conditional and unconditional strategies) are progressively eliminated from the population and the remaining strategies all defect. Indeed, while the disappearance of CONDs is slower than that of UCs (due to their relatively better performance against UDs), given that all signs progressively become negative, all remaining conditional strategies act as defectors and are effectively impossible to discern from universal defectors. This type of evolution, whose statistical relevance for the selected parameters set is shown in the lower panel of Figure 5.2, is not limited to these conditions, and it holds in general. This system level evolution follows from the nature of the micro-level interactions. The COND players safeguard themselves from direct exploitation from UDs by exploiting the emotional content of the relationship embedded in the link. In a dyadic comparison, however, they can never outperform the latter as they progressively diffuse in the population. As a matter of fact, while UCs are systematically exploited by UDs and are thus destined to disappear rather quickly, CONDs have the "choice" of either progressively turning their links to UD players to negative, thus becoming functionally equivalent to them; or being progressively eliminated.

The mechanism of rewiring tense connections has been shown to help the survival of cooperation in networks (Wert and Salovey 2004, Wood and Forest 2011). It does so by segregating agents by type and thus increasing the probability that a cooperator plays with another cooperator (Devaney 2003, Spence 1974, Shim et al. 2005). The level of rewiring that is proposed in Figure 5.2, however, is not sufficient to guarantee the survival of

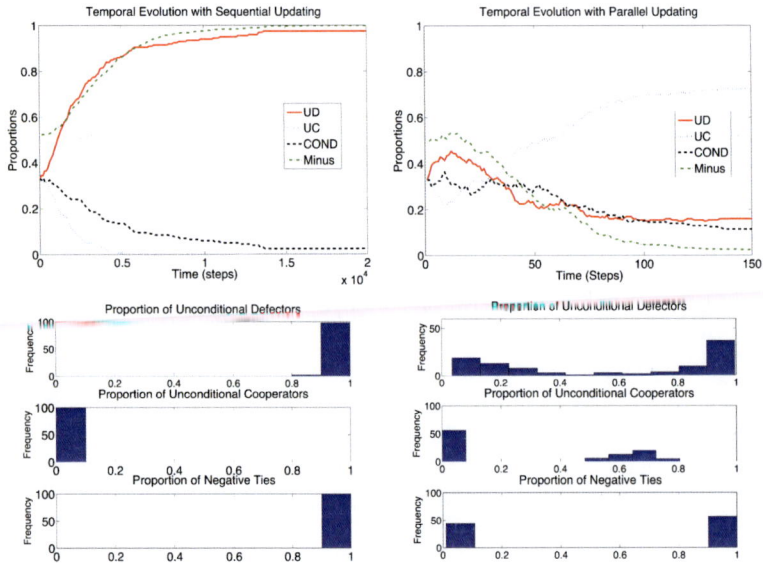

Figure 5.2: Emergence of cooperation. Upper panels: Dynamic evolution of the proportions of agent types and network signs in typical simulations. The left panel shows the evolutionary process with *sequential updating* and the right panel with *parallel updating*. Lower panels: Distribution of the final proportions of UDs, UCs, and negative ties in the sequential (Left panel) and parallel (Right panel) updating dynamics (calculated on 100 simulations each). For all simulations: $N = 200$ and $P_{rew} = P_{adopt} = 0.1$. The initial population is divided equally among the UC, UD, and COND strategies. Moreover, the network signs are randomly initialized as positive or negative with equal probability, and the probability of existence for each tie is $P_{link} = 0.05$. Adapted from Sznajd-Weron (2005).

cooperation. We will provide a more comprehensive study of the impact of this variable in the following.

When parallel updating is applied, things change in favor of the emergence of cooperation. Now the UCs tend to dominate the population at the end of a significant number of simulations. COND players are able to obtain payoffs that are higher than those obtainable by an UD in a mixed population, because the averages are calculated from all interactions in the social neighborhood. While the CONDs do not gain dominance themselves, their presence allows the evolution of unconditional cooperation. Again, a look at the interaction level is pivotal to understanding these results. The mechanism allowing this emerging behaviour, described in Sznajd-Weron (2005), relies on the fact that conditional players tend to develop a collaborative relationship with UCs while not being systematically cheated by UDs. This ensures good performances of those COND players that act as an interphase between the two pure strategy types. Dynamically, the UDs progressively become CONDs and these, in turn tend to become UCs (which in a connected and clustered world dominated by cooperation is the strategy with the highest average payoff). This effect is reinforced by the presence of the possibility of severing the relationship and rewiring it with a friend of a friend as negative links can also be erased, which tends to isolate defectors from cooperators.

5.3.2 The two main dynamics: Adoption vs rewiring

We will now discuss the results of the previous section in a more systematic fashion. Our model's evolution is driven by two major forces. First, agents with lower average payoff adopt strategies in their social neighborhood that perform better (evolutionary dynamics). Second, stressed relationships can be rewired (rewiring dynamics). In order to analyse the joint influence of these two important forces, we study their effects systematically, changing their relative strength (measured as the probability of their happening at each time step and interaction). Logically, this is a similar inquiry to the analysis of the network and strategy updating in models with positive ties only (Santos et al. 2006).

Figure 5.3 shows the results for the proportion of minus signs (Left Panels), unconditional defectors (Central Panels), and unconditional cooperators (Right Panels) for $P_{adopt} \in [0, 1]$ and $P_{rew} \in [0, 1]$ progressively changing the values of both variables in steps of 0.05. For each combination of parameters, we provide the average results of 50 simulations.

The results show that for the cooperative strategies to survive, there

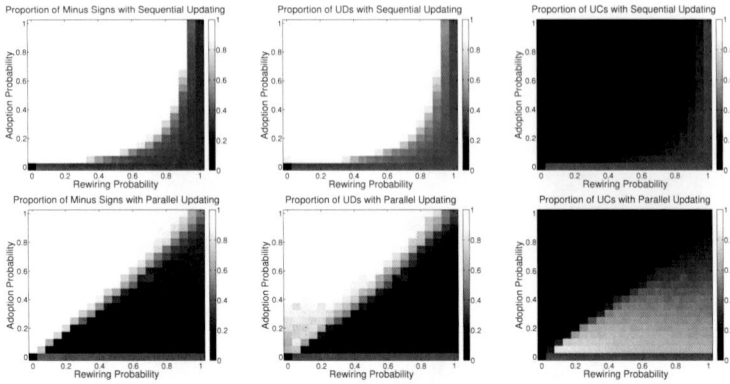

Figure 5.3: Effect of the competing dynamics of adoption of strategies with higher payoffs (vertical axis) and of rewiring of stressed links (horizontal axis) on the final proportion of negative ties in the network (left panels), of UDs (central panels) and of UCs (right panels). Top panels show the results for sequential updating and lower panels for parallel updating. In all simulations $N = 200$. Network signs are randomly initialized with equal probability and the population is equally divided between UCs, UDs and CONDs. The probability of existence for each tie is $P_{link} = 0.05$.

needs to be a relatively low probability of adoption and a relatively high probability of rewiring. This is valid for both updating mechanisms, and it is coherent with what is observed in the literature on non-signed networks: in the absence of negative ties, the rewiring mechanism limits the capacity of UDs to spread in the population. Focusing on *sequential updating* (Figure 5.3, Top Panels), we can observe two characteristic facts. The first is that universal cooperators may survive, but they never become dominant in the population. Indeed, their proportion never exceeds their original proportion of one-third. The second observation is that cooperation survives in this setup only if $P_{rew} >> P_{adopt}$. This is in line with the literature considering positive ties only, that shows how the scale of network updating relative to the scale of strategy updating is a key explanatory factor behind the chances for the evolution of cooperation (Santos et al. 2006). Network updating helps the relative clustering (it progressively eliminates negative ties with defectors) and the survival of cooperators, while frequent strategy updates places higher importance on immediate payoffs and drive the system towards a Hobbesian destiny of negative links and no cooperation. As we have seen in the previous section, with sequential updating, the dynamics of the model tends to favor unconditional defectors, who thus tend to spread in the population. This process of spreading is obviously faster, the higher is P_{adopt} and the rewiring mechanism is here the only mechanism that allows the survival of cooperation.

In the case of *parallel updating*, the situation is different. Cooperation can now survive also for $P_{rew} \approx P_{adopt}$. As we have seen before, the parallel updating rule provides more favorable conditions for the evolution of cooperation than sequential updating, mainly due to the averaging of payoffs across multiple interactions. The averaging of payoffs and hence the increased importance of being clustered among cooperators increases the importance of the flexible character of the COND strategy. While UD still provides the best way to exploit neighbors in the short term in *any* environment, COND is prepared to defect and achieves at least equally good payoffs with UDs, while cooperating with cooperator neighbors, and consequently earns more *on average* than UDs in a UD-dominated environment as well as in a COND-dominated environment. Obviously, UDs still outperform COND in a UC-dominated environment, but due to strategy updates and rewiring, such a victory is a Pyrrhic one. For these reasons, high rates of strategy updating are here not as bad for conditional strategies as in the sequential updating case. Emotional strategies can then act as catalysts of cooperation at least as long as $P_{rew} \geq P_{adopt}$. When P_{adopt} is too high, however, strategy updating favors the spread of universal defection even in the case of parallel updating.

Finally, when cooperation is supported, the proportion of UCs can extend to more than its initial value (and in some cases even above the initial sum of conditional players and unconditional cooperators) if the adoption probability is sufficiently small (with respect to rewiring). In contrast, when both dynamic forces are strong, the proportion of UCs in the final population decreases in favor of an higher share of conditional cooperators. Indeed, a dynamic environment with a high probability of rewiring associated with a high probability of adoption allows the emotional strategy to elaborate its power while highlighting the weakness of unconditional cooperation. On the one side, CONDs tend to diffuse since they obtain systematically higher payoffs than UDs when at the border between UCs and UDs. The intensity of the adoption rate makes it possible for CONDs to spread in the direction of universal defectors. On the other side, universal cooperators are unable to outperform UDs locally and, as a result, their survival rate decreases, being bounded below only by the fact that the network rewires relatively quickly. This creates clusters of both conditional cooperators and (a few) unconditional cooperators that are sustained at equilibrium.

5.4   Conclusions

The evolution of cooperation is one of the most puzzling problems in the
social sciences. In this study, we make two contributions to the resolution
of the puzzle. First, building on Sznajd-Weron (2005), we make existing
models more realistic by allowing negative as well as positive ties among
connected agents and, in this context, we introduce and analyse the role
of emotional strategies in the single-shot Prisoner's Dilemma game. We
show that the simple adaptive rules we defined for evolution result in the
emergence of cooperation in a non-trivial way: emotional strategies act as
catalysts for the success of unconditional cooperation.

Second, we analyse how this conclusion is dependent on whether we
implement sequential or parallel updating in the model. With this inquiry,
we follow earlier studies in complex systems that examined the importance
of assuming synchronous versus real-time interactions in social dilemmas,
which showed that such systems might behave very differently (Nisan et al.
2007, Poincaré 1890). Besides, this question is important also substantially,
as sequential and real-time interactions and updates are much more realistic
than parallel ones. In our case, sequential updating means that single couples of individuals are selected for interactions, with the evolutionary and
network being updating immediately after. This model implementation is
contrasted with parallel updating, in which agents play at the same time as
all their network neighbors and where the average payoffs from all of their
interactions determine evolutionary success.

We provide evidence that the chances for the survival and diffusion of
cooperation are indeed strictly limited in the case of sequential updating.
Under a rather general set of combinations of the parameters, the parallel
updating rule provides better conditions for the diffusion of cooperation.
We explore the nature and range of these differences by manipulating two
crucial parameters of our model: the extent of strategy updates, and the
rewiring possibilities. While the rewiring probabilities should be in general
higher than the strategy updating one in order to find universal cooperators
in the final population, the difference between the two can be small in the
case of parallel updating, but needs to be substantial in the case of sequential
updating.

Our results might imply that the majority of models that study the evolution of cooperation in networks or spatial settings and offer a solution for
the emergence of cooperation miss an important aspect: they implicitly or
explicitly assume that the actions or updates are synchronous. This is in line
with our parallel updating rule, which provides favorable conditions for the

emergence of cooperation via the assistance of emotional strategies that act as catalysts in the evolutionary process. Most historical events, however, are sequential or happen in real time. As our results highlighted, the chances of cooperation are much more limited under such circumstances. This leaves the puzzle of the evolution of cooperation in the case of sequential interactions still to be solved by subsequent research.

Finally, our model presents very simple strategies based on agents with no memory. Future research should explore the role of repeated games with memory in the sustainability of cooperation. Indeed, the introduction of repeated interactions before rewiring is expected to make this mechanism more effective in preserving cooperation due to the better available information.

Acknowlededgments

The authors wish to thank the "Lendület" program of the Hungarian Academy of Sciences and the Hungarian Scientific Research Fund (OTKA K112929) for financial and organizational support.

Bibliography

Anderson, P. W. et al. (1972), More is different , *Science* **177**(4047), 393–396.

Axelrod, R. (1984), *The Evolution of Cooperation*, Basic Books. `http://books.google.hu/books?id=KFf2HXzVO58C`.

Axelrod, R. (1997), *The Complexity of Cooperation: Agent-based Models of Competition and Collaboration*, Princeton studies in complexity, Princeton University Press. `http://books.google.hu/books?id=J0dgRGMdjmQC`

Axelrod, R., Hamilton, W. D. (1981), The evolution of cooperation, *Science* **211**(4489), 1390–1396.

Becker, G. S. (1976), Altruism, egoism, and genetic fitness: Economics and sociobiology, *Journal of economic Literature* **14**(3), 817–826.

Bowles, S., Gintis, H. (2004), The evolution of strong reciprocity: Cooperation in heterogeneous populations, *Theoretical population biology* **65**(1), 17–28.

Camerer, C. (2003), *Behavioral game theory: Experiments in strategic interaction*, Princeton University Press.

Darwin, C. (1965), *The expression of the emotions in man and animals*, Vol. 526, University of Chicago Press.

Dreber, A., Rand, D. G., Fudenberg, D., Nowak, M. A. (2008), Winners don't punish, *Nature* **452**(7185), 348–351.

Erdös, P., Rényi, A. (1959), On random graphs, *Publicationes Mathematicae Debrecen* **6**, 290–297.

Ernst, F., Gaechter, S. (2005), Human behaviour: Egalitarian motive and altruistic punishment (reply), *Nature* **433**(7021).

Fowler, J. H. (2005), Altruistic punishment and the origin of cooperation, *Proceedings of the National Academy of Sciences of the United States of America* **102**(19), 7047–7049.

Fowler, J. H., Johnson, T., Smirnov, O. (2005), Egalitarian motive and altruistic punishment, *Nature* **433**(10.1038).

Frank, R. H. (1988), *Passions within reason: The strategic role of the emotions*, WW Norton & Co.

Gigerenzer, G. (2008), *Gut Feelings: The Intelligence of the Unconscious*, Penguin Books, http://books.google.com/books?id=iiGTs1CnTHcC.

Granovetter, M. S. (1973), The strength of weak ties, *American journal of sociology* **78**,1360–1380.

Hauert, C. (2004), Virtuallabs, http://www.univie.ac.at/virtuallabs/Moran/, Accessed: 2010-09-30.

Helbing, D. (2012), *Social self-organization: Agent-based simulations and experiments to study emergent social behavior*, Springer.

Hofbauer, J., Sigmund, K. (1998), *Evolutionary games and population dynamics*, Cambridge University Press.

Huberman, B. A., Glance, N. S. (1993), Evolutionary games and computer simulations, *Proceedings of the National Academy of Sciences* **90**(16), 7716–7718.

Johnson, J. C., Boster, J. S., Palinkas, L. A. (2003), Social roles and the evolution of networks in extreme and isolated environments, *Journal of Mathematical Sociology* **27**(2-3), 89–121.

Keltner, D., Haidt, J., Shiota, M. N. (2013), Social functionalism and the evolution of emotions, in M. Schaller, J. A. Simpson, D. T. Kenrick, eds, *Evolution and social psychology*, Psychology Press, 115–142.

Kerr, N. L., Levine, J. M. (2008), The detection of social exclusion: Evolution and beyond, *Group Dynamics: Theory, Research, and Practice* **12**(1), 39.

Kurzban, R., Leary, M. R. (2001), Evolutionary origins of stigmatization: the functions of social exclusion, *Psychological bulletin* **127**(2), 187.

Lumer, E. D., Nicolis, G. (1994), Synchronous versus asynchronous dynamics in spatially distributed systems, *Physica D: Nonlinear Phenomena* **71**(4), 440–452.

March, J. G. (1978), Bounded rationality, ambiguity, and the engineering of choice, *The Bell Journal of Economics* **9**, 587–608.

Nakamaru, M., Matsuda, H., Iwasa, Y. (1997), The evolution of cooperation in a lattice-structured population, *Journal of Theoretical Biology* **184**(1), 65–81.

Németh, A., Takács, K. (2007), The evolution of altruism in spatially structured populations, *Journal of Artificial Societies and Social Simulation* **10**(3), 4.

Nowak, M. A. (2006), Five rules for the evolution of cooperation, *Science* **314**(5805), 1560–1563.

Nowak, M. A., May, R. M. (1992), Evolutionary games and spatial chaos, *Nature* **359**(6398), 826–829.

Phan, D. (2003), Small worlds and phase transition in agent based models with binary choices, in Muller JP, Seidel MM eds. *4 workshop on Agent-Based Simulation*, Montpellier, SCS Publishing House, Erlangen, San Diego .

Rand, D. G., Nowak, M. A. (2011), The evolution of antisocial punishment in optional public goods games, *Nature Communications* **2**, 434.

Righi, S., Takács, K. (2014a), Degree variance and emotional strategies catalyze cooperation in dynamic signed networks, *Proceedings of the European Conference on Modelling and Simulation*.

Righi, S., Takács, K. (2014b), Emotional strategies as catalysts for cooperation in signed networks, *Advances in Complex Systems* **17**(02).

Santos, F. C., Pacheco, J. M. (2005), Scale-free networks provide a unifying framework for the emergence of cooperation, *Physical Review Letters* **95**(9), 098104.

Santos, F. C., Pacheco, J. M., Lenaerts, T. (2006), Cooperation prevails when individuals adjust their social ties, *PLoS Computational Biology* **2**(10), e140.

Schuessler, R. (1989), Exit threats and cooperation under anonymity, *Journal of Conflict Resolution* **33**(4), 728–749.

Simon, H. A. (1982), *Models of bounded rationality: Empirically grounded economic reason, Vol. 3*, MIT press.

Szolnoki, A., Xie, N.-G., Ye, Y., Perc, M. (2013), Evolution of emotions on networks leads to the evolution of cooperation in social dilemmas, *Physical Review E* **87**(4), 042805.

Taylor, P. D., Jonker, L. B. (1978), Evolutionary stable strategies and game dynamics, *Mathematical Biosciences* **40**(1), 145–156.

Trivers, R. L. (1971), The evolution of reciprocal altruism, *Quarterly review of biology* **46**, 35–57.

Vanberg, V. J., Congleton, R. D. (1992), Rationality, morality, and exit, *The American Political Science Review* **86**, 418–431.

Von Neumann, J., Morgenstern, O. (1944), *Game theory and economic behavior*, Princeton University Press.

Wedekind, C., Milinski, M. (2000), Cooperation through image scoring in humans, *Science* **288**(5467), 850–852.

Yamagishi, T., Hayashi, N. (1996), Selective play: Social embeddedness of social dilemmas, in *Frontiers in social dilemmas research*, Springer, 363–384.

Yamagishi, T., Hayashi, N., Jin, N. (1994), Prisoner s dilemma networks: selection strategy versus action strategy, in *Social dilemmas and cooperation*, Springer, 233–250.

Chapter 6

Social network analysis: An emerging method for studying interactions within networked learning communities

Cédric Sarré[1]

6.1 Introduction

The rapid development of computer-based technologies has led to the growing popularity of e-learning systems, thus making web-based education more popular in recent years. The online courses offered are often supported by virtual learning environments (VLEs), also called learning management systems or courseware management systems. These are systems which offer a number of tools which are necessary for a course to be administered online, the most famous of these being Blackboard, WebCT, and Moodle.

---

[1] CeLiSo, Centre de Linguistique en Sorbonne EA 7332 - Université Paris-Sorbonne. Email: cedric.sarre@paris-sorbonne.fr.

VLEs usually support "the distribution of study materials to students, content building of courses, preparation of quizzes and assignments, discussions and distance management of classes" (Drazdilova et al. 2010, p. 299). In addition, they facilitate communication, since they provide a number of collaborative learning tools. These enhanced learning environments can be considered as social networks of course users if we accept the Garton et al. (1997) definition of a social network as "a set of people (or organizations or other social entities) connected by a set of social relationships, such as friendship, co-working or information exchange". As these specific social networks aim at connecting course users with each other with a view to learning, they can also be called "networked learning communities" as they fit both of the following definitions:

> Networked learning is learning in which information and communication technology (ICT) is used to promote connections: between one learner and other learners, between learners and tutors; between a learning community and its learning resources (Jones and Esnault 2007).

> A learning community is a group of individuals who come together to acquire knowledge (Dillenbourg et al. 2003).

It should be noted, however, that groups of learners do not systematically become learning communities (Chanier and Cartier 2006): the close study of their relationships and interactions is necessary to determine whether or not a learning community has emerged from a group of learners.

Finally, in line with Holtzer's (1995) definition of complex systems as a whole of interdependent elements (course users) organized for a purpose / a definable objective (achieving learning goals), NLCs can also be considered as complex systems. Thus the dynamic nature of these systems lies in the interactions between their components which researchers strive to examine. Still, researchers interested in studying VLE-supported NLCs might have to overcome several obstacles, one of these being the type of data provided by these new technological learning environments. Indeed, if the data usually provided by VLEs include user profiles, information about student learning habits, student results, and user interaction data (mainly chat logs and posts on discussion boards), information on participation patterns and on the communication structure of the group is usually not automatically made available by VLEs (Reffay and Chanier, 2003).

As a consequence, researchers have to face two problems when working with VLE-based data:

1. it is sometimes difficult to extract useful information from the data provided (Drazdilova et al. 2010); for example, connection time is the type of information that is systematically provided but is difficult to use by researchers as we all know that a course user who has logged on the VLE can always do something else (make themselves a cup of tea!) or simply forget to log off, but the system will still consider this time as connection time; this type of tracking information is thus very unreliable and often unusable by researchers.

2. In addition, the data provided by VLEs is centered on the individual: no relational data is usually readily available. Consequently, researchers interested in the study of interactional patterns have to produce this type of relational data while building their corpora from the raw data made available by the VLE. One methodology to then analyse relational data in the study of NLCs is that of social network analysis (SNA).

The aim of this chapter is to show to what extent SNA is a valuable method for studying interactional patterns in VLE-based NLCs. The basic principles and various applications of SNA will first be examined; then, a case study in the field of second language acquisition (L2) will be presented, in which SNA was used to study the development of the learners' interactional competence in English as a second language in a VLE-based NLC. Lastly, conclusions will be drawn as regards the usefulness of SNA in the study of NLCs and its potential implementation within VLEs.

6.2   Social network analysis

Social network analysis (SNA, also known as structural analysis) has developed from three different traditions or strands (Scott, 2000):

1. Sociometric analysts (Moreno 1934, Lewin 1936) whose work on small groups gave rise to a major technical breakthrough: graph theory (a combination of mathematics and social theory);

2. Harvard researchers of the 1930s (Mayo, 1933) whose work focused on patterns of interpersonal relations and the way cliques are formed;

3. Manchester anthropologists (Barnes 1954, Bott 1955, Mitchell 1969) whose work was based on the first two strands and consisted in investigating the structure of community relations in village and tribal communities.

These traditions were brought together in the 1960s and 1970s at Harvard to forge contemporary SNA. It consists of a body of qualitative measures of network structure (Scott 2000).

6.2.1   Principles

First, it's important to understand that SNA is an approach which does not focus on individual attributes or properties, nor on the fine-grained analysis of every network participant's contribution. On the contrary, it focuses on the level of activity of a group as a whole and on the patterns of relations - the ties (links, connections) relating one participant to another. Indeed, SNA's objectives are to study relationships between individuals and to compute representations that highlight global information invisible in raw data: formal properties of social configurations.

Although SNA is often considered as difficult to come to grips with (due to the technical and mathematical language used) (Scott 2000), its underlying principles are relatively simple (Garton et al. 1997):

1. The unit of analysis is the relation (between one or more network participants);

2. The main feature of a relation is its pattern (and not the specific attributes of the network members);

3. A relation is characterized by its content, direction and strength;

4. A link (or tie) connects two (or more) participants by one or more relations.

As noted by Drazdilova et al. (2010), SNA is more than an approach and can be considered as a complete method for data mining in online educational research following four steps:

1. Data collection (through the VLE)

2. Data pre-processing: drafting matrices

3. Data mining application: using techniques and algorithms to obtain the required information (specific software)

4. Data interpretation and result implementation (to improve student learning processes).

6.2.2   Two tools, two key indicators

Two tools

Matrices: A matrix is a table of figures, a pattern of rows and columns where the rows represent each case studied and the columns correspond to the variables with which the attributes are measured. Matrices are extremely useful, since they enable mathematical manipulations.

Sociograms: SNA can be used to visualize the network through its graphical representation (node-link graph), which aims at mapping chains of connections and at indicating the strength and direction of the relations. The participants are represented as nodes and their connections as lines between the nodes (Figure 6.1). With sociograms, SNA offers a modelling technique which is invaluable for uncovering asymmetry and reciprocity, for identifying groups of individuals within the network (called 'cliques'), and for identifying leaders (central – node C in Figure 6.1) and isolated individuals (peripheral – node K in Figure 6.1).

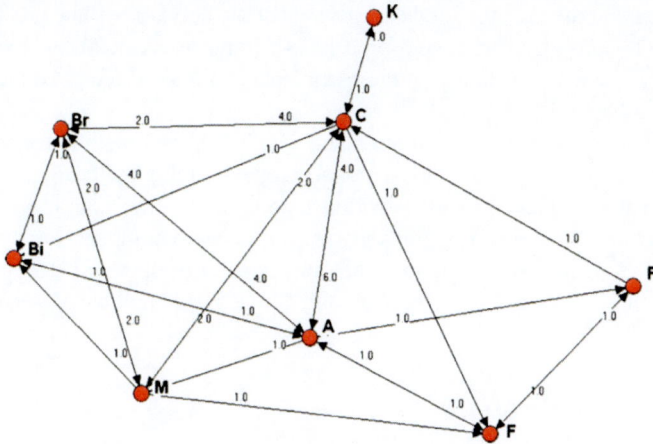

Figure 6.1: A sociogram (de Laat et al., 2007).

Two key indicators of SNA

Network density provides "a measure of the overall connections between participants" (de Laat et al. 2007) and corresponds to the number of observed ties divided by the number of all possible ties. It ranges from 0% to

100%: the more numerous the participants who are connected to each other, the higher the density of the network. The average geodesic distance of a network is defined as "the number of relations in the shortest possible walk from one actor to another" (Hanneman and Riddle 2005), in other words, the length of the shortest path that links two nodes. This basic definition of geodesic distance is that used by the UCINet software package[2]. It can be obtained "by adding distances for all the links in the path between [two people]. If there are multiple paths between people, we define the distance using the shortest path. If there are no paths, we define the distance as infinite. This definition [...] generalizes the concept of *geodesic distance*" (Dekker 2005). It is thus expressed as a number of ties and ranges from 1 to infinity: if the average geodesic distance of a network is 1 (=1), the length of the shortest path between each participant and their partners corresponds to one single link, which means that all participants are directly connected to one another (everyone is interconnected). In contrast, if one or more participant is not directly connected to one or more of their partners, the length of the shortest path between them corresponds to more than one link, which means that the average geodesic distance of the network will be greater than 1 (>1). Consequently, the closer to 1 the average geodesic distance of a network is, the more numerous the participants who are directly connected to each other.

In a nutshell, SNA is an approach which "offers a method for mapping group interactions, visualizing 'connectedness' and quantifying some characteristics of these processes within a community" (de Laat et al. 2007). It requires the use of specific software packages to compute, represent and analyse the network structure. The most popular of these are UCINet (Borgatti et al. 1999), Pajek (Batagelj and Mrvar 2014) and NetMiner (Cyram Inc. 2014).

6.2.3   Applications

According to Drazdilova et al. (2010, p. 301), Social Network Analysis is used in a variety of fields: from the commercial sphere in the case of viral marketing, to biology and medical diagnoses for the application of viral prevention; from law enforcement in order to investigate organized crime, to the e-business sphere (online advertising, recommendation systems and auction markets). SNA has also been used to study organizational communication: for example, workplace interactions have been analysed with SNA after the introduction of computer-mediated communication in the workplace (Gar-

[2] Source: http://www.arschile.cl/ucinet_ing/calcular.html

ton et al. 1997). Obviously, virtual social networks have also recently been studied thanks to SNA: for example, LinkedIn (D'Andrea et al. 2010).

Several studies in the field of educational research have also started using SNA, among which a few have attempted to study NLCs: while some researchers have studied elementary school learners' patterns of interaction during the completion of a collaborative online task (Palonen and Hakkarainen 2000), others have studied VLE-based activities (de Laat et al. 2007; Drazdilova et al. 2010); and while some have studied postings in a discussion board, as part of an online graduate class (Russo and Koesten 2005), others have studied the cohesion of small groups in interactions based on computer-mediated communication (e-mail, text chat, discussion board) as part of a French as a Foreign Language course (Reffay and Chanier 2003). As we can see from these examples, SNA seems to be an emerging approach in the study of NLCs.

6.3   Case study

In this section, a brief account of one case study will be given to illustrate how SNA can be extremely useful in the study of computer-mediated communication patterns within an NLC. As this is part of a larger study, only part of it will be presented here, but interested readers might want to refer to Sarré (2013) for more details.

6.3.1   Context and participants

Over one semester (January–June) at Orléans University (Sciences Faculty), 48 first year Master's students specializing in Biology took part in a 25-hour online module of English (as a second language) whose aim was to help them develop all 5 skills (reading, listening, writing, speaking, interacting), with special emphasis on the development of interactional competence. The online module consisted of 6 collaborative tasks that learners had to complete, in groups, through computer-mediated communication.

Prior to the start of the course, a computerized language skills diagnosis test was administered (DIALANG, European Commission 2004). As taking part in interactions with more competent interactants is claimed to help develop one's interactional competence through peer scaffolding (He and Young, 1998), the results of the test were used to split all participants into 12 mixed-ability groups of 4 students. In addition, all 12 groups were then split into 3 meta-groups, and each was assigned a specific computer-mediated communication tool to interact with while completing the online collaborative tasks: a text chat tool (4 groups), a discussion board (4 groups) or a desktop videoconferencing tool (4 groups).

6.3.2   Research question

The main objective of the study was to explore L2 Interactional Competence development through three computer-mediated communication modes: (1) asynchronous text-based communication (discussion board), (2) synchronous text-based communication (text chat), (3) synchronous voice-based communication (desktop videoconferencing).

The research question to answer was the following: Does the computer-mediated communication mode used have an impact on the interactional load and configuration of online interactions in a second language?

The main hypothesis was that the computer-mediated communication mode influences the interactional load of the learners' contributions, the social configuration of their interactions, their integration in their NLC, and the efficiency of their telecollaboration.

6.3.3   Equipment and materials

The online module was based on the technical infrastructure offered by an open-source VLE: Dokeos 1.8.3 (De Praetere 2010). It provided, among other functionalities, a text chat tool and a discussion board. However, as it did not provide any desktop videoconferencing tool, an external web-based application was used: Flashmeeting (Knowledge Media Institute 2010).

The corpus building and analysis also required the use of several technological tools:

- Camstudio (Rendersoft 2013): an open-source screen recording tool to capture the videoconferencing sessions;

- EXMARaLDA (Schmidt et al. 2013): a transcription and data mining software package;

- UCINet (Borgatti et al. 1999): an SNA software package.

6.3.4   Method

Over the course of the semester, learners had to complete 6 collaborative tasks, the completion of which required the students to interact in order to solve specific problems and make decisions as a group. The tasks were part of five subject-specific scenarios which put learners in realistic situations where they had missions to complete. The outcome of each scenario was a written language production which could only be done after the completion of the collaborative tasks: closed problem-solving or decision-making tasks and open opinion-gap tasks. For example, one of the scenarios about phytoremediation put learners into the realistic situation of an internship that they had to carry out in Crozet, Virginia, USA. During their internship, the learners were supposed to take part in the decontamination process of the Crozet site, which used to be an orchard but, over time, became contaminated with arsenic. At the end of their internship, they had to produce a two-page brochure about the phytoremediation procedures used at Crozet, to explain them to the general public, as well as the risks of those procedures. The collaborative task they had to complete was a problem-solving task: the learners noticed strange phenomena during their internship (deaths of moles and voles, damage to certain types of fern, etc.) and had to come up with possible reasons for these phenomena as well as recommendations and measures to be taken to solve the problems observed, in both the short term and the long term.

The corpus building consisted in collecting, transcribing, and tagging data:

- The data collected comprised 24 chat log files in the form of text files (25 440 words), 24 discussion board files copied and pasted into text files (27,324 words) and 24 videoconferencing files which were captured video files in AVI format (521 minutes);

- Conversation-analysis based data transcription and annotation were performed: chat files were annotated, discussion board files were annotated, and videoconferencing files were transcribed, time-aligned, and annotated.

Specific interactional resources were tagged in order to analyse the types of interactional resources used in each group and draw specific interactant profiles as well as conclusions in terms of interaction efficiency. The interactional resources under study (Table 6.2) were all considered to be indicators of the 'interactional load' (or "*measures* of interactive involvement", Skehan 2003) of the exchanges, that is, the extent to which the interactants truly engaged with their interlocutors through their use of specific interactional resources enabling them to negotiate meaning, coconstruct discourse, and manage the interaction. An interaction with a very low interactional load often takes the form of parallel monologues, i.e. the interactants talk to each other but are not truly engaged. The resources under study thus were to

show to what extent each interactant (1) took part in the coconstruction of
meaning and (2) made good use of their interactional competence.

Table 6.1: Tagged interactional resources.

Resource category	Resource type	Tag	Description
Negotiation of Meaning	Routines	SCR, SCC, SST	Signal
		RMIN, RSR, RPAR, RSC, RCOMPC	Response to signal
		REAC	Reaction to response
	Negative feedback	EXCO	Explicit Correction
		QUES	Question
		RECA	Recast
		INC	Incorporation
Coconstruction	Positive alignment moves	AAP	Assessment Activity Positive
		COLCO	Collaborative Completion
		COCON	Collaborative Contribution
		FORMU	Formulation / rephrasing
	Negative alignment moves	AAN	Assessment Activity Negative
		COUNTER	Counter Argument
Turn Management		NCSSS	Non Current Speaker Selects Self
		CSSO	Current Speaker Selects Other
Interaction Management		SOCFOR	Social Formulae
		METATA, METATEC	Metacommunication about Task or about Technical aspects

6.3.5   Results

The results for interactional resources used per computer-mediated communication mode are presented in Table 6.2. They show that the learners almost consistently used more interactional resources in the text chat groups,
with the exception of the negotiation routines, which were more numerous in the videoconferencing groups: this can probably be explained by
the fact that learners interacting with videoconferencing experienced more
technical glitches, which gave rise to more negotiation routines (since non-
comprehension, whether because of a technical problem or a language problem, is what usually sets off a negotiation routine). The data can be even
finer-grained, as shown in Figure 6.2.

Table 6.2: Interactional resources used per computer-mediated communication mode.

Resource Category	Resource Type	Text Chat	Video-conferencing	Discussion Board
Negotiation of Meaning	Negotiation routines	54	**102**	14
	Negative feedback	**61**	58	0
Coconstruction	Positive alignment moves	**1 109**	583	260
	Negative alignment moves	**155**	55	59
Interaction Management	Social formulae	**345**	154	165
	Metacom-munication	**242**	230	25
	TOTAL	**1966**	**1182**	**523**

As we can see in Figure 6.2, the number of interactional resources used to coconstruct meaning follows the global pattern mentioned above: text chat groups consistently used more than the other groups. We could, in turn, examine the other types of interactional resources under study with such detail. Still, however interesting they may be, these quantitative analyses do not account for how the various interactional resources are used within the NLCs. This is where SNA comes into play to provide a more qualitative analysis of the data. As previously mentioned, the first thing we need in order to perform SNA is relational data.

Figure 6.3 represents the extraction of the annotated data and shows that each participant is potentially assigned three separate tiers: tiers coded [v] (v for verbal) correspond to the orthographic transcription of the exchange, the other tiers [NOM] (for Negotiation of Meaning) and [INTERAC] (for Interactional Resources) are devoted to the tagging of the specific interactional resources under study. During the tagging phase, the interpersonal links between participants were also coded on resource-specific tiers. For example, on line 2, when LAU asks MAM to clarify what she just said, this particular interactional resource (a clarification request, coded SCR on the [NOM] tier since this is a signal used in the negotiation of meaning routines) was tagged as well as the participant it was addressed to (the SCR tag

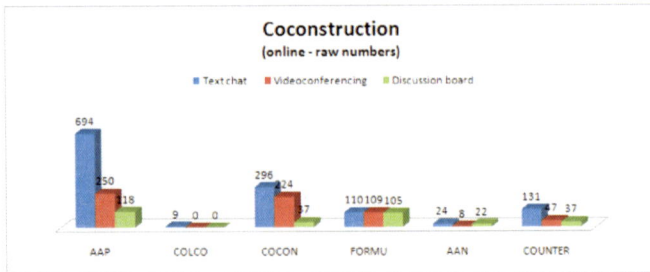

Figure 6.2: Coconstruction resources used per CMC mode.

is followed by (MAM) to identify the participant). Another example can be seen on line 3, when COR says she agrees with MAM: on the [INTERACT] tier, the AAP tag has been used to identify the type of interactional resource used (a Positive Assessment Activity), as well as the (MAM) tag to indicate whom the link was made with. Thanks to this specific tagging of interpersonal link from the raw data, which is a way of producing relational data, matrices were drafted.

The matrix shown in Figure 6.4 represents all the links made by the members of Group 1 during the completion of their 6 collaborative tasks. Column A and line 1 show sets of three letters which are used to identify the participants, the convention commonly used being to indicate the origin of the link on the line and the destination of the link in the column. Each cell presents the number of connections made between the different participants as coded on the [NOM] and [INTERACT] tiers. For example, line 3 column B shows that JUM made direct interactional contact with HAY 18 times. As the connections between participants can be two-way (reciprocal ties) and as the intensity of each tie does matter (the number of connections made), the sociograms drafted were both valued and directed. Thanks to these matrices, the density and geodesic distance measurements were made using the UCINet software package. The results are presented in Table 6.3.

In addition, weighted (or valued) directed graphs were produced using UCINet to show who interacted with whom during the completion of the online tasks (interactional patterns and communication configuration). An additional attribute was added to the graphs: the amount of participation (number of turns) was represented by the nodes themselves (the bigger the diameter of a node, the more the learner participated in the interactions).

[1]

MAM [v]	so in few months i think we will determine a new biosafety
MAM [INTERAC]	NCSSS
MAM [NOM]	TCONT

[2]

MAM [v]	leveldo you agree ?		
MAM [INTERAC]	CSSO (ALL)		
MAM [NOM]			
LAU [v]		can you explain please ?	
LAU [INTERAC]		CSSO (MAM)	
LAU [NOM]		SCR (MAM)	
ADE [v]			yes mistake like
ADE [INTERAC]			NCSSS AAP (MAM)

[3]

MAM [v]			a man who
MAM [INTERAC]			NCSSS
MAM [NOM]			RPAR (LAU)
ADE [v]	that should not occur		
ADE [INTERAC]	COCON (MAM)		
COR [v]		yes, i agree with you MAM.	
COR [INTERAC]		NCSSS AAP (MAM)	

[4]

MAM [v]	walk in the corridor without safety with dangerous cultures i
MAM [INTERAC]	
MAM [NOM]	

[5]

MAM [v]	think there is a big problem
MAM [INTERAC]	
MAM [NOM]	

Figure 6.3: Extraction of tagged data.

	A	B	C	D	E
1		HAY	JUM	KHE	CEL
2	HAY	0	14	11	44
3	JUM	18	0	15	25
4	KHE	13	10	0	25
5	CEL	47	27	17	0

Figure 6.4: Group 1 Matrix.

Table 6.3: Density and average geodesic distance.

		Density	Average Geodesic Distance
Text Chat	Group 1	22.1667	1
	Group 2	26.7500	1
	Group 3	36.1667	1
	Group 4	29.9167	1
Videoconferencing	Group 5	7.1667	1
	Group 6	14.5000	1.083
	Group 7	27.9167	1
	Group 9	4.5833	1.250
Discussion Board	Group 14	12.4167	1
	Group 15	9.3333	1.111
	Group 16	3.8333	1.167
	Group 17	2.0833	1.167

6.4   Discussion and conclusions

In all three computer-mediated communication modes (Figures 6.5 to 6.8), SNA has made it possible to identify one (or more) key player(s), the "virtuosos" (Perkins and Newman, 1996) who are "highly skilled practitioner[s] of e-discourse", in other words, a participant who "serves as a guide, gentle teacher and exemplar" (ibid., p. 163).

SNA also showed that text chat interactions are the most symmetrical ones, as their characteristics include: (1) balanced participation and strength of ties, (2) reciprocal ties only (geodesic distance = 1), (3) interactional load is high and well distributed (high density).

At the other end of the scale, the sociometric analysis also highlighted the fact that discussion board interactions are the most asymmetrical ones, as their characteristics include: (1) unbalanced participation and weak ties, (2) few reciprocal ties (geodesic distance >1), (3) one (or more) peripheral participant(s), the "lurkers" (Perkins and Newman 1996) who are participants who do not actively take part in the exchanges but simply read/listen to other participants' contributions, (4) the interactional load is low and unevenly distributed (low density): there are many "parallel monologues" (House 2002).

As for desktop videoconferencing interactions, SNA showed that they

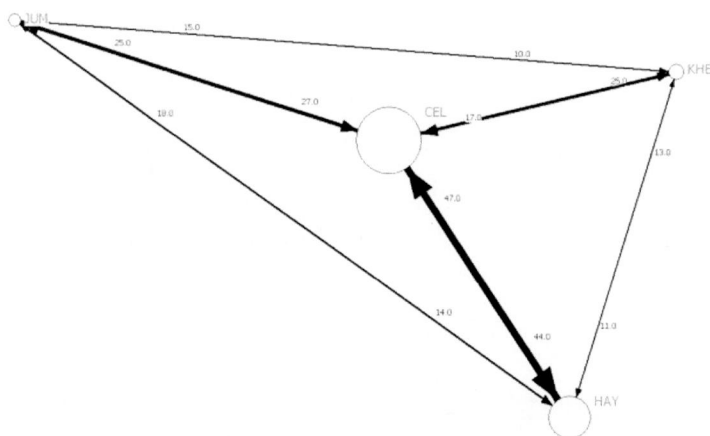

Figure 6.5: Text chat (group 1).

are the most difficult ones to map, as half seem fairly symmetrical, and half asymmetrical. It is hypothesized that this may be due to the videoconferencing tool itself: indeed, Flashmeeting, like most desktop videoconferencing applications to date, does not allow multiple speakers to speak at the same time when using their webcam (audio and video feeds), which means that a queuing system has to be used to be given the floor. This probably explains why certain participants rush and say everything they need to say without really engaging with their interlocutors, as they are afraid they might not get to talk again later in the exchange. Whatever the interpretations, this type of qualitative analysis would not have been possible without using a method like SNA. As previously mentioned, this sociometric analysis is part of a larger study that includes in-depth quantitative analyses and qualitative micro-analyses based on the tagged interactional resources presented here in order to uncover the interactional patterns at work when participants truly engage in a computer-mediated exchange and display their interactional competence in a second language.

Clearly, there is a growing need today for new ways of analysing NLCs, since social interactions are now central to online learning communities and not "simply scaled-up individuals and ties" (Garton et al. 1997). This is where the SNA approach provides real added value, especially when studying computer-mediated communication, as it (1) is a valuable complementary analytical tool in NLC research, (2) can be an answer to the need for data triangulation, (3) has a role to play in mixed-method research. It seems

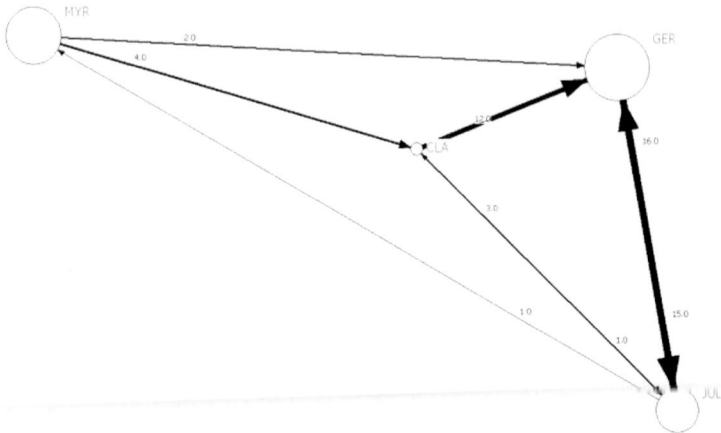

Figure 6.6: Videoconferencing (group 9).

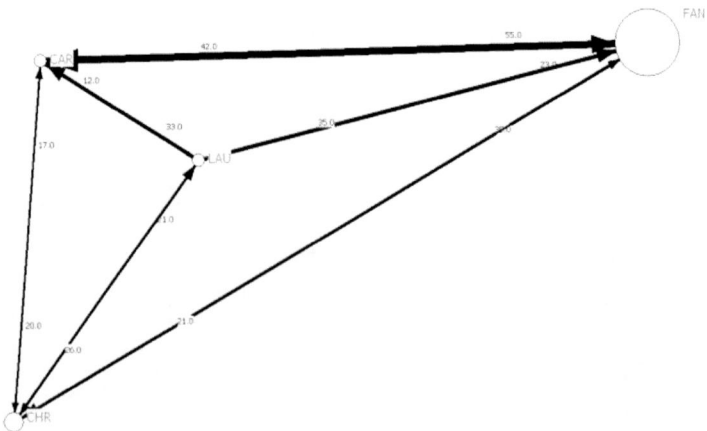

Figure 6.7: Videoconferencing (group 7).

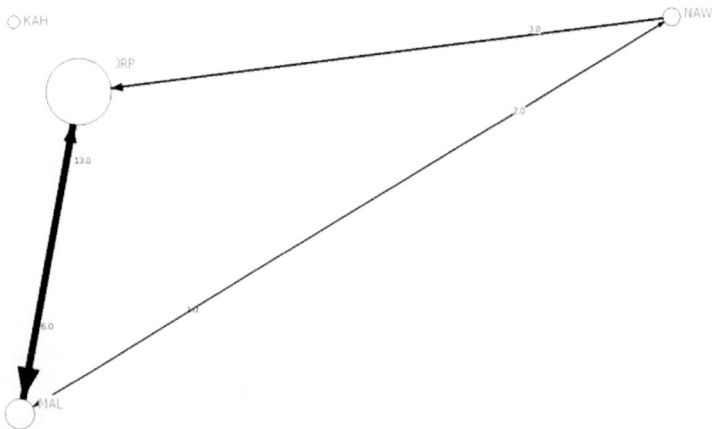

Figure 6.8: Discussion board (group 17).

necessary indeed to study NLCs as complex environments in rich ecolog-ical settings, both from a quantitative and a qualitative point of view. For all these reasons, SNA is definitely a promising approach in the study of networked learning communities.

Finally, in an ideal world, we could go so far as to imagine that future VLEs will make the most of SNA and start implementing monitoring func-tionalities which will automatically make relational data available to tutors and researchers, since these are often a lot more relevant and interesting than connection time, for example!

Bibliography

Barnes, J. A., 1954, Class and committee in a Norwegian island parish. *Human Relations*, **7**, 39–58.

Batagelj, A., and A. Mrvar, 2014, Pajek (version 4.01) [computer software]. University of Ljubljana, Slovenia. http://pajek.imfm.si/

Borgatti, S. P., M. G. Everett, and L. C. Freeman, 1999, UCINet 6.0 (Version 1.00) [computer software]. Natick, Mass., Analytic Technologies.

Bott, E., 1955, Urban families: Conjugal roles and social networks. *Human Relations*, **8**, 345–384.

Chanier, T., and J. Cartier, 2006, Communauté d'apprentissage et communauté de pratique en ligne – Le processus réflexif dans la formation des formateurs. *Revue Internationale des Technologies en Pédagogie Universitaire*, **3**(3), 64–82.

Cyram, Inc., 2014, Netminer 4 (version 4.2.1) [computer software]. Seoul, South Korea: Seoul University Research Park. http://www.netminer.com

D'Andrea, A., F. Ferri, and P. Grifoni, 2010, An overview of methods for virtual social network analysis. In A. Abraham, A. E. Hassanien, and V. Snasel, eds. *Computational Social Network Analysis—Trends, Tools and Research Advances*, 3–25, Berlin: Springer-Verlag.

de Laat, M., V. Lally, L. Lipponen, and R.-J. Simons, 2007, Investigating patterns of interaction in networked learning and computer-supported collaborative learning: A role for social network analysis. *Computer-Supported Collaborative Learning*, **2**(1), 87–103.

De Praetere, T., 2010, Dokeos (version 1.8.3) [computer software]. Brussels, Belgium: Dokeos S.A. http://www.dokeos.com.

Dekker, A., 2005, Conceptual distance in social network analysis. *Journal of Social Structure*, **6**(3).

Dillenbourg, P., C. Poirier, and L. Carles, 2003, Communautés virtuelles d'apprentissage: e-jargon ou nouveau paradigme? In A. Taurisson and A. Sentini, eds., *Pédagogies.net*, pp. 1-26, Montréal, Quebec, Canada: Presses Universitaires du Québec.

Drazdilova, P., G. Obadi, and K. Slaninova, 2010, Analysis and visualization of relations in eLearning. In A. Abraham, A. E. Hassanien, and V. Snasel, eds., *Computational Social Network Analysis—Trends, Tools and Research Advances*, 291–318, Berlin: Springer-Verlag.

European Commission, 2004, DIALANG [computer software]. Lancaster, England: University of Lancaster. http://www.lancaster.ac.uk/researchenterprise/dialang/about.htm

Garton, L., C. Haythornthwaite, and B. Wellman, 1997, Studying online social networks. *Journal of Computer Mediated Communication* **3**(1).

Hanneman, R., and M. Riddle, 2005, *Introduction to Social Network Methods*. University of California at Riverside. http://faculty.ucr.edu/~{}hanneman/nettext/

He, A., and R. Young, 1998, Language proficiency interviews: A discourse approach. In R. Young and A. He, eds., *Talking and Testing: Discourse Approaches to the Assessment of Oral Proficiency*, 1–24, Amsterdam: John Benjamins.

Holtzer, G., 1995, *Autonomie et didactique des langues : le Conseil de l'Europe et les langues étrangères*. Besançon, France: Presses Universitaires de Franche-Comté.

House, J., 2002, Developing pragmatic competence in English as a lingua franca. In K. Knapp and C. Meierkord, eds., *Lingua Franca Communication*, 245–267. Frankfurt, Germany: Peter Lang.

Jones, C., and L. Esnault, 2007, The metaphor of networks in learning: Communities, collaboration and practice. In S. Banks, P. Goodyear, V. Hodgson, C. Jones, V. Lally, D. McConnell, and C. Steeples, eds., *Networked Learning 2004: Proceedings of the Fourth International Conference on Networked Learning 2004*, pp. 317-323, Lancaster, England: Lancaster University and the University of Sheffield.

Knowledge Media Institute, 2010, Flashmeeting [computer software]. Milton Keynes, England: The Open University. http://flashmeeting.open.ac.uk/home.html

Lewin, K., 1936, *Principles of Topological Psychology*. New York: McGraw-Hill.

Mayo, E., 1933, *The Human Problems of an Industrial Civilisation*. Cambridge, Mass.: Macmillan.

Mitchell, J. C., 1969, The concept and use of social networks. In J. C. Mitchell, ed., *Social Networks in Urban Situations*, Manchester, England: Manchester University Press.

Moreno, J. L., 1933, Psychological organization of groups in the community. *57th Yearbook of Mental Deficiency*, 3–25

Palonen, T., and K. Hakkarainen, 2000, Patterns of Interaction in Computer-Supported Learning: A Social Network Analysis. In B. Fishman and S. O'Connor-Divelbiss, eds., *Fourth International Conference of the Learning Sciences*, 334–339, Mahwah, NJ: Erlbaum.

Perkins, J., and K. Newman, 1996, Two archetypes in a discourse: Lurkers and virtuosos. *Journal of Educational Telecommunications*, **2**(2), 155–170.

Reffay, C., and T. Chanier, 2003, How social network analysis can help to measure cohesion in collaborative distance learning. In: B. Wasson, S. Ludvigsen, and U. Hoppe, eds., *Designing for Change in Networked Learning Environments*, 343–352, Dordrecht, the Netherlands: Kluwer.

Rendersoft, 2013, Camstudio (version 2.7.2) [computer software]. Southend-on-Sea, UK, http://camstudio.org.

Russo, C., J. Koesten, 2005, Prestige, centrality and learning: A social network analysis of an online class. *Communication Education*, **54**(3), 254–261.

Sarré, C., 2013, Technology-mediated tasks in English for Specific Purposes (ESP): Design, implementation and learner perception. *International Journal of Computer-Assisted Language Learning and Teaching*, **3**(2), 1–16.

Schmidt, T., K. Wörner, T. Lehmberg, H. Hedeland, 2013, EXMARaLDA (Partitur Editor version 1.5.2) [computer software]. Hamburg, Germany: University of Hamburg, http://www.exmaralda.org.

Scott, J., 2000, *Social Network Analysis: A Handbook*, 2nd ed. Los Angeles, Calif.: SAGE Publications.

Skehan, P., 1998, *The cognitive approach to language learning*. Oxford University Press.

Chapter 7

Link prediction in large-scale multiplex networks

Manisha Pujari[1], Rushed Kanawati[2]

7.1   Introduction

Complex networks and their characteristics have gained considerable attention in the field of computational analysis of social networks. A complex
network can be any real-world network which has an abstract form, without any predefined structure or pattern of evolution. At times, they can be
highly dynamical in nature, evolving or changing constantly. Also, starting
with a tiny form, in this era of big data, they take the form of huge networks.
Analysing these dynamic large-scale networks is a major challenge for social network researchers. These networks are often heterogeneous in nature.
This means that a real network may have different types of nodes and also
different types of links. Multiplex networks are a category of heterogeneous
complex networks, which have different kinds of links between the same

---

[1]Laboratoire d'Informatique de Paris Nord (LIPN), Universite Sorbonne Paris Cite. Email:
manisha.pujari@lipn.univ-paris13.fr

[2]Laboratoire d'Informatique de Paris Nord (LIPN), Universite Sorbonne Paris Cite. Email:
rushed.kanawati@lipn.univ-paris13.fr

nodes. They can be represented as a set of simple networks, each having the same nodes but different sets of links. A common example of a multiplex network is a scientific collaboration network. Researchers or authors of research papers are an important part of these networks. They can be linked if they have co-published some articles or if they have published their articles in the same conferences or if they have the same research domain. They can also be linked if they have referred to the same publications in their articles. Figure 7.1 shows a diagrammatic representation of a scientific collaboration network and the multiplex structures present within it. In panel (b) of Figure 7.1, it is shown how an author network can be represented by multiplex layers, each having the same nodes but different types of links or edges.

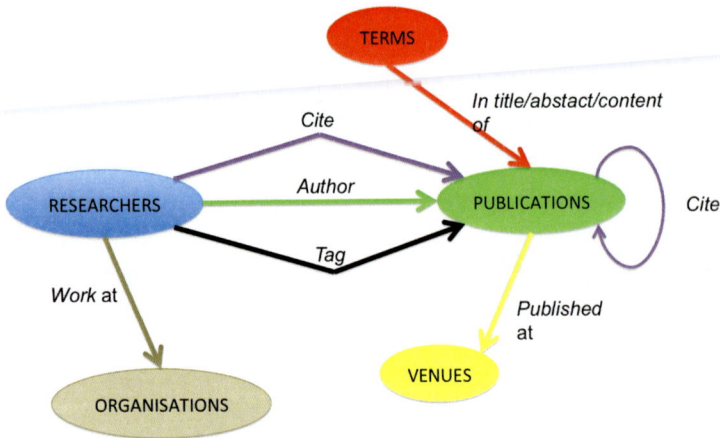

(a) Scientific collaboration network

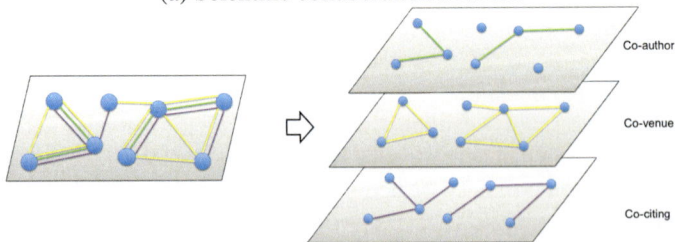

(b) Multiplex layers in author (researcher) network

Figure 7.1: Multiplex structure in a scientific collaboration network.

One of the most important problems in the analysis of complex networks is the problem of *link prediction* (Weibull 1995, Uhl-Bien 2006). It consists in estimating the likelihood of the existence or appearance of an edge between two unlinked nodes, based on the observed links and attributes that contain information about the nodes, edges or the entire graph. It has important applications in many fields, including social, biological and information systems, etc. Link prediction has been widely used in biological networks, such as protein interaction networks (Becker 1964, Mincer 1974), metabolic networks, food webs, etc., to find missing links and thereby help in reducing the experimental cost if its predictions are accurate. In social interaction and scientific or commercial collaboration networks they can play an important role in predicting new associations (new links) (Taubman and Wachter 1986, Damasio 1994, Parker 2012). This further has a usefulness in the recommendation task: a service provided by almost all social networks and mainly used in e-commerce networks (Poincaré 1890). Link prediction can also be helpful in finding hidden criminal links (Nisan et al. 2007, Maynard-Smith 1974), which is another critical field of research.

There are two basic types of link prediction: *structural* and *temporal*.

- *Structural link prediction* refers to the problem of finding *missing links* which probably exist in a partially observed network or graph (Stephan and Maiano 2007, Taskar et al. 2003, Yin et al. 2011, van Knippenberg et al. 2004). Structural link prediction is obviously useful for finding missing or hidden information in a network. It is directly applicable to finding unobserved patterns of genes, proteins and yeast interactions for medical research involving various diseases, such as cancer, HIV, Alzheimer's syndrome, etc. (Becker 1964, Mincer 1974). It can also help to find existing criminal links which often remain hidden from the public eye.

- *Temporal link prediction* refers to the problem of finding *new links* by studying the temporal history of a network (Jaoul-Grammare and Nakhili 2010, Kopel 2005, Lehrer 2009, Srivastava and Beer 2005, Damasio 1994, Schultz 1961). So here we have information about the network till time t and the goal will be to predict a new link that may appear at some point of time in the future, say $t + k$. It has its application primarily in recommendation systems, which are being used widely in e-commerce websites for product recommendations, in any search engines to help users with probably relevant terms they might be searching for, for recommending tags in social resource-sharing

websites like Flickr[3], YouTube[4], De.li.ci.ous[5], etc., and are very commonly used for recommending friends in many social networks, like
Facebook[6] and Twitter[7]. It has another significant use in predicting
future collaboration between researchers for academic or scientific
purposes. It also has an important use in order to identify probable
upcoming criminal associations.

The problem has been keenly studied by many researchers over a long
period of time. Many link prediction approaches has been proposed in recent
years. They can be classified as *node-features based approaches* or *topological approaches*, based on whether they use the node-features or only
structural information of the graph for prediction. In node-features based
approaches, apart from the structure of the graph, we also have some extra
information regarding the properties or characteristics of the nodes. This
extra information can be helpful in predicting links when the nodes are very
sparsely connected in the graph. One such approach is the local probabilistic model proposed by Wang et al. (2007). Topological approaches refer
to those which involve only the exploitation of the graph structure of the
network. They compute scores for pairs of unconnected nodes, based only
on the graphical features of the network structure, and without any extra
information about the features of the nodes. They observe mainly how the
connections have been established between the nodes and how they change
over time. Based on the former, they try to predict a missing link; based on
the latter, they predict a new link.

Node features are very useful when the network graph is very sparsely
connected and not much can be learned from the graph topology, whereas
topological approaches are very efficient in the absence of information about
the content of the features. Both have their own utility, and at times a combination of both can make a very good predictor. This kind of approach can
be termed a *hybrid* approach.

The topological (graph based) link prediction approaches can be further categorized as *temporal* or *non-temporal / static* based on whether they
take into consideration the dynamic aspects of the network or not. Another
way to classify them is as *dyadic* or *structural* approaches, based on the
type of score calculation involved. The approaches using a link score are
dyadic approaches, whereas those looking for mining rules of the evolution

---

[3]http://www.flickr.com
[4]http://www.youtube.com
[5]http://www.delicious.com/
[6]http://www.facebook.com
[7]http://www.twitter.com

of sub-graphs can be considered as structural approaches. Link prediction approaches can also be classified as *supervised* or *unsupervised*. Supervised approaches generate a model using many topological scores for unlinked node pairs to predict links, whereas unsupervised approaches use a single type of score for the node pairs and mostly use ranking to predict new links.

In Section 7.3, we give a brief account of the traditional methods of link prediction in a complex network, focusing mainly on dyadic topological approaches. In Section 7.4, we describe our approach, based on supervised rank aggregation. In Section 7.5, we present our new approach of link prediction in multiplex networks, which uses multiplex link information.

7.2   Formal definitions and notation

Before going further to describe our methods of link prediction, we would like to give a formal definition of link prediction and the mathematical notation used in the paper. In topology based link prediction approaches, only the structural properties of the underlying graph are used to implement statistical relational learning and to find a model that will be used to predict links. Suppose we have a social network graph $G = < V, E >$ where V is the set of nodes or vertices and E is the set of edges present in the graph. The goal of link prediction is to find pairs (u, v) such that $u, v \in V$ and $(u, v) \notin E$.

To predict new links at a certain point of time t_{n+1}, given network information till time t_n, the network will be presented as a sequence of graphs representing different snapshots of the network at different points of time $< t_0, t_1, ..., t_n >$. Suppose the temporal sequence of graphs is $G = < G_0, G_1, ..., G_n >$, so the goal of link prediction here will be to find the structure of the graph G_{n+1}. In other words, we will try to find pairs (u, v) such that $u, v \in V$ and $(u, v) \notin E$ where $V = \bigcup_{i=0}^{n} V_i$ and $E = \bigcup_{i=0}^{n} E_i$.

In any network G, $\Gamma(v)$ is the set of neighbours of the node v. The degree of a node is given by

$$k_i = \begin{cases} |\Gamma(v_i)| & \text{if } G \text{ is simple graph} \\ \sum w_j \; \forall \; x_j \in \Gamma(v_i) & \text{if } G \text{ is weighted graph} \end{cases}$$

In machine learning terms, the unlinked pairs of nodes are called *examples* or *instances*. If the time aspects of the network are to be considered also, then the examples can be generated as follows. Let $G = < G_1, ..., G_n >$ be a temporal sequence of an evolving graph. The whole sequence is divided into two parts: *training* and *testing*. Each part is then again divided into two phases: one to generate examples, and the other to label these examples. Thus, for example, in training, we shall have *learning* and *labeling* phases, resulting in the graphs G_{learn} and G_{label} generated by taking the union of the temporal sequences of the graphs for three corresponding time slots. The training data is constructed as follows. An example will be a couple of nodes (x, y) that are not linked in G_{learn} but both belong to the same connected component. The class is obtained by checking whether this couple of nodes is indeed connected in G_{label}. If such a connection exists, then it will be a *positive* example in the supervised learning task, but if no connection exists, it will be a *negative* example (Jevons 1871). Thus, examples are generated from these graphs for both training and testing. These examples are also characterized by a given number of topological attributes computed on the learning (or test) graphs. Figures 7.2 and 7.3 illustrate the process.

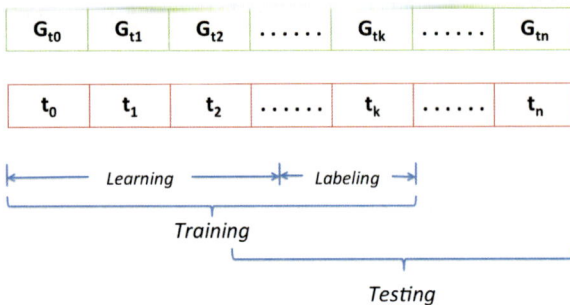

Figure 7.2: Generation of examples.

7.3   Link prediction approaches

The basic and most simple approach for predicting links using network graphs is to compute similarity scores for the unlinked pairs of nodes, and based on this score, to decide the presence or appearance of a link between them. In the scientific literature, we find many ways of computing this score. They can be *neighborhood-based*, *distance-based* or *an aggregation of node properties*. These approaches are mostly unsupervised. Below we list few of the important methods that have been used for link prediction.

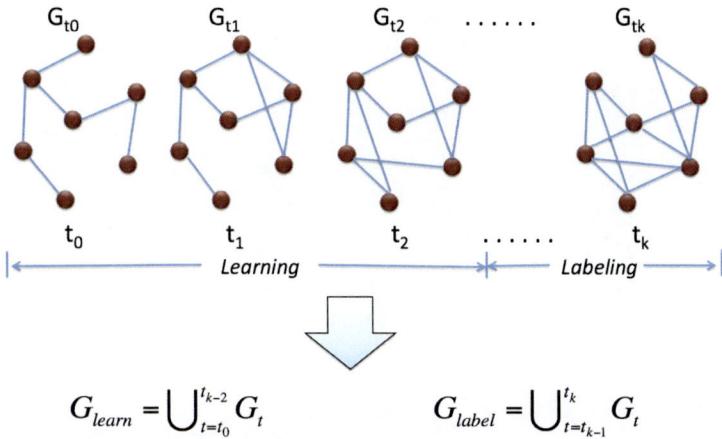

$$G_{learn} = \bigcup\nolimits_{t=t_0}^{t_{k-2}} G_t \qquad\qquad G_{label} = \bigcup\nolimits_{t=t_{k-1}}^{t_k} G_t$$

Figure 7.3: Construction of learning and labeling graphs.

Neighborhood based features

Common neighbors: Common neighbors counts the number of nodes (i.e. neighbors) that are connected to both the nodes under observation. Wert and Salovey (2004) used this quantity for studying collaboration networks, while Spence (1974) used it while analysing large-scale social networks.

$$CN(x,y) = \mid \Gamma(x) \cap \Gamma(y) \mid$$

Jaccard coefficient: The Jaccard coefficient calculates the ratio of the number of common neighbors to that of the total number of neighbors of the two nodes (Shapiro and Stiglitz 1984).

$$JC(x,y) = \frac{\mid \Gamma(x) \cap \Gamma(y) \mid}{\mid \Gamma(x) \cup \Gamma(y) \mid}$$

Adamic Adar coefficient: This metric proposes to weight the common neighbors based on their connectivity while computing the score. It gives more weight to less connected neighbors, increasing their contribution in the score (Stephan and Maiano 2007).

$$AA(x,y) = \sum_{z \in \Gamma(x) \cap \Gamma(y)} \frac{1}{log \mid \Gamma(z) \mid}$$

It is based on the coefficient proposed by Barrouillet and Camos (2008) to find the similarity between two web pages. For two web pages x and y, sharing a set of features z, this coefficient is computed as

$$\sum_{z:feature\ shared\ by\ x\ and\ y} \frac{1}{log(frequency(z))}$$

Resource allocation: This metric is based on the resource allocation dynamics in complex networks (Wood and Forest 2011). Like the Adamic Adar coefficient, this index also depresses the contribution of high-degree common neighbors.

$$RA(x,y) = \sum_{z \in \Gamma(x) \cap \Gamma(y)} \frac{1}{\mid \Gamma(z) \mid}$$

Path based features

Distance based

Shortest path length: The number of edges in the shortest path between x and y can also serve to predict links. It is also known as the distance between nodes. The longer the distance, the less is the similarity between the nodes, and the less is the chance of having a link between them. This metric captures the fact that the path between two nodes in a social network can affect the formation of a link between them, due to the fact that a friend of a friend can be a friend in a social network.

Katz's index: One of the well known scoring indexes, commonly known as the *Katz index*, has been proposed by Shim et al. (2005). It is based on the paths between nodes in a graph. It sums over a collection of paths and is exponentially damped by the length, to give shorter paths more weight. Mathematically, it is defined by

$$Katz(x,y) = \sum_{l=1}^{\infty} \beta^{\ell} \times \mid path_{x,y}^{(\ell)} \mid$$

where $path_{x,y}^{(\ell)}$ is the set of paths between x and y of length ℓ and β is a positive parameter (i.e. a damping factor) which favors the shortest paths.

The same can be presented using the adjacency matrix

$$Katz(x, y) = \beta A_{xy} + \beta^2 (A^2)_{xy} + \beta^3 (A^3)_{xy} + \ldots$$

Here, A_{xy} is the adjacency matrix, i.e. its entries are either 1 or 0 based on whether x and y are directly connected. $(A^2)_{xy}$ is the matrix showing the number of paths of length 2 between x and y, and so on. A very small β leads to a score close to the number of common neighbors, because long paths contribute very little. The matrix showing the Katz scores between all pairs of nodes is

$$K = (I - \beta A)^{-1} - I$$

β must be lower than the reciprocal of the largest eigenvalue of the matrix A to ensure the convergence of above equation (Uhl-Bien 2006).

Random walk based

Matrix forest index: The matrix forest index computes the similarity between two nodes as the ratio of the number of spanning rooted forests where the two nodes are in the same tree rooted at one of the nodes, to all the spanning rooted forests of the network. It can be computed as $M = (I - L)^{-1}$, I being the identity matrix and $L = D - A$ the Laplacian matrix of the network, where D is the degree matrix and A is the adjacency matrix (Mariotte 1717). This index was used for the collaborative recommendation task in Orlean (1995).

Hitting time and commute time: The hitting time is a random walk based feature that counts the time required by a random walker to go from node x to node y in a graph. It is defined as the expected number of steps required for a random walker to walk from one node to the other. A smaller hitting time can indicate that the nodes are similar and have a higher chance of linking in the future. As this metric is not symmetrical, often for undirected graphs, the average commute time is used instead. If $HT(x, y)$ is the hitting time to reach node y from node x, the average commute time is given by

$$CT(x, y) = HT(x, y) + HT(y, x)$$

A negated value of hitting or commute time can be used as a score for predicting links.

Rooted Pagerank: Pagerank denotes the importance of a node x by summing up the importance of all the other nodes linked to x. This importance can also be represented by the stationary distribution weight of a node. This feature can be altered to find a similarity score between two nodes and is termed the *rooted pagerank* in Stephan and Maiano (2007). The similarity between two nodes x and y is measured as the stationary probability of y in a random walk that returns to x with probability $1 - \alpha$ at each step, moving to a random neighbor with probability α. The rooted pagerank for all node pairs can be computed as follows.

$$RPR = (1 - \alpha)(I - \alpha N)^{-1}$$

where D is the diagonal degree matrix and $N = DA^{-1}$ is the adjacency matrix with row sums normalized to 1.

PropFlow: PropFlow captures the probability that a restricted random walk starting from one node x ends at another node y in l or less steps when using the link weights as the transition probabilities. The restriction is that a walk terminates on reaching y or on revisiting any node including x. The walk selects links based on their weights, which produces a score to estimate the likelihood of new links. This measure is a more localized measure of propagation and is insensitive to topological noise far from the source node (Sznajd-Weron 2005).

Aggregation of node features

Preferential attachment: Preferential attachment combines the degrees of the two concerned nodes and can be used as a score for predicting links. Here, the probability of appearance of a new link is directly proportional to the degree of the observed nodes (Green and Zhu 2010).

$$PA(x, y) = \mid k_x \times k_y \mid$$

For a simple un-directed and un-weighted graph, the degree of a node is equal to the number of neighbors, i.e. $k_x = \Gamma(x)$.

Sum of neighbors: In Damasio (1994), the sum of neighbors is used as a topological feature for characterizing an unlinked node pair. Formally, it can be defined as $\Gamma(x) + \Gamma(y)$.

Aggregation of clustering coefficients: As described in Chapter 1, the clustering coefficients of a node quantify the probability of the neighbors of the node to get connected to each other.

$$cf(x) = \frac{3 \times \#Triangles\ adjacent\ to\ x}{\#Possible\ triples\ adjacent\ to\ x}$$

This property can also be used for link prediction by taking an aggregation (sum or product) of the clustering coefficients of two unconnected nodes. The similarity score for any two nodes x and y will be

$$CC(x, y) = cf(x) \times cf(y) \quad or \quad CC(x, y) = cf(x) + cf(y)$$

In the seminal Weibull (1995), it was shown that simple topological measures representing relationships between pairs of unlinked nodes in a complex network can be used for predicting the formation of new links. Consider the case of employing *common neighbors* as a topological measure. Let $\mathcal{L}$ be the list of pairs of unlinked nodes (belonging to the same connected component). We have $\mathcal{L} = \{(x, y)\}$. Let $\Gamma(x)$ be the function returning a set of direct neighbors of node x in the graph. The common neighbors function of two nodes x, y is then defined by

$$CN(x, y) = |\Gamma(x) \cap \Gamma(y)| \tag{7.1}$$

The list $\mathcal{L}$ is sorted according to the values obtained by applying the common neighbors function to the couples of unlinked nodes. The top k couples of nodes are then returned as the output of the prediction task. The assumption here is that the more a couple of unlinked nodes share common neighbors, the more they are likely to have a link in future. In Weibull (1995), k is equal to the number of really appearing links. Other types of topological measures can be applied for the same purpose.

Many other publications have focused on how to combine different topological metrics in order to enhance the prediction performance. One widely applied approach is based on expressing the problem of link prediction as a problem of binary classification. The idea is to compute, for each unlinked couple of nodes in $\mathcal{L}$, a set of topological measures. Then with each element in $\mathcal{L}$, associate one of the labels Linking (positive) or Not-linking (negative) based on the status of the graph at a future step. The dataset computed this way (topological features with classes) can then be used to learn a model for discriminating the *linking* class from the *not-linking* one, using classical supervised machine learning approaches (Damasio 1994, Jevons 1871).

Another main category is that of the matrix based approaches. In van Knippenberg et al. (2004), the authors use a supervised matrix factorization approach for link prediction. The model learns latent features from the structure of a graph. The authors combine these latent features with explicit node features and also with the outputs of other models to make better predictions. They propose a new approach to deal with the class imbalance problem, by directly optimizing a ranking loss. The model is optimized with stochastic gradient descent and also scales to large graphs. Another model for temporal link prediction based on matrix factorization is given in Pierce (1980), and exploits multiple information sources in the network to predict link occurrence probabilities as a function of time. They propose a unique model combining global network structure, content information of nodes and local proximity information. To combine the temporal information of the network, they use a weighted exponentially decaying model to build an aggregate weighted link matrix over a set of T time slices.

Other approaches include probabilistic models, stochastic block models, hierarchical models, etc. A more detailed survey of link prediction and approaches can be found in Devaney (2003) and Uhl-Bien (2006).

7.4 Link prediction based on supervised rank aggregation

None of the previous research has attempted to combine the predictive power of individual topological measures by applying computational social choice algorithms (also known as rank aggregation methods Marois and Ivanoff (2005)). *Rank aggregation* can be defined as a process of combining a number of ranked lists or rankings of candidates or elements to get a single list and with the least possible disagreement with all the experts or voters who provided these lists. These methods are a part of social choice theory and are mostly applied to political and election related problems (*Mémoire sur les élections au scrutin* 1781, Young and Levenglick 1978, Black et al. 1998).

These techniques are designed to ensure fairness between experts, while combining their rankings, and hence all experts are given equal weight. Expressing the link prediction problem in terms of a vote is straightforward: the candidates are examples (pairs of unconnected nodes), while the voters are topological measures computed for these pairs of unlinked nodes. Then we have a voting problem with a quite huge set of candidates and rather a reduced set of voters. These settings are very similar to those encountered when considering the problem of ranking documents in meta-search engines, where voting schemes have also been applied with success (Miller 1956, Ericsson et al. 1993, Waylen et al. 2004).

In our settings, prediction performance can be boosted by weighting differently the applied topological measures, as functions of their individual performance in predicting new links. We propose here two different weighting schemes. Weights are used in two different weighted rank aggregation methods: the first one is based on the classical Borda count approach (*Mémoire sur les élections au scrutin*, 1781), while the second is based on the Kemeny aggregation rule.

The latter is known to compute the *Condorcet* winner of an election (if it exists): the candidate that wins each duel with all the other candidates is the winner.

Before describing the approaches based on *supervised rank aggregation*, which refers to the same process of combining rankings but giving different weights to the voters and these weights are learned in a due process of training, here is a brief description about two of the well known classical rank aggregation methods.

- **Borda's method** The *Mémoire sur les élections au scrutin* (1781) is a truly positional method, as it is based on the absolute positioning of the ranked elements rather than their relative rankings. A Borda score is calculated for each element in the lists. Based on this score, the elements are ranked in the aggregated list. It is primarily applicable to complete lists of ranked elements. Complete ranked lists are lists with exactly the same elements but in different orders or rankings. Also, in a list L_k, the rank of an element x is represented by $L_k(x)$. For a set of complete lists $L = [L_1, L_2, L_3, ..., L_n]$, Borda's score for an element x and a list L_k is given by

$$B_{L_i}(x) = \{count(y)|L_i(y) < L_i(x) \& y \in L_i\} \qquad (7.2)$$

The total Borda' score for an element is

$$B(x) = \sum_{t=1}^{n} B_{L_i}(x) \qquad (7.3)$$

Borda's method is mostly applicable to complete lists, and is not very suitable for partial lists, where the lists can have some different elements. Borda's method, being a method based on the absolute position, is incapable of correctly treating elements which are not present in all the lists taken into consideration. One option is to assign equal scores to all unranked elements, as suggested in Yaniv and Kleinberger (2000). This is a separate topic of research and outside the scope of this paper.

- **Kemeny optimal aggregation**, proposed in Mazari and Recotillet (2013), makes use of Kendall Tau distance to find the optimal aggregation. Kendall Tau distance counts the number of pairs of elements that have opposite rankings in the two input lists, i.e. it calculates the pairwise disagreements.

$$K(L_1, L_2) = |\ (x, y)\ s.t.\ L_1(x) < L_2(y)\ \&\ L_1(x) > L_2(y)\ | \quad (7.4)$$

The first step is to find an initial aggregation of the input lists, using any standard method. The second step is to find all possible permutations of the elements in the initial aggregation. For each permutation, a score is computed, which is equal to the sum of the distances between this permutation and the input lists. The permutation having the lowest score is considered to be the optimal solution. For example, for a collection of input rankings $\tau_1, \tau_2, \tau_3, ..., \tau_n$ and an aggregation π, the score is given by

$$SK(\pi, \tau_1, \tau_2, \tau_3, ..., \tau_n) = \sum_{i \in n} K(\pi, \tau_i) \quad (7.5)$$

The speciality of Kemeny optimal aggregation is that it complies with the *Condorcet principle*, which is not the case with positional methods like Borda's algorithm (Young and Levenglick 1978). The Condorcet principle states that if there exists an item that defeats every other item in simple pairwise majority voting, then it should be ranked above all others.

In spite of all its advantages, Kemeny optimal aggregation is computationally hard to implement. So while looking for an alternative solution that gives a similar kind of aggregation but is computationally feasible, we were led to another approach, named *Local Kemenization* (Mazari and Recotillet 2013). A full list π is a locally Kemeny optimal aggregation of partial lists $\tau_1, \tau_2, \tau_3, ..., \tau_n$ if there is no full list π' that can be obtained from π by performing a single transposition of a single pair of adjacent elements for which

$$SK(\pi', \tau_1, \tau_2, \tau_3, ..., \tau_n) < SK(\pi, \tau_1, \tau_2, \tau_3, ..., \tau_n)$$

In other words, it is impossible to reduce the total distance of an aggregation by flipping any adjacent pair of elements in the aggregation.

Looking into the work based on rank aggregation techniques, we can say that not much has been explored when it comes to the application of rank aggregation to link prediction. Moreover, this literature applies mostly unsupervised rank aggregation algorithms, giving equal weight to all the experts who provided the ranked lists. One of the well known methods is the weighted majority algorithm proposed in Turner (1984), using weights for predictors, all having equal weights in the beginning. There is a master predictor which makes the final prediction based on the class which corresponds to a maximum of the total weights of the predictors. If the final prediction is wrong, then the weights of all the predictors who disagreed with that label are increased by a factor β such that $0 \leq \beta < 1$, thus reducing the effect of unworthy predictors at each iteration. This approach has a limitation, in that the performance of the master predictor can be at most equal to the best performing predictor. In contrast, the use of rank aggregation can provide even better prediction at times. This may be due the fact that in these algorithms, the *"likes"* of the majority of the predictors is given higher preference. At the same time, the *"dislikes"* are given less preference. So these algorithms are much more resistant to spam and noise.

A significant contribution to supervised rank aggregation has been made by Turner et al. (1987), proposing supervised aggregation by a Markov chain to enhance the ranking results of meta-searches. However, it has been shown that local Kemenization improves on approaches based on Markov chains (Mazari and Recotillet 2013).

In the very recent Subbian and Melville (2011), supervised rank aggregation is used to find influential nodes and future links. Those authors propose their own supervised Kemeny aggregation method based on quick sort and apply it to Twitter and citation networks. However, their method is mostly based on the topological features of the nodes, whereas the method to be presented in this chapter is based on the features of a couple of nodes (edges) with the use of a merge sort algorithm to find supervised local Kemeny aggregation. The reason why we use merge sort is that it is seemingly more stable than quick sort.

In the next part (subsection 7.4.1) there will be presented a description of our work on link prediction using rank aggregation. We contribute in three ways: first we provide a way to generate weights for the topological measures; second, we propose a new way of introducing weights to approximate Kemeny aggregation; and third, we use supervised or weighted rank aggregation for link prediction in complex networks. Our approach is evaluated in the context of a link prediction task applied to academic co-authorship networks. The experiments were conducted on real networks extracted from the now well-known DBLP bibliographical server.

7.4.1   Link prediction by supervised rank aggregation

Each attribute of an example has the capacity to provide some unique information about the data when considered individually. The training examples
are ranked based on the attribute values. So, for each attribute we will get
a ranked list of all examples. Considering only the top k ranked examples and with the assumption that when we rank the examples according
to their attribute values, the positive examples should be ranked on the top,
we compute the performance of each attribute. This performance is measured in terms of either *precision* (maximization of identification of positive
examples) or *false positive rate* (minimization of identification of negative
examples) or a combination of both. Based on the individual performances,
a weight is assigned to each attribute.

For validation, we use examples obtained from the validation graph
characterized by the same attributes and try to rank all examples based on
their attribute values. So for n different attributes, we shall have n different
rankings of the test examples. These ranked lists are then merged using a *supervised rank aggregation* method and the *weights of the attributes* obtained
during the learning process. The top k ranked examples in the aggregation
are taken to be the predicted list of positive examples. Using this predicted
list, we calculate the performance of our approach. In this case, k is equal
to the number of positive examples in the validation graph.

Computation of the weights

We propose to compute the voters' (topological measures) weights based on
their ability to identify the correct elements in the top k positions of their
rankings. The weights associated to the applied topological measures are
computed based on the following criteria:

- **Maximization of positive precision:** Based on the maximization of
  the identification of positive examples, the attribute weight is calculated as

  $$w_i = n * Precision_i \qquad (7.6)$$

  where n is the total number of attributes and $Precision_i$ is the precision of attribute i based on the identification of positive examples. Recall that the precision is defined as the fraction of retrieved instances
  that are relevant.

- **Minimization of false positive rate:** By minimizing the identification of negative examples, we get a weight as below:

$$w_i = n * (1 - FPR_i) \qquad (7.7)$$

where n is the total number of attributes and FPR_{a_i} is the false positive rate of attribute a_i based on the identification of negative examples. The false positive rate is defined as the fraction of non-relevant instances that are retrieved as relevant.

Supervised rank aggregation

First we define some basic functions used later in defining weighted aggregation functions. Let L_i be a ranked list of n candidates (a vote). $L_i(x)$ denotes the rank of element x in the list L_i. The top ranked element has the rank 0. The basic individual Borda score of an element x for a voter i is then given by

$$B_i(x) = n - L_i(x)$$

Let x and y be two candidates. We define the local preference function as follows.

$$Pref_i(x, y) = \begin{cases} 1 & \text{if } B_i(x) > B_i(y) \\ 0 & \text{if } B_i(x) < B_i(y) \end{cases} \qquad (7.8)$$

Introducing weights in the Borda aggregation rule is rather straightforward.

Let $(w_1, w_2, \ldots, w_r)$ be the weights for r voters providing r ranked lists on n candidates. The weighted Borda score for a candidate x is then given by

$$B(x) = \sum_{i=1}^{r} w_i * B_i(x) \qquad (7.9)$$

To approximate Kemeny aggregation (Mazari and Recotillet 2013), we introduce weights into the definition of the non-transitive preference relationships between candidates. This is modified as follows. Let w_T be the sum of all computed weights, i.e. $w_T = \sum_{i=1}^{r} w_i$. For each couple of candidates x, y we compute a score function as follows: $score(x, y) = \sum_{i=1}^{r} w_i * Pref_i(x, y)$ The weighted preference relation ($\succ_w$) is then defined as follows:

$$x \succ_w y : score(x, y) > \frac{w_T}{2}$$

This new preference relation is used to sort an initial aggregation of candidates in order to obtain a supervised Kemeny aggregation. The initial aggregation can be any of the input lists or an aggregation obtained by applying any other classical aggregation method, like Borda. In our algorithm, we have applied merge-sort for the time being.

7.4.2   Experimentation

We evaluated our approach using data obtained from DBLP [8] databases. DBLP is a scientific bibliography website containing a large database of articles mostly related to computer science. Our network consists of authors as nodes and they are linked if they have co-published at least one paper during the observed period of time.

The data comes from 1970–1979. We created three datasets. Following the procedure described previously, we generated examples for each dataset. Table 7.1 provides information about the training or test graphs while Table 7.2 summarizes information about the examples generated.

Table 7.1: Statistics on the training or test graphs.

Years	Properties	Co-Author
1970–1973	*Nodes*	91
	Edges	116
	Density	0.028327
1972–1975	*Nodes*	221
	Edges	319
	Density	0.013122
1974–1977	*Nodes*	323
	Edges	451
	Density	0.008673

Table 7.2: Examples from co-authorship graph.

Years		# Positive	# Negatives
Train/Test	Labeling		
1970–1973	1974–1975	16	1810
1972–1975	1976–1977	49	12141
1974–1977	1978–1979	93	26223

We applied our approach to the complete datasets. For rank aggregation, we have used the supervised Borda and supervised Kemeny methods.

---

[8]http://www.dblp.org

We compare our approach with link prediction approaches using basic machine learning algorithms like Decision tree, Naive Bayes, and the k-Nearest neighbors algorithm. We named our approaches Supervised Borda 1 and Supervised Borda 2, based on how the attribute weights are computed. 1 indicates that the weights were computed based on the maximization of positive precision, and 2 indicates that the weights were computed based on the minimization of false positive rates. We will follow the same convention with regard to the supervised Kemeny algorithms. We selected the following topological attributes: number of common neighbors (CN), Jaccard coefficient (JC), preferential attachment (PA) (Poincaré 1890), Adamic Adar coefficient (AA) (Barrouillet and Camos 2008), resource allocation (RA) (Zhou and Zhang 2009) and shortest path length (SPL).

Table 7.3: Results in terms of F1-measure.

	Learning: 1970–1973 Test: 1972–1975	Learning: 1972–1975 Test: 1974–1977
Decision tree	0.0357	0.0168
Naive Bayes	0.1032	0.0070
Kemeny	0.2449	0.0860
Supervised Kemeny 1	0.4286	0.2581
Supervised Kemeny 2	0.4286	0.2258

Table 7.3 summarizes the results obtained in terms of the F1-measure. The F1-measure is the harmonic mean of *precision* and *recall*. *Recall* is the proportion of correctly predicted links out of the total number of actual new links and *precision* is the proportion of correctly predicted links out of the total number predictions made. Formally, it is given by

$$\text{F1-measure} = \frac{2 * Precision * Recall}{Precision + Recall} \qquad (7.10)$$

While k-nearest neighbors and our methods based on Borda and supervised Borda failed to provide any substantial results (due to which we have not listed them in the table), our approximate Kemeny and supervised Kemeny based methods outperform the decision tree and naive Bayes algorithms for both datasets. This shows the validity of our approach.

Although it is still early to say that rank aggregation based methods perform better than the other approaches to link prediction, the preliminary results do show that rank aggregation, especially with the Kemeny method, indeed adds some new information which may enhance the results of a prediction. This is quite encouraging for us to continue this work further. Still, the fact remains that rank aggregation methods, especially the Kemeny method, has a high computational complexity, which puts into question its applicability for link prediction in large scale networks. To cope with this, we will be working on an application of top-k rank aggregation. Much work needs to be done in this regard.

7.5   Link prediction using multiplex links

All the work that we have presented till now addresses the link prediction only in simple networks having homogeneous links. In this section, we explain how the prediction of links can be done in a multiplex scenario and how the prediction performance can be enhanced using multiplex information.

As far as our knowledge extends, not much has been explored in this regard. Although there have been a few recent publications proposing methods for predicting links in heterogeneous networks (networks which have different types of nodes as well as edges (Sun et al. 2011), and there has been a little work on extending simple structural features like the degree, path, etc., to the context of multiplex networks (Harsanyi and Selten 1992, Lorenz 1964), nothing has been attempted to use them for link prediction. We propose a new approach for exploring multiplex relations to predict future collaboration (co-authorship links) between authors. The applied approach is based on supervised machine learning, where we attempt to learn a model for link formation based on a set of topological attributes describing both positive and negative examples. While such an approach has been successfully applied in the context of simple networks, different options can be used to extend it to the context of multiplex networks. One option is to compute topological attributes in each layer of the multiplex. Another is to compute directly new multiplex-based attributes quantifying the multiplex nature of dyads (potential links). Both approaches will be discussed in the next section.

7.5.1   Our approach

Our approach includes computing simple topological scores for unconnected node pairs in a graph. Then we extend these attributes to include information from other dimension graphs. This can be done in three ways: first we compute the simple topological measures in all dimensions; second is to take the average of the scores; and third we propose an entropy based version of each topological measure, which places importance on the presence of a non-zero score of the node pair in each dimension. In the end, all these attributes can be combined in various ways to form different sets of vectors of attribute values characterizing each example or unconnected node pair. Formally, we have a multiplex graph $G = < V, E_1, \ldots, E_m >$, which in fact is a set of graphs $< G_1, G_2, \ldots, G_m >$, and we want to compute a topological attribute X for it. For any two unconnected nodes u and v in G_i (where we want to make a prediction), the attribute $X(u, v)$ computed on G_i will be a *direct* attribute, and the same one computed on all other dimension graphs will be *indirect* attributes. The second category computes an average of the attribute over all the dimensions, i.e.

$X_{average} = \frac{\sum_{\alpha=1}^{m} X(u,v)^{[\alpha]}}{m}$ for $u, v \in V$ and $(u, v) \notin E_i$. where m is the number of types of relations in the graph (dimension or layer). In the third category, we propose a new attribute called *product of node degree entropy* (PNE) which is based on the *degree entropy*, a multiplex property proposed by Harsanyi and Selten (1992). If the degree of node u is $k(u)$, the degree entropy is given by $E(u) = -\sum_{\alpha=1}^{m} \frac{k(u)^{[\alpha]}}{k_{total}} \log(\frac{k(u)^{[\alpha]}}{k_{total}})$ where $k_{total} = \sum_{\alpha=1}^{m} k(u)^{[\alpha]}$ and we define the *product of node degree entropy* as

$$PNE(u, v) = E(u) * E(v)$$

We also extend the same concept to define the entropy of a simple topological attribute, say X_{ent}

$$X_{ent}(u, v) = -\sum_{\alpha=1}^{m} \frac{X(u,v)^{[\alpha]}}{X_{total}} \log(\frac{X(u,v)^{[\alpha]}}{X_{total}})$$

where $X_{total} = \sum_{\alpha=1}^{m} X(u,v)^{[\alpha]}$.

The entropy based attributes are more suitable for capturing the distribution of the attribute value over all dimensions. A higher value indicates a uniform distribution of the attribute's value across the multiplex layers. We refer to the average and entropy based attributes as *multiplex attributes*.

7.5.2   Experiments

We evaluated our approach using data obtained from DBLP[9] databases, from which we created three datasets, each one corresponding to a different period of time. Table 7.4 summarizes the information about the graphs

---

[9]http://www.dblp.org

of each dataset. Each graph has four years for learning or training. The next two years are used to label the examples generated from the learning graphs. Examples are unconnected node pairs and they are labeled as *positive* or *negative* based on whether they are connected during the labeling period or not. Table 7.5 shows the number of examples obtained for each dataset.

Table 7.4: Graphs.

Years	Properties	Co-Author	Co-Venue	Co-Citation
1970–1973	*Nodes*	91	91	91
	Edges	116	1256	171
1972–1975	*Nodes*	221	221	221
	Edges	319	5098	706
1974–1977	*Nodes*	323	323	323
	Edges	451	9831	993

Table 7.5: Examples from co-authorship graph.

Years		# Positive	# Negatives
Train/Test	Labeling		
1970–1973	1974–1975	16	1810
1972–1975	1976–1977	49	12141
1974–1977	1978–1979	93	26223

We use the same attributes that were used in the previous section for the supervised rank aggregation based approach, i.e. the number of common neighbors (CN), Jaccard coefficient (JC), preferential attachment (PA) (Poincaré 1890), Adamic Adar coefficient (AA) (Barrouillet and Camos 2008), resource allocation (RA) (Zhou and Zhang 2009), and shortest path length (SPL). For any attribute XX

- XX_{aut}: Value of attribute was computed on co-authorship graph during learning period

- XX_{ven}: Value of attribute was computed on co-venue graph

- XX_{cit}: Value of attribute was computed on co-citation graph

- $AvgXX$: Average of the attribute value over the different relation graphs in our case $m = 3$ as we are using co-authorship, co-venue and co-citation graphs.

- PNE: Product of node degree entropy. If the degree of node i is $k(i)$, the entropy for node i is calculated as

$$E_i = -\sum_{\alpha=1}^{m} \frac{k(i)^{[\alpha]}}{k_{total}} \log(\frac{k(i)^{[\alpha]}}{k_{total}})$$

where $k_{total} = \sum_{\alpha=1}^{m} k(i)^{[\alpha]}$ and

$$PNE(i,j) = E_i * E_j$$

- $XXent$: Entropy value of the corresponding attribute (based on the entropy equation proposed for the node degree in Harsanyi and Selten (1992))

$$XXent(i,j) = -\sum_{\alpha=1}^{m} \frac{XX(i,j)^{[\alpha]}}{XX_{total}} \log(\frac{XX(i,j)^{[\alpha]}}{XX_{total}})$$

We apply the decision tree algorithm to one dataset to generate a model and then test it on another dataset. We used the data mining tool Orange[10] for that. We used four types of combinations of the attributes, creating five different sets: Set_{direct} (attributes computed only in the co-authorship graph); $Set_{direct+indirect}$ (attributes computed in co-authorship, co-venue and co-citation graphs); $Set_{direct+multiplex}$ (attributes computed from co-authorship graph, entropy based attributes and also with average attributes obtained from three dimension graphs); Set_{all} (attributes computed in co-authorship, co-venue and co-citation graphs, with average of the attributes, and also entropy based attributes); and $Set_{multiplex}$ (average attributes and entropy based attributes). Table 7.6 shows the results obtained in terms of the F1-measure and area under the ROC curve (AUC). We can see that there is an improvement in the F1-measure when we use multiplex attributes. The AUC is better for all the sets that include multiplex and indirect attributes for both datasets.

The results clearly show that the inclusion of different types of links surely affects the prediction of new links for the better. Though marginal, the enhanced results in preliminary experiments do seem to validate our approach and encourage further experimentations.

---

[10]http://orange.biolab.si

Table 7.6: Results of decision tree algorithm.

Attributes	Learning: 1970–1973 Test: 1972–1975		Learning: 1972–1975 Test: 1974–1977	
	F1-measure	AUC	F1-measure	AUC
Set_{direct}	0.0357	0.5263	0.0168	0.4955
$Set_{direct+indirect}$	0.0256	0.5372	0.0150	0.5132
$Set_{direct+multiplex}$	0.0592	0.5374	0.0122	0.5108
Set_{all}	0.0153	0.5361	0.0171	0.5555
$Set_{multiplex}$	0.0374	0.5181	0.0185	0.5485

7.6 Conclusions

In this paper we present the problem of link prediction in complex networks and multiplex networks. We present here a brief overview of the state of the art of various link prediction approaches, focusing mainly on dyadic topological approaches. The unsupervised methods involve the computation of scores for unlinked pairs of nodes. While neighborhood based scores are easy to compute, some path based measures, like Katz, commute time, and rooted pagerank, can be really time consuming. The same is the case for other matrix based approaches, which have issues of computation time and memory when applied to real large scale networks. This makes them difficult to be employed for evolving real networks. So some approximate solutions for these measures, such as truncated Katz and others can be a good choice.

In the supervised approaches, especially machine learning based methods, an attempt is made to combine the effect of various topological attributes to generate a model which is then used to predict links on a test graph. The same is done in our proposed approach, based on supervised rank aggregation. While machine learning methods have been in use for a long time and have given reliable performance in various contexts, the supervised rank aggregation method is quite new and requires much work to establish its applicability in real applications. Also the fact remains that as they involve the use of aggregation methods, like approximate Kemeny aggregation, they have a computational complexity of $O(rn \log(n))$ where r is the number of attributes used and n is the number of examples in each input ranked list provided. But the preliminary results we get on the DBLP datasets validate the approach and encourage us to explore the method further.

A major challenge faced with these types of supervised approaches is the well known *extreme class skewness or class imbalance problem*. The number of actual new links is very small as compared to the number of possible links. As we can see, in the DBLP datasets we have used, the ratios of positive vs negatives links are 1:113, 1:248 and 1:282 in the three datasets, respectively. Also note that this imbalance increases with the size of the graphs used for experimentation. This makes it more difficult for an algorithm to generate a good model and make a good inference from the test data. Although very few of the negative examples have actual predictor value as positive examples, the model ends up giving a large number of raw false positives. Also, in the presence of a large class skew, the information carried by the positive examples gets diluted in the vast negative class. Moreover, unlike the classical machine learning context, in link prediction, the correct classification of positive examples is more important. The most common solution to this problem, as suggested by the existing research, is the sampling of negative examples. This can be done by random methods or by using some filters by distance, node degree, etc. Another way, on which we are working, is to use a filter based on a community detection algorithm. The assumption here is that two nodes that do not belong to the same community tend to remain unlinked for a longer period than those belonging to the same community. Thus they can have more meaning as negative examples during the learning of the model. Each method has its own advantages and disadvantages, but some some can be fairer than others. The sampling of the data is mostly done on the learning data. Sometimes it is required to sample test data, as in the case when an extremely large number of test examples places unreasonable demands on processing resources and storage. If for any reason this has to be done, proper care should be taken based on the context where the link prediction is to be done. More details about the class imbalance problem can be found in Devaney (2003), Tajfel (1978). In Tajfel (1978), there is a detailed description about how the predictor performance changes with the sampling of test data. They also provide valuable information about which performance measure is to be used for evaluating different link prediction techniques.

Last, but not least, we have presented, at the end of this paper, how to extend the traditional supervised machine learning based link prediction approach to predict links in multiplex networks. We proposed new attributes that capture multiplex information. By applying them to the prediction of co-authorship links, we showed that the use of multiplex attributes improves the prediction result. The same method can be used to predict links in any of the multiplex layers. With the preliminary results, we are really excited and hopeful that the multiplex information can prove to be very useful for different tasks in the analysis of a network.

Bibliography

Adamic, L. A., Buyukkokten, O., Adar, E. (2003), A social network caught in the Web, *First Monday* **8**(6).

Adamic, L., Adar, E. (2003), Friends and neighbors on the Web, *Social Networks* **25**(3), 211–230.

Airoldi, E. M., Blei, D. M., Fienberg, S. E., Xing, E. P., Jaakkola, T. (2006), Mixed membership stochastic block models for relational data with application to protein-protein interactions, in *Proceedings of the International Biometrics Society Annual Meeting*.

Al Hasan, M., Chaoji, V., Salem, S., Zaki, M. (2006), Link prediction using supervised learning, in *SIAM Workshop on Link Analysis, Counterterrorism and Security with SIAM Data Mining Conference*, Bethesda, MD.

Al Hasan, M., Zaki, M. J. (2010), A survey of link prediction in social networks, in C. C. Aggarwal, ed., *Social network Data Analysis*, Springer, chapter 9, 243–275.

Aslam, J. A., Montague, M. (2001), Models for metasearch, in *Proceedings of the 24th annual international ACM SIGIR conference on Research and development in information retrieval*, SIGIR '01, ACM, New York, NY, USA, 276–284, http://doi.acm.org/10.1145/383952.384007.

Barabási, A.-L., Albert, R. (1999), Emergence of scaling in random networks, *Science* **286**(5439), 509–512.

Battiston, F., Nicosia, V., Latora, V. (2013), Metrics for the analysis of multiplex networks, http://arxiv.org/abs/1308.3182.

Benchettara, N., Kanawati, R., Rouveirol, C. (2009), Calcul de recommandation par prédiction de liens dans un graphe biparti, in *Actes de l'atelier sur l'apprentissage et graphes pour les systèmes complexes (plate-forme AFIA 2009)*, Hammemt, Tunisie.

Benchettara, N., Kanawati, R., Rouveirol, C. (2010a), Apprentissage supervisé pour la pr'ediction de nouveaux liens dans des réseaux sociaux bipartie, in *Actes da la 17ième Rencontre de la société francophone de classification (SFC'2010)*, St. Denis, La réunion, 63–66.

Benchettara, N., Kanawati, R., Rouveirol, C. (2010b), Supervised machine learning applied to link prediction in bipartite social networks, in *International Conference on Advances in Social Network Analysis and Mining*, ASONAM 2010'.

Berlingerio, M., Bonchi, F., Bringmann, B., Gionis, A. (2009), Mining Graph Evolution Rules, in W. L. Buntine, M. Grobelnik, D. Mladenic, J. Shawe- Taylor, eds, 'ECML/PKDD (1)', Vol. 5781 of *Lecture Notes in Computer Science*, Springer, 115–130.

Berlingerio, M., Coscia, M., Giannotti, F., Monreale, A., Pedreschi, D. (2011), Foundations of Multidimensional Network Analysis, in *International Conference on Advances in Social Networks Analysis and Mining (ASONAM)*, IEEE, 485–489, http://dx.doi.org/10.1109/asonam.2011.103.

Black, D., Newing, R., McLean, I., McMillan, A., Monroe, B. (1998), *The Theory of Committees and Elections by Duncan Black, and Revised Second Editions Committee Decisions with Complementary Valuation by Duncan Black*, 2nd edn, Kluwer Academic Publishing.

Chebotarev, P., Shamis, E. (1997), The matrix-Forest Theorem and measuring Relations in small Social Groups, *Automation and Remote Control* **58**(9), 1505–1514.

Chevaleyre, Y., Endriss, U., Lang, J., Maudet, N. (2007), A short introduction to computational social choice, *SOFSEM 2007: Theory and Practice of Computer Science*, 51-69, http://www.springerlink.com/index/768446470RPLJ120.pdf.

Clauset, A., Moore, C., Newman, M. E. J. (2008), Hierarchical structure and the prediction of missing links in networks, *Nature* **453**(7191), 98–101.

Dwork, C., Kumar, R., Naor, M., Sivakumar, D. (2001a), Rank aggregation methods for web, in *Proceedings of the 10th international conference on World Wide Web, WWW '01*, ACM, Hong Kong, 613–622, http://doi.acm.org/10.1145/371920.372165.

Dwork, C., Kumar, R., Naor, M., Sivakumar, D. (2001b), Rank aggregation, spam resistance, and social choice, in *WWW '01: Proceedings of 10th international conference on World Wide Web*, 613–622.

Eronen, L., Toivonen, H. (2012), Biomine: predicting links between biological entities using network models of heterogeneous databases, *BMC bioinformatics* **13**(1), 119.

Fire, M., Puzis, R., Elovici, Y. (2013), Link prediction in highly fractional data sets, in *Handbook of Computational Approaches to Counterterrorism*, Springer New York, 283–300.

Fouss, F., Yen, L., Pirotte, A. , Saerens, M. (2006), An Experimental Investigation of Graph Kernels on a Collaborative Recommendation Task, in *Sixth International Conference on Data Mining (ICDM'06)*, IEEE, 863–868, http://dx.doi.org/10.1109/icdm.2006.18.

Fu, Y., Xiang, R., Liu, Y., Zhang, M., Ma, S. (2007), Finding Experts Using Social Network Analysis, in *Web Intelligence*, IEEE Computer Society, 77–80.

Gao, S., Denoyer, L., Gallinari, P. (2011), Temporal link prediction by integrating content and structure information, in *Proceedings of the 20th ACM international conference on Information and knowledge management - CIKM '11*, ACM Press, New York, USA, 1169, http://dblp.uni-trier.de/db/conf/cikm/cikm2011.html#GaoDG11.

Huang, Z., Li, X., Chen, H. (2005), Link prediction approach to collaborative filtering, in M. Marlino, T. Sumner, F. M. S. III, eds, *JCDL*, ACM, 141–142.

Huang, Z., Lin, D. K. J. (2008), The Time-Series Link Prediction Problem with Applications in Communication Surveillance, *INFORMS Journal on Computing*, **21**(2), 286–303.

Jaccard,P. (1901), Étude comparative de la distribution florale dans une portion des Alpes et des Jura., *Bulletin de la Société Vaudoise des Sciences Naturelles* **37**, 547–579.

Katz., L. (1953), A new status index derived from sociometric analysis., *Psychmetrika*, **18**(1), 39–43.

Kossinets, G. (2006), Effects of missing data in social networks, *Social Networks* **28**(3), 247–268.

Liben-Nowell, D. (2005), An Algorithmic Approach to Social networks, *PhD thesis*, M.I.T.

Liben-Nowell, D., Kleinberg, J. (2003a), The link prediction problem for social networks, in *Proceedings of the twelfth international conference on Information and knowledge management, CIKM '03*, ACM, New York, USA, 556–559, http://doi.acm.org/10.1145/956863.956972.

Liben-Nowell, D., Kleinberg, J. M. (2003b), The link prediction problem for social networks, in *CIKM*, ACM, 556–559.

Liben-Nowell, D., Kleinberg, J. M. (2007), The link-prediction problem for social networks, *JASIST* **58**(7), 1019–1031.

Lichtenwalter, R. N., Lussier, J. T., Chawla, N. V. (2010), New perspectives and methods in link prediction, in *Proceedings of the 16th ACM SIGKDD international conference on Knowledge discovery and data mining - KDD '10*, ACM Press, New York, USA, 243, http://dblp.uni-trier.de/db/conf/kdd/kdd2010.html#LichtenwalterLC10.

Lichtenwalter, R., Chawla, N. (2012), Link prediction: Fair and effective evaluation, in *International Conference on Advances in Social Networks Analysis and Mining (ASONAM)*, IEEE/ACM ,376–383.

Littlestone, N., Warmuth, M. K. (1989), Weighted majority algorithm, *IEEE Symposium on Foundations of Computer Science*.

Liu, Y.-T., Liu, T.-Y., Qin, T., Ma, Z.-M., Li, H. (2007), Supervised rank aggregation, in *Proceedings of the 16th international conference on World Wide Web, WWW '07*, ACM, New York, USA, 481–490, http://doi.acm.org/10.1145/1242572.1242638.

Lü, L., Zhou, T. (2011), Link prediction in complex networks: A survey, *Physica A: Statistical Mechanics and its Applications* **390**(6), 1150–1170. http://dx.doi.org/10.1016/j.physa.2010.11.027.

Mémoire sur les élections au scrutin (1781).

Menon, A. K., Eklan, C. (2011), Link prediction via matrix factorization, in D. Gunopulos, T. Hofmann, D. Malerba, M. Vazirgiannis, eds, Machine Learning and Knowledge Discovery in Databases, Vol. 6912 of *Lecture Notes in Computer Science*, Springer Berlin Heidelberg, 437–452.

Montague, M., Aslam, J. A. (2002), Condorcet fusion for improved retrieval, in *Proceedings of the eleventh international conference on Information and knowledge management, CIKM '02*, ACM, New York, USA, 538–548, http://doi.acm.org/10.1145/584792.584881.

Newman, M. E. J. (2004), Coauthorship networks and patterns of scientific collaboration, *Proceedings of the National Academy of Science of the United States (PNAS)* **101**, 5200–5205.

Ou, Q., Jin, Y. D., Zhou, T., Wang, B. H., Yin, B. Q. (2007), Power-law strength-degree correlation from resource-allocation dynamics on weighted networks, *Phys. Rev. E* **75**, 021102, http://dx.doi.org/10.1103/PhysRevE.75.021102.

Sculley, D. (2007), Rank aggregation for similar items, in *Proceedings of the Seventh SIAM International Conference on Data Mining (SDM)*.

Subbian, K., Melville, P. (2011), Supervised rank aggregation for predicting influence in networks, in *Proceedings of the IEEE Conference on Social Computing (SocialCom-2011)*, Boston.

Taskar, B., Wong, M. F., Abbeel, P., Koller, D. (2003), *Link Prediction in Relational Data*, in S. Thrun, L. K. Saul, B. Schölkopf, eds, NIPS, MIT Press.

Wang, C., Satuluri, V., Parthasarathy, S. (2007), Local Probabilistic Models for Link Prediction, in Y. Shi, C. W. Clifton, eds, *IEEE International Conference on Data Mining (ICDM)*.

Yin, D., Hong, L., Davison, B. D. (2011), Structural link analysis and prediction in microblogs, in *Proceedings of the 20th ACM international conference on Information and knowledge management - CIKM '11*, ACM Press, New York, USA, 1163, http://dblp.uni-trier.de/db/conf/cikm/cikm2011.html#YinHD11.

Sun, Y., Barber R., Gupta M., Aggarwal, C.C., Han, J. (2011), Co-Author Relationship Prediction in Heterogeneous Bibliographic Networks, in *Advances on social network Analysis and mining (ASONAM)*, Kaohsiung, Taiwan.

Young, H., Levenglick, A. (1978), A consistent extension of condorcet's election principle, *SIAM Journal on Applied Mathematics* **35**(2).

Zhou, T., Lu, L., Zhang, Y.-C. (2009), Predicting missing links via local information, *The European Physical Journal B*, **71**(4), 623–630.

Chapter 8

The social functions of gossip and the Leviathan model

Sylvie Huet[1]

8.1   Introduction

In everyday conversation, 70% of our time is spent gossiping (Emler, 1990; Foster, 2004; Wert and Salovey, 2004)! Our peers are our preferred subject of discussion whenever we meet each other. This choice is so natural and generic to human societies that it is an issue of great interest for social scientists. Indeed social scientists commonly wonder about the purpose and the impact of such a frequent activity. A body of literature has proposed many theories explaining, in terms of social functions, why we gossip (Foster, 2004; Conein, 2011; Beersma and Van Kleef, 2012) but the question remains open to debate.

   To participate in the debate, we studied the Leviathan model. It considers agents gossiping and having an opinion of each other (Deffuant, Carletti, and Huet, 2013). The question of the social impact of gossip is then relevant

[1]Irstea, UR Laboratoire d'ingénierie des systèmes complexes, 9 avenue Blaise Pascal, F-63178 Aubière, France, E-mail: sylvie.huet@irstea.fr

for this model and our conclusions can feed the debate in social psychology even if they are only really shown to be significant for this model.

The Leviathan model has been recently proposed. It brings a new and unique insight into the relation between agents' respective evaluations and group structure. What is the essence of this model? It is a theory explaining how people structure themselves from the agent's need to form an opinion of the others, including themselves. It considers agent interactions through meeting in pairs. Motivated by the need to be held in high esteem (Hobbes, 1651), agents act in self-defence, applying a process called vanity. They protect themselves from being despised by sanctioning the despiser, or favour a compliment by rewarding the compliment giver. They also gossip about their peers, influencing each other with regard to what they think of them. Various structures and leadership styles emerge from the meeting dynamics. The result could be an absolute dominance, a very hierarchical society, or a crisis in which everyone hates each other, including themselves. Egalitarianism and elitism are also forms of power structures emerging from the Leviathan model.

The power of such a virtual world model is the understanding of the impact the agents' behaviour has on the structure. The choice to gossip or not can change the whole way people see each other. The intensity of the gossip can change the equality of the structure and the leadership style. Even if the Leviathan model is a very simply constructed world, we can experiment with this world. We can observe and understand how agents' direct interactions and practices of gossip build the structure. We can demonstrate the emergence of a leadership style from chosen agent dynamics. That is the reason why it is the relevant tool to feed the debate, and possibly inspire new experiments on the social functions of gossip.

In the Leviathan model, gossip occurs during dyadic interactions. That is the most common situation for gossip accordingly to Conein (2011). However, the model also considers gossip varying in several ways. It can vary in intensity, from its absence to a high number of discussed peers: the more people a speaker talks about, the more intense is the gossiping during a meeting. The impact of gossiping is considered according to various levels of openness of people. This openness corresponds to a parameter controlling how high a speaker should be held in esteem to influence the listener. Very open-minded agents are influenced whatever the level of esteem in which they hold the speaker. Very narrow-minded agents are only influenced by speakers held in high esteem. The strength of gossip is also ruled by a propagation coefficient. This coefficient and the openness are also used to control how much two talkers influence each other.

In this paper, we are interested in understanding the relation between gossip and two social phenomena: consensus and the positivity bias. These two properties of a population together seem somewhat counterintuitive: a population needs some consensus to act as a group, at the same time the positivity bias is said to be quite universal and it means people diverge. This paradox can perhaps be solved by the understanding of its links to gossip and its social functions (Foster, 2004).

Deffuant, Carletti, and Huet (2013) have shown that the Leviathan model is able to exhibit these two social phenomena. They emerge from the individual's need to form a valuation of themselves (i.e. self-valuation), as well as defining the value of others, through direct interaction and gossip. The particular role of gossip in their emergence and maintenance has not been exhaustively investigated in this model.

That is the purpose of this paper, which starts from four hypotheses: gossip leads to consensus which increases with its intensity; gossip decreases the strength of the positivity bias and can suppress it; positivity bias and disagreement are linked to each other; the positivity bias and bias to negativity occurring in the Leviathan model appear conjointly whatever the level of gossip (they have been conjointly diagnosed in the first investigations of Deffuant et al. 2013). Overall, our hypotheses are confirmed. We especially find that consensus is almost never reached without gossip. We also show how an asymmetrical level of openness to the influence of others, depending on how high they are held in esteem, is important for the positivity bias, as well as for the bias to negativity. The number of peers discussed during a meeting is also shown to be essential for consensus.

While the next section is dedicated to a short review of the literature, the following one presents the Leviathan model as well as our experimental design. A section presenting the results of our analysis comes next. A final section is entirely focused on synthesizing and discussing our conclusions.

8.2   Literature review

Firstly, we briefly review the body of literature related to gossip and the properties of populations. Secondly, we sum up what we already know from the Leviathan model which is likely to help our study. This literature review will be used in section 8.3.2 to argue about our hypotheses.

8.2.1   Gossip, reputation and positivity bias in the social literature

The social functions of gossip

In his review, Foster (2004) mentions that gossip has been poorly studied in the past. From the research it reviews, it reports four major social functions of gossip: "information (Dunbar and Duncan, 1997), entertainment, friendship, and influence". Among the numerous functions proposed by the review, we can cite the following ones. "From gossip, the individual gets a map of his social environment ... can benefit from an elevated social status. It allows building norms ... reinforcing friendship through sharing norms..., controlling cheating, ... ; it hold communities together against the forces of social entropy". Similarly, Kniffin and Wilson (2005) suggest that "gossip can be used within groups to enforce norms". Overall, it seems a lot of social functions have been proposed for gossip, without being always supported by experiments, or hierarchized.

Reputation

A reputation is a "consensus among knowledge informations as to the attributes of targets" (Emler, 1990; Moscovici, 2000; Arrow and Burns, 2004). Emler describes reputations as "social representations" (Moliner (2001), describing Moscovici's concept) of the self: "reputations are social constructions, created collectively through processes of social communication, and are not to be confused with one individual's perception of another". He also explained that reputation emerges from interactions between individuals who are motivated to form a social identity: "I also want to suggest that defining yourself in terms of a particular social identity is largely a matter of persuading others to so define you. ... Social identity are conferred or agreed by the collective, not merely assumed by the individual. ... they imply shared definitions. ... they are overtly expressed or communicated claims on the individual's part. ... those claims are negotiated with others." This negotiation and sharing is made through direct interactions between individuals and gossip, as is again explained by Emler (1990): "... that people do regularly exchange information about their mutual acquaintances (Emler and Fisher, 1981). In a series of small-scale exploratory surveys of the informal conversations of students and teachers, we discovered that the most common topic of conversation after self-disclosure was named acquaintances, ... Gossip of this kind does seem to be a pervasive human activity (Paine, 1967), and given the frequency of such conversations (Emly and Grady, 1986), it seems a reasonable inference that people do in fact engage in extensive information gathering about reputations. But in these conversations we also exchange a great deal of information about ourselves. There are grounds for believing these aspects of conversation concern our own reputations."

Complementary to Emler's point of view, we have to cite the process of self-categorization through which an individual socially defines himself as a group member. The category, that we can also call a stereotype, can be seen as a reputation. This process is the core of social identity building and the formation and maintenance of a group (Tajfel, 1978; Turner, 1984; Turner et al., 1987; Hogg, Turner, and Davidson, 1990).

Positivity bias

By contrast to consensus, the positivity bias comes from someone's self-judgement that they are better than the others. But similarly to the consensus, it is also a social phenomenon, not a fact about an individual. Indeed, it is diagnosed in an experiment when a majority of participants in the experiment say they are better than the average (Hoorens, 1993). Positivity bias, which can also be called confirmatory bias or illusory superiority (Hoorens, 1993) is strongly supported empirically. It is diagnosed especially when, to the classical question "do you assess yourself better than the average in a particular domain", a more or less large majority of people answers "yes". It is well established that people tend to rate themselves as better than average across many domains. Only the size of the majority of people answering "yes" seems to change over the domain. For example, it is well known that drivers rate themselves better than the average, whatever their level of training and experience (Waylen et al., 2004). The same authors indicate "Grayson (1998) found that, in a sample of more than 1000 motorists, 80% considered themselves to have better than average driving ability". Many other examples of what it is also called self-serving bias are given in (Myers et al., 2009), such as "90% of business managers rate their performance as superior to the average". This bias has been shown to be quite universal (Myers et al., 2009). Gossip, the universality and importance of which has also been shown (Foster, 2004; Conein, 2011; Beersma and Van Kleef, 2012) is said to be a source of self-suffering as well as a way to punish cheaters. It probably affects the positivity bias since it seems able to decrease someone's self-opinion (i.e. the opinion people have of themselves).

8.2.2   The Leviathan model, reputation, and positivity bias

The modelling approach

The Leviathan model was designed according to a classical approach in the field of social simulation or in sociophysics. It consists in making a few simple assumptions about the rules of interactions between agents and the way they change their opinions. The aim is to study the emerging behaviours obtained. In some of these models, the agents have binary opinions (Sznajd-Weron, 2005; Galam, 2008), while in others the opinions are continuous (Deffuant et al., 2000, Fortunato et al., 2005; Deffuant, 2006; Lorenz and Urbig, 2007; Urbig and Malitz, 2007; Huet and Deffuant, 2008; Gargiulo and Huet, 2010; Huet and Deffuant, 2010; Gargiulo and Huet, 2012) (see Castellano et al., 2009 for a review). For some parameters, our model is also close to ones which include a set of affinities between agents, leading to emerging networks (Carletti and Righi, 2011).

This model is called Leviathan model in reference to Hobbes (1651), who pointed out that the feeling of being undervalued is a major source of violence. In practice, the basic Leviathan model (Deffuant et al. 2013) assumes that each agent can have a continuous opinion about every other agent, truncated if necessary to remain between -1 and +1. In the initial state, the agents don't have an opinion about the others. The agents inter-act in randomly chosen pairs and two different processes apply. The first one supposes that during any interaction, each agent propagates its opinions about herself, about its interlocutor and about several randomly chosen other known agents. In this propagation, highly valued agents are more in-fluential. The second process represents a vanity effect: an agent likes to be highly valued by the others, thus it increases its opinion of those who value its well. In contrast, it decreases its opinion of those who undervalue her. These assumptions are inspired by Hobbes (1651), but also by more recent experiments and observations from social psychologists (Fein and Spencer, 1997; Buckley et al., 2004; Srivastava and Beer, 2005; Leary et al., 2006; Stephan and Maiano, 2007; Wood and Forest, 2011). Moreover, we suppose that access to the opinions of others is not perfect: people may not express exactly what they think, and the listener may misinterpret these expressions. We consider such a defect as distorting the opinion of the speaker. To model this distortion, we add or suppress a random noise to the opinion to be prop-agated.

The behaviour of the model

From its first study and despite its simplicity, the model shows a surprising variety of dynamic structures when changing the parameters. The following structures have been identified up to now.

- Equality. Each agent **has a positive opinion about herself**; it is **connected by strong positive mutual opinions** with a small set of agents and has very negative opinions about all the others. All agents have a similar number of positive (and negative) links, called friends (respectively, foes). For some parameters, the network of positive links shows the characteristics of small world networks.

- Elite. The pattern shows two categories of agents: the elite agents and second category agents. The **elite agents have a positive self-opinion** and **are strongly supported by a friend**, but they have a very negative opinion of all the other elite agents and of all the second category agents. The second category agents have a very negative self-opinion, they have a very negative opinion of all the other second category agents, and their opinion about the elite agents is moderate.

- Hierarchy. **All agents share a similar opinion about every other agent** (called reputation) and the reputations are widely spread between -1 and +1. There are more agents of low reputation than of high reputation: this gives the image of a classical hierarchy with a wide base and progressively shrinking when going up to the top.

- Dominance. As in the hierarchy pattern, **all agents share a similar opinion about every other agent,** but a single agent has a high reputation while most of the other agents have a very low reputation.

- Crisis. Each agent **has a very negative opinion of all the others and of herself.**

One can notice that these structures exhibit some properties that we relate to gossip in the previous section: agreement about opinions, positivity bias, ... The first study has mainly pointed out the importance of the propagation coefficient of the influence[2] compared to the vanity one for explaining how these structures emerge. The particular impact of gossip has not been deeply investigated.

---

[2] both in direct meetings and via gossip

Two various forms of positivity bias have been identified in the first study of the Leviathan model and linked to each other.

The first case of positivity bias shown in Deffuant et al. (2013) is associated to a value of the influence coefficient higher than 0 while the vanity coefficient is zero. This bias is not significant in terms of the number of people held in low esteem, but in terms of the average self-opinion over time compared to the average reputation (the average of all the opinions) over time. A small bias in favour of self-opinion compared to reputation was noticed by Deffuant et al. (2013). For these particular parameter values, someone's self-opinion is very close to the opinion others have of the agent. The bias is due "to propagation coefficient ruling the influence. The higher is someone's self-opinion, the larger is its influence propagation coefficient" (since the other's opinion of its is very close to this self-opinion). This larger influence is due to the asymmetry of the propagation coefficient computation ruled by σ giving more influence to an agent held in high esteem. Because of this difference of influence, when an agent's self-opinion is higher than its reputation, the others have less influence on the self-opinion than when the self-opinion is lower than the reputation (everything else being equal). However, the effect of this average difference between self-opinion and reputation depends on the value of the agent's reputation: ... Hence, highly valued agents tend to lead the others' opinions and, with the statistical bias for a self-opinion higher than the reputation, they tend to increase their reputation. This is in contrast to the case of poorly valued agents, who tend to naturally decrease their self-opinion, only by the effect of the propagation coefficient.

Below a threshold of self-opinion, reputations and self-opinions tend to be biased towards negative valuations, and above that threshold, towards positive valuations. The value of this threshold depends on σ, determining the propagation coefficient, and also on k, the number of agents about whom the opinions are propagated during the encounters. Indeed, this number has an impact on how the agents propagate their opinion about the others.

These observations are useful for understanding the processes when the vanity is higher than zero. First note that the vanity process enhances the tendency of self-opinion to be higher than reputation. Indeed, the small statistical positive bias for self-opinion that is due to the opinion propagation leads, on average, the agents to consider themselves as more or less undervalued by the others, thus they devalue them due to their vanity. This explains the tendency of agents of the Leviathan model to obtain negative opinions.

Considering a larger vanity leads us to the second case of positivity bias shown in Deffuant et al. (2013) even if it has not been identified as such. It is associated to a very low propagation coefficient for a large vanity one. The reference paper (Deffuant et al., 2013) does not explicitly talk about positivity bias in this case but it is obvious from the analytical demonstration regarding how to compute the average number of agents positively assessed by an agent. Indeed, for these particular values of the parameters, the matrix of opinion is symmetrical around the diagonal representing the self-opinion of individuals. An agent can only be highly positively valued or highly negatively valued. The agents maintain themselves with a good self-opinion in a dynamic relational equilibrium between a few friends who flatters them and who are flattered in return, and a large number of foes who punish them and who are punished in return. For these particular parameter values, the number of friends and foes is similar for every agent. Foes are agents held in low esteem while friends are agents held in higher esteem than oneself. The number of foes can be computed analytically, as shown in Deffuant et al. (2013). This number gives an indication of the strength of the positivity bias: having more foes than one-half of one less than the size of the population indicates a positivity bias. From their demonstration, the authors conclude that the effect of the main parameters, for this particular set of parameter values (a coefficient very close to zero for influence and a large coefficient for vanity):

- "Increasing δ (noise parameter) decreases the final equilibrium value of the self-opinion, and hence the final number of friends of each agent;

- Increasing σ (ruling the slope of the sigmoid function of the influence parameter) will decrease the difference of weights between friends and foes and therefore decrease the equilibrium value for a given number of friends. This will thus result in a larger number of friends at the stable state;

- Increasing the number of agents in the population decreases the valuation by the self-opinions for the same number of friends. Therefore, it will result in a larger number of friends at the stable state.

- Increasing ρ increases the fluctuations around the equilibrium self-opinion and hence will decrease the number of friends at the stable state. "

Regarding the "gossip" parameters we are interested in, we retain from the first case of positivity bias that increasing σ decreases the strength of the positivity bias while increasing ρ increases it. In the second case, we retain σ and k or an increase of vanity can change the observed dynamics. More-over, from the previous work, we also know that for these parameter values, and due to vanity, the whole population is biased toward the negative. We also notice here the presence of a global bias of the population toward the negative. This global bias to negativity seems to be a characteristic of the dynamics of the Leviathan model since it can come both from vanity and influence.

Overall, Deffuant et al. (2013) proposes that the positivity bias is due to some particular *equilibria* associated to pairs of the influence and vanity coefficients. However, a possibly much more important role of gossip in the emergence and the disappearance of the positivity bias deserves a specific investigation.

8.3   Material and methods

8.3.1   The model [3]

We consider a set of N agents. Each agent i is characterized by its list of opinions about the other agents and about herself: $(a_{i,j})$, $\ 1 \leq i, j \leq N$. We assume $a_{i,j}$ lies between -1 and +1, or it is undefined (equal to nil) if the agent i never met j and nobody has yet talked to i about j. At initialisation, we suppose that the agents have never met, therefore all their opinions are undefined. When opinions change, we always keep them between -1 and +1, by truncating them to -1 if their value is below -1 after the interaction, or to +1 if their value is above +1.

The agents interact in uniformly and randomly drawn pairs (i, j) and at each encounter, we apply two processes: the face-to-face management, implying influence attempts and vanity between the two agents meeting each other; and the gossip, consisting in influence trials about people they know (sometimes including themselves). Someone knows someone else if they have already met or been heard about. In this case, the opinion of one about the other is different from nil.

We follow the people's interactions considering a time range called it-eration. We assume one iteration, i.e. one time step $t \rightarrow t + 1$, has $N/2$

---

[3]In this paper, the functions of the Leviathan model have been presented in a different order, allowing distinguishing the direct experience of meeting a peer from the indirect contact with a peer through gossiping. In practice, this means that the events do not occur at exactly the same time during the meeting and that the number of times the bounding feature of the opinion is checked is smaller. The model remains unchanged despite this small difference.

random pair interactions (each agent interacts N times on average during one iteration).

The following describes the processes occurring during a pair's meeting, with an incremental approach allowing understanding how they are coupled to each other. We start with gossip, which allows us to explain how agents influence each other. Then we continue with the management of the face-to-face meetings during which not only influence but also vanity plays a role. We finally present the global iterative loop during which pair meetings occur.

Gossip: Agents discuss their peers

Let us assume that agents i and j have been drawn. During an encounter, we suppose that agent j propagates to i its opinions about itself (j), about i, and about k agents randomly chosen from its acquaintances. Moreover, we suppose that if i has a high opinion of j, then j is more influential.

This hypothesis is implemented by introducing a propagation coefficient, denoted p_{ij}, which is based on the difference between the opinion of i about j (a_{ij}) and the opinion of i about itself (a_{ii}). It uses the logistic function with parameter σ. If a_{ij} = nil (j is unknown to i), we assume that i has a neutral opinion about j and we set a_{ij} to 0. Let us also observe that, at the initialisation, an agent has no opinion about herself (a_{ij}= nil), before it takes part in a first encounter, thus we also set a_{ij}= 0. Then we compute the propagation coefficient p_{ij}, which rules the intensity of the opinion propagation from j to i:

$$p_{ij} = \frac{1}{1 + \exp\left(-\frac{a_{ij}-a_{ii}}{\sigma}\right)}$$

The parameter σ defines the slope of the function close to $a_{ij} - a_{ii}$. Figure 8.1 represents the value of p_{ij} when the difference $a_{ij} - a_{ii}$ varies (between -2 and +2), for three different values of σ. One can observe that p_{ij} tends to 1 when $a_{ij} - a_{ii}$ is close to 2 (i values j more highly than herself), and tends to 0 when it is close to -2 (i values j lower than herself). Indeed, when σ is small, p_{ij} rapidly changes from 0 to 1. When σ is large, this change is progressive.

A parameter ρ controls the impact of the coefficient p_{ij}. The agent i modifies its opinion about the agent z that i talked about applying the product of the influence coefficient ρ by the propagation coefficient to the difference between what j said about z and what it thinks of z. However, i has no direct access to the opinion of j and can misunderstand j: for example,

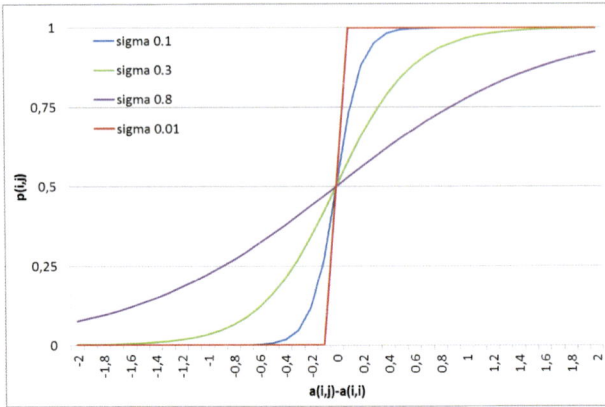

Figure 8.1: Examples of variations of the propagation coefficient p_{ij} when $a_{ij} - a_{ii}$ varies, and for four values of the parameter σ when σ decreases, this function tends towards a threshold function returning 0 for negative entries and 1 for positive entries.

j may not express exactly what they think and/or i may misinterpret these expressions. To take this into account in the model, the propagated opinions are distorted by noise. This noise is identical for every conversation and can also be seen as a difficult context not allowing the two partners to perfectly understand each other.

To take into account this difficulty, we consider the perception of i as the value a_{jz} more or less a uniform noise drawn between $-\delta$ and δ (δ is a model parameter). This random addition then corresponds to a systematic error the agents make regarding the others' opinions. More formally, the process can be written in pseudo-code as follows:

Gossip(i, j)
  Repeat k times:
    Choose randomly z taking into account $a_{jz} \neq$ nil, z /neq j.
    If $a_{jz} = nil$, $a_{jz} \leftarrow 0$
    $a_{iz} \leftarrow a_{iz} + \rho p_{ij}(a_{jz} - a_{iz} + \text{Random}(-\delta, +\delta))$

Random $(-\delta, \delta)$ returns a uniformly distributed random number between $-\delta$ and $+\delta$, which can be seen as a noise that distorts the perception that i has about j's opinions. The parameter δ rules the amplitude of this noise.

The face-to-face meeting activates influence and vanity

During their first meeting, i and j don't know each other and their opinions are nil. Then, they instantaneously become 0, which is the neutral opinion. This initiates the meeting dynamics and allows influence and vanity to act.

Indeed, when agents i and j meet, they talk about themselves: i talks about itself and j, while j talks about itself and i. This direct exchange implies two processes occurring at the same time: the influence of each of them on what they think about themselves and the other, and a vanity process applied only by the listener to the talker. This vanity process expresses the fact that agents tend to reward the agents that value them more positively than they value themselves and to punish the ones that value them more negatively than they value themselves. Then, added to the influence i received from j regarding what it thinks about j, agent i compares her self-opinion a_{ii} to the opinion j transmits about its a_{ij}. If the perceived opinion of the other (j) is higher than its self-opinion, i increases her opinion of j (reward). Otherwise, i decreases its opinion of j (punishment).

The parameter ω rules the importance of the vanity process. The modification of i's opinion of j is assumed to depend simply on the difference between the opinion of i about itself and the opinion of j about i (slightly modified, at random).

The face-to-face meeting can be formally described in pseudo-code as follows.

Face-to-face meeting(i,j)
   If $a_{ii} = nil$, $a_{ii} \leftarrow 0$
   If $a_{ij} = nil$, $a_{ij} \leftarrow 0$
   $a_{ii} \leftarrow a_{ii} + \rho p_{ij}(a_{ji} - a_{ii} + \text{Random}(-\delta, +\delta))$
   $a_{ij} \leftarrow a_{ij} + \rho p_{ij}(a_{jj} - a_{ij} + \text{Random}(-\delta, +\delta)) + \omega(a_{ji} - a_{ii} + \text{Random}(-\delta, +\delta))$

During the interaction, face-to-face meeting (i, j) and face-to-face meeting (j, i) are successively applied.

Summary

The model has 7 parameters:

- N, the number of agents;

- δ, the maximum intensity of the noise when someone is alluded to;

- σ, the reverse of the sigmoidal slope of the propagation coefficient;

- ρ, the parameter controlling the intensity of the coefficient of the influence process (applied to the propagation coefficient);

- k, the number of acquaintances an agent talked about during a meeting – they are randomly chosen from among its acquaintances; and

- ω, the coefficient of the vanity process.

The following algorithm describes one iteration: $N/2$ random pairs of agents are drawn, with reinsertion, and we suppose that each agent influences the other during the encounter.

```
Repeat N/2 times:
    Choose randomly a couple (i, j)
    Save the opinions which are going to change in temporary variables to ensure the update
    during the i and j meeting is synchronous
    Face-to-face meeting (i, j)
    Face-to-face meeting (j, i)
    Gossip(i, j)
    Gossip(j, i)
```

The update is synchronous: every opinion change occurring during a meeting is computed on the same value of opinions taken at the beginning of a pair meeting.

8.3.2   Hypothesis and methods

This section describes our hypothesis as well as the experimental design and the measured indicators. A later subsection describes the results.

Hypotheses

From our literature review, we formulated four hypotheses to test.

H1:   In the Leviathan model, the gossip parameters have a strong influence on the level of consensus of the population. In the previous study of the model, the appearance of consensus has been analyzed in terms of a rapport between the power of vanity and the power of the propagation of opinion. On the other hand, the social psychology literature explains that consensus emerges and is maintained through direct interaction and gossip: this would be relevant to explore and debate the specific role of gossip by considering it in the Leviathan model;

H2:   In the Leviathan model, the gossip parameters have a strong influence on the strength of the positivity bias. Previous study has mentioned some effect of the openness of people and sometimes of the number of discussed peers without really investigating them. Moreover, in the literature, gossip is at the same time considered as dangerous for someone's self-suffering from a bad reputation, and explained as a social way to control cheating. That would mean gossip probably decreases the strength of the positivity bias. Perhaps it can lead a majority of people to undervalue themselves. Once again, the Leviathan model can provide elements to discuss;

H3:   Positivity bias and consensus level of the population are related to each other: it seems indeed counterintuitive that the positivity bias and consensual reputation can be present simultaneously. How can a population that judges themselves higher than the average agree on someone's value? We hypothesize there is a negative correlation between positivity bias and consensus;

H4:   The simultaneous occurrence of positivity bias and a bias to negativity does not depend on gossip. The two biases have been jointly observed in previous study of the Leviathan model.

Three parameters of the model are involved in the control of gossip: ρ, σ and k. While the impact of ρ has been already properly studied in regards to ω, only two values of k and σ have been tested (2 and 10, respectively, 0.3 and 0.5). Then, in order to draw conclusions about the impact of gossip in the model, we have to vary these two parameters more.

To test our hypotheses, we elaborated the following experimental design.

Experimental design

The model includes 7 parameters, and it is difficult to make an exhaustive study in the complete parameter space. We aim to study the impact of gossip. In the model, the intensity of gossip is controlled through three parameters: the coefficient of propagation, the number of agents someone talks about, and the degree of openness to influence. While the first study (Deffuant et al. 2013) varied a lot the influence and vanity coefficients, only two values have been tested for the other two parameters. To study the impact of gossip, we have now to focus our experimental design on these two parameters: k, ruling the gossip intensity in terms of number of agents one agent talks about during a meeting; and σ, the level of openness to influence, depending on the level of esteem for the talker. We also decided to vary the noise δ, but not as much as k and σ. Also ρ and ω vary little, since their co-variation has been studied in detail. We fix N, the number of agents to 40, in order to make our investigation tractable.

In practice, we vary the other parameters as follows:

- k, the number of discussed acquaintances, takes the values 0, 1, 2 , 3, 4, 5, 10, 15, 20, 25 and 30;

- σ, ruling the slope of the logistic function determining the propagation coefficients, takes the values 0, 0.01, 0.02, 0.06, 0.1, 0.2, 0.3, 0.4, 0.5, 0.6, 0.7, 0.8, 0.9, 1.0;

- ρ, ruling the intensity of the overall influence by being applied to the propagation coefficient, takes on three values: 0.05, 0.5, 1;

- ω, ruling the intensity of vanity, 0.05, 0.5, 1;

- δ, the intensity of noise disturbing the evaluation of others' opinions, takes the values 0.1 and 0.3.

For each set of parameter values, we ran the model for 201,000 iterations (one iteration corresponding to $N/2$ random pair interactions), and we repeated this 15 times.

Measuring indicators

From iteration 30,000 to 200,000, we measured every 10,000 iterations a group of values allowing us to draw some conclusions about the impact of gossip. The measures, averaged over number of times of a run and over the 15 replicas, gives us indicators. The indicators we built to test our hypothesis are the following.

The average disagreement. This indicates the level of consensus. It is computed as the average over the population of the maximum difference of opinion about someone, considering everyone except the agent herself. The average measure is finally called the average disagreement. The disagreement varies from 2 to about δ (recall that we set δ equal to 0.1 and 0.3). The value 2 corresponds to the maximum difference between the minimum (-1) and the maximum (+1) opinion and δ to the average "noise" around the opinion of someone when everyone agrees on this opinion.

The average number of underestimated agents. This indicates the level of the positivity bias.

It is the average number of agents that each agent of our population assesses as lower than herself. Our experimental design considers a population of 40 agents. This means that if an agent considers that more than (40-1)/2=19.5 agents are lower than herself, it assesses itself as better than the average. Then, we decide to diagnose the existence of a positivity bias when the average number of underestimated (equal to "assessed as lower") agents is higher than 19.5. Indeed, this indicates that more than 50% of the agents assess themselves as higher than the average. When the average number of underestimated is close to 39, this means that almost everyone believes that they are the highest valued agent in the population.

The average share of positive opinion. This indicates a bias to negativity of the whole population.

We measure the number of positive opinions and compute the relative share (ie number of positive opinions/total number of opinions) to assess the bias to negativity of the population. Indeed, this bias increases when the average share decreases. It is 0 when the average share of positive opinions is 0.5 and maximal when the average share of positive opinions is 0.

8.4   The results of the simulations

This section presents the influence of gossip in the Leviathan model on various aspects and the results of the tests of our four hypotheses. The first subsection shows its influence on the level of consensus. The second deals with the strength of the positivity bias in regard to gossip. The third shows how the level of consensus, the positivity bias, and the bias to negativity, identified in Deffuant et al. (2013), relate to each other.

8.4.1   Gossip and the level of consensus

Our purpose is to understand the influence of gossip on the level of consensus in the population on the valuation of each agent. To measure the level of consensus, we compute the average level of disagreement, which varies from 2 to about δ (see section 8.3.2 for more details).

Figure 8.2 shows that disagreement decreases with the intensity of gossip in terms of the number of people someone talked about, k (on the top). Disagreement is particularly high in the absence of gossip (i.e. $k = 0$). However, in the bottom panel of Figure 8.2, we can observe that consensus can be reached even in the absence of gossip for some particular sets of parameter values. This occurs in two different cases: for $\sigma < 0.2$ and ρ=0.05

for $\omega = 0.5$ or 1 and $\rho = 0.5$ for $\omega = 1$ in which the disagreement is zero; for $\sigma \geq 0.2$ and $\rho = 0.5$ for $\omega = 0.05$ and $\rho = 1$ for $\omega = 1$ in which the disagreement roughly corresponds to the average value of tested δ (0.1 and 0.3). On the right part of Figure 8.2, we also observe that depending on whether the vanity is low or high, an increase in the level of openness can increase or decrease the level of disagreement.

The top of Figure 8.3 shows how these parameters change the average disagreement. For $\rho \geq 0.5$, the average disagreement tends to increase with ρ and ω and is not highly sensitive to σ. In contrast, for $\rho = 0.05$, the average disagreement becomes very sensitive to σ when ω is larger than ρ (ω equal to 0.5 or 1). We observe a particular "threshold" behaviour occurring for the smaller value of ρ and larger values of ω (0.5 and 1), and depending on σ: a low disagreement for $\sigma < 0.2$, and a high disagreement for $\sigma \geq 0.2$. We know this threshold behaviour is not explained by k since it always occurs. But even for this low $\rho = 0.05$, we can see on the bottom of Figure 8.3 that the average disagreement remains sensitive to gossip since the results for $k=0$ are different from the results for $k > 0$ whatever the σ. Also, on the lower panel of Figure 8.3 we can observe, even in the conditions of high disagreement due to a low value of ρ and a high value of σ (red bar), an increase in the intensity of gossip. This means again an increase in k decreases the average disagreement.

When the influence coefficient ρ is low (0.05) combined with a larger vanity ($\omega > 0.05$), accepting to be influenced in face-to-face meeting or in gossip by people held in lower esteem than oneself (i.e. larger values of σ) is a source of disagreement between people. For these conditions, the behaviour is very singular since we observe a threshold: the level of disagreement is mainly driven by σ and depends on whether σ is lower than a threshold around 0.2. Under the threshold, the disagreement is low, but it is higher above the threshold. In any case, the average disagreement is decreased by an increase in k, but if the influence parameter is low regarding the vanity, this can require a very high intensity of gossip k. This particular behaviour is partly due to the characteristics of our indicator. Indeed, as shown at the bottom panel of Figure 8.4, we can observe only for our critical small $\rho=0.05$, low in regard to ω (=1 in the figure) that the values of σ for which the average disagreement is the lowest lie in a zone where the average number of agents holding themselves in the highest esteem (i.e. they think they are the best) is equal to the population size. In this particular case, the average disagreement is zero since it corresponds to the average maximum difference of opinion on someone without taking into account their self-opinion. Then, as everyone is valued as the worst, the difference is

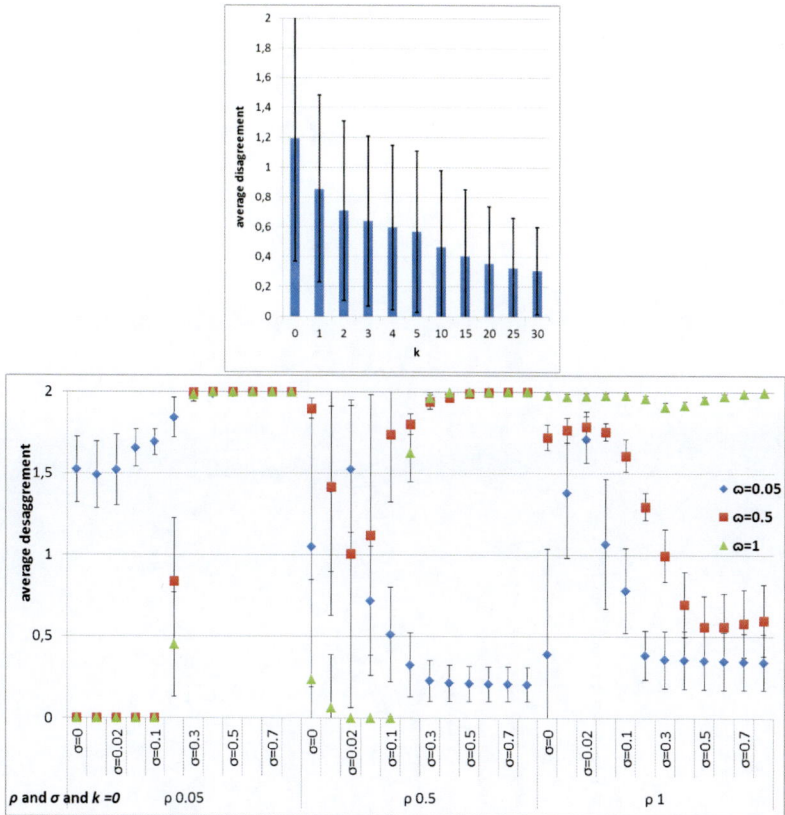

Figure 8.2: Average disagreement for various values of k (in abscissa of the graph on the top), and for $k = 0$ various values of ρ and σ) (in abscissa of graph at the bottom) and $\omega($ = 0.05 blue diamonds; = 0.5 red squares; = 1 green triangles). Each bar corresponds to the average over all results obtained from all tested values for all the parameters considered in the experimental design. Error bars corresponds to one standard deviation over or under these results.

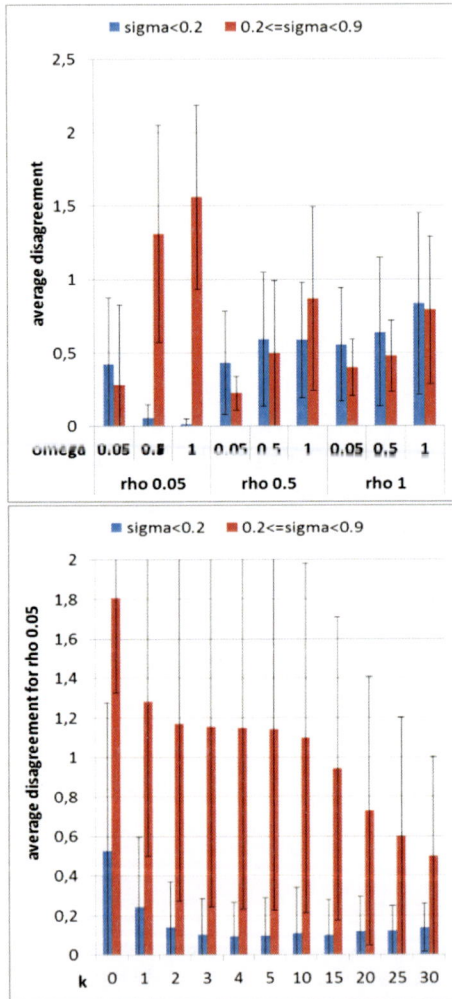

Figure 8.3: Average disagreement for, $\sigma < 0.2$ (blue bars) and $0.2 \le \sigma < 0.9$ (red bars): on the top, for various ρ and for each value of ρ, various ω on the bottom, for various k (in abscissa) and ρ=0.05. Each bar corresponds to the average over all results obtained from all tested values for all the parameters considered in the experimental design.

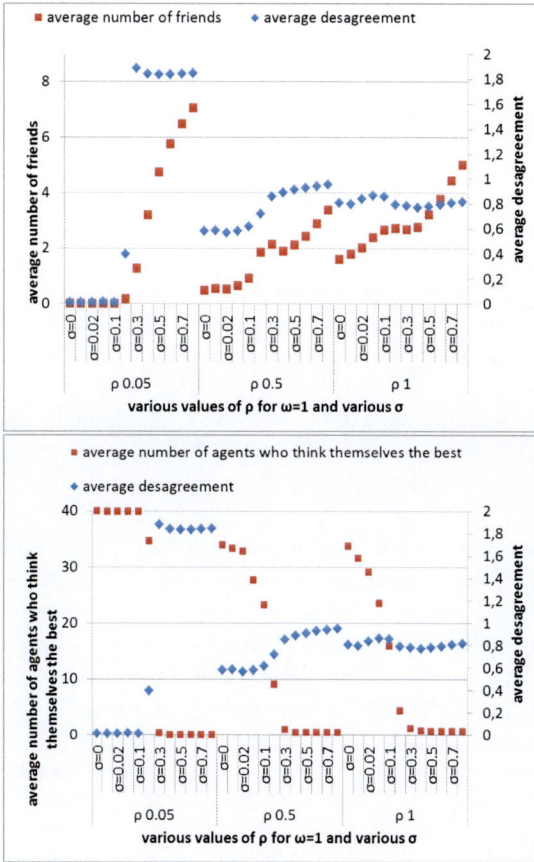

Figure 8.4: Average disagreement and average number of friends (on the top), and average number of agents who think they are the best (at the bottom) for $\omega=1$, various values of ρ and σ. Each dot corresponds to the average over all results obtained from all tested values for all the parameters considered in the experimental design.

zero, and the level of disagreement is assessed to be zero. This corresponds
to the structure "equality" in which the agents have no friends.

For larger values of σ, we observe on the top panel that the number of
friends (people of whom one has a positive opinion) increases dramatically
compared to the measures for the larger values of ρ. This set of parameter
values corresponds to a particular equilibrium exhibiting the positivity bias
identified in Deffuant et al. (2013) and called equality. This equilibrium
between "friends and foes" (see section 8.2.2 for more explanations) only
occurs when ρ is very close to being null and the vanity coefficient ω is large
in comparison. It is considered as a population of agents having friends held
in high esteem while they have the worst possible opinion of all the other
agents. Friends have mutually positive opinions of each other, and reward
each other. That is the way they are able to resist a large set of "hated"
agents who punish them and try to destroy the good opinion they have of
themselves. For this equilibrium, the disagreement is close to the maximum,
since friends are close to having the best possible positive opinion and foes,
the worst. Such a difference of opinions between people is not observable
elsewhere in the parameter space. Overall, that explains why we observe a
"threshold" behaviour depending on σ for these parameters of ρ=0.05 and
ω=0.5 and 1.

The main results of this section are: **no gossip, almost no agreement;
increasing gossip decreases the disagreement**. A consensus can hardly
be obtained when there is no gossip. It can emerge from direct meetings
only when the influence parameter is high for a very low vanity coefficient
and agents are open enough to each other. With the absence of gossip under other conditions, the disagreement remains. When gossip is introduced,
the disagreement decreases. More generally, the more people someone talks
about, the more consensual is someone's value. **This confirms our hypothesis H1**: gossip has a strong influence on the consensus level of the
population.

8.4.2   Gossip and the positivity bias

In order to test hypothesis H2, this section aims to determine whether and
how gossip participates in the emergence of the agents' positivity bias. To
diagnose a positivity bias in the model, we compute the average number
of agents that each agent of our population assesses as lower than herself.
When this average number is higher than 19.5 (for our experimental design
with 40 agents in the population), we diagnose a positivity bias (see section
8.3.2 for more details).

Figure 8.5 presents how the average number of underestimated agents, indicating the strength of the positive bias, is sensitive to k, σ, ρ and ω. **The bias is almost always observable for $k = 0$** (see the two panels) whatever ρ or σ except for $\omega = 0.05$ and $\rho=0.5$ or 1 (at the bottom) except for very large values of σ (see Figure 8.7, $\sigma=100$). It is not observable for the same value leading to a consensus in the previous section (see Figures 8.2 and 8.5 at the bottom). On the other hand, we notice that **an increase of σ always decreases the strength of the positivity bias whatever the other parameters until it disappears** (a large increase in σ suppresses all the positivity bias, as illustrated in Figure 8.7). **The bias always occurs for lower values of σ.**

Regarding k, when k is higher than zero, the impact is very complex and depends on the overall dynamics: from dynamics mainly driven by ω over ρ to dynamics mainly driven by ρ over ω. We can sum up the sensitivity to k in three cases. Figure 8.6 shows these three cases. To represent directly the strength of the positivity bias, we compute the average number of underestimated agents $(N - 1)/2$ with N the size of the population of agents. This "new" indicator is zero when there is no bias and 19.5 when the bias is a maximum.

When ρ is small (0.05), two cases should be distinguished:

1. $\rho = \omega$ (0.05): **an increase in k decreases the strength of the bias until it disappears** or remains at the same level (Figure 8.6 on the top left panel).

2. $\rho \ll \omega$ (0.5 and 1): **an increase in k increases, and then decreases the strength of the bias** (Figure 8.6, top right panel).

When ρ is larger (0.5 and 1 in our experimental design), two regimes exist for a the same couple of parameter values (ρ, ω), depending on the value of σ:

1. for large values of σ, **an increase in k decreases the strength of the bias until it disappears** or remains at the same level (Figure 8.6, bottom right panel) ;

2. for low values of σ, **an increase in k at first decreases, and then increases the strength of the bias** (Figure 8.6, bottom right panel and detail with a zoom in the bottom left panel).

The vanity coefficient ω plays an important role in the absence of gossip (see on the right of Figure 8.5). It is **particularly strong when ρ is low**

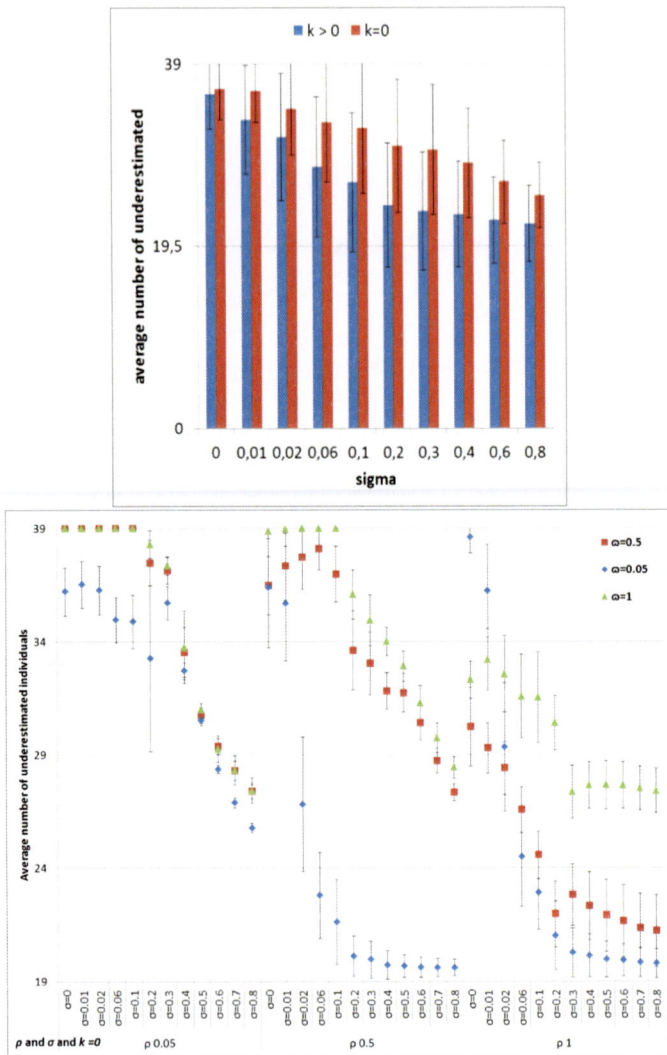

Figure 8.5: Average number of underestimated below oneself for various values of k: on the top, for various values of σ (in abscissa), $k = 0$ (red bars) and $k > 0$ (blue bars) and all values of all other parameters; at the bottom for $k = 0$ and the various ρ and σ values and ω =0.5 (red squares), ω =0.5 blue diamonds, ω =1 (green triangles. Each dot corresponds to the average overall results obtained from every tested value for all the parameters considered in the experimental design. Error bars correspond to one standard deviation over these results.

in comparison to ω and leads to the largest positivity bias which can be observed in the model. This is the largest positivity bias identified in Deffuant et al. (2013) and explained as associated to a particular equilibrium between "friends and foes" (see section 8.2.2 for explanations) called equality.

The parameters corresponding to equality are the only ones in our experimental design requiring a larger value of σ (than those of the experimental design) to suppress the positivity bias. Indeed, the left panels of Figure 8.6 have their top right corners totally empty, indicating there is no positivity bias in these zones (in contrast to the top right figures, describing the case "equality"). However we are sure that a σ close to infinity suppresses the bias. This is illustrated by Figure 8.7, showing there is no more positivity bias whatever the parameters for $\sigma = 100$. Nevertheless, the particular equilibrium of equality makes the bias very robust to gossip.

Moreover, the sensitivity to k for these values of ρ and ω is very singular (see the top right panel of Figure 8.6). An increase in k at first increases the strength of the bias and then as k continues to increase, the strength of the bias decreases. **This particular behaviour is probably linked to the probability of a peer's being the subject of discussion during a meeting**. This probability is ruled by k but a small population composed of a large set of foes against a small set of friends is very sensitive to such a probability. As the first step, an increase of k increases the probability of "friends" being discussed by everyone, then agents who have a very strong negative opinion of them make those of them held in lower esteem more likely to be held in low esteem. This decreases the average number of agents held in low esteem, and thus the positivity bias. In the second step, a new increase in k is going to increase the number of evocation of "friends" by the minority of "friends" itself and "reinforce" them, decreasing the positivity bias at the same time. However, the bias remains, since the difference of opinion between foes and friends is kept very large by the vanity process. This particular sensitivity was not identified in Deffuant et al. (2013).

For larger values of ρ, the impact of vanity is not as high and the sensitivity of the bias to the intensity of the gossip in terms of the value of k depends on σ (see the bottom panels of Figure 8.6). Although an increase in k for large values of σ decreases and suppresses the bias, it decreases and increases the bias for small values of σ. **We have been unable to explain such a result thus far.** The second type of positivity bias identified in Deffuant et al.2013) and occurring in the absence of vanity corresponds to the one observed for these larger ρ values. It was assessed as weak in Deffuant et al. (2013) since it was measured for a sufficiently large value of σ. However, we have shown that the positivity bias always occurs whatever ρ and ω

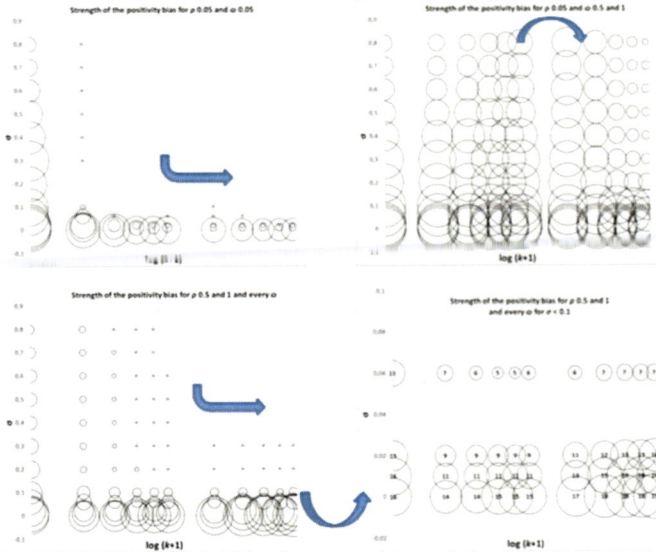

Figure 8.6: Strength of the positivity bias for: ρ=0.05 and ω=0.05 (on the top left); ρ=0.05 and ω=0.5 and 1 (on the top right); ρ=0.5 and 1 and every ω (on the bottom left) with a focus on small values of σ (on the bottom right). The size of the circle represents the strength. This strength is indicated in the centre of the circle in the zoom on the lower right panel. The blue arrows simply represent the way the strength changes with the increase of the value of k. For the lower left panel, there are two arrows since there are two different ways of changing, depending on σ.

for low values of σ. **This makes our knowledge of the emergence of the positivity bias more generic.**

To sum up, the positivity bias always occurs for lower values of σ. **An increase of** σ **always decreases the strength of the positivity bias until it disappears, whatever the other parameters are**. This is easy to understand, since an agent held in low esteem has less chance to become held in high esteem than an agent held in high esteem to become held in low esteem. Indeed, the strength of the influence of the agent held in high esteem is very high, while that of an agent held in low esteem is close to zero, especially for lower σ and large ρ. In the case of little disagreement between people, it remains possible that people held in high esteem think themselves lower and influence strongly the listener in this way. In contrast, agents held in low esteem thinking themselves higher have little chance to change the listener's mind. When σ increases, the difference of influence between those held in high and low esteem decreases.

In the absence of gossip (k=0), the positivity bias is very robust to the value of σ for large values of ω, and a very large value of σ is required to suppress the bias. When gossip is introduced ($k > 0$) and σ is large enough, the positivity bias can be suppressed extremely quickly, especially when ρ and ω are low (see the disappearance of the bias in the top left Figure 8.7 (from k=0 to k=1 for $\sigma \geq 0.1$)).

All these results confirm our hypothesis H2 that the gossip parameters have a large influence on the positivity bias. Indeed, in the Leviathan model, the bias can be suppressed if agents are open enough to the influence of people held in low esteem, and not only to the influence of agents held in high esteem. In the absence of gossip, except for ω= 0.05 and ρ= 0.5 or 1, agents have to be open to the influence of almost everyone with the same sensitivity to suppress the positivity bias. Then, only gossip is able to suppress the bias when the hypothesis of a difference of openness to agents held in low esteem compared to agents held in high esteem is maintained in the model. The intensity of gossip, in terms of the number of people someone talks about, changes the strength of the positivity bias. However, the way this intensity changes the strength of the bias differs a lot depending on the coefficient ρ, σ and ω.

8.4.3   Positivity bias, level of disagreement, and bias to negativity

We know from the two previous sections that the level of disagreement is driven by gossip controlled by k, the number of discussed peers during a meeting, and the positivity bias is driven by σ, the openness to others depending on the level of esteem for them. We try now to assess our hypotheses 3 and 4 regarding on the one hand the link between agreement and positivity bias and on the other hand, the link between the positivity bias and the bias to negativity diagnosed in the Leviathan model.

We hypothesized there is a global negative correlation between positivity bias and consensus. But we know now they are not driven by the same parameters: while the consensus is mainly driven by k and ρ, the biases, particularly here the positivity one, are mainly driven by σ and ω! Thus this means in practice they can be both present as well as one present and the other absent. In Figure 8.7, we can observe that disagreement (green circles) tends to slightly increase in step with σ whatever k, while the positivity bias (red squares) represented by the share of underestimated agents in the population tends to decrease. However, we have seen previously the first lowest step for small values of σ is only explained by the lowest value of ρ (section 4.1 and the left panel of Figure 8.3) and the characteristics of our indicator. Indeed, for ρ=0.05 and ω=1, the average disagreement is very low for small σ because all the agents hold themselves in the highest esteem and have the poorest opinion of the others. As our indicator does not consider the opinion of oneself for its computation, such a pattern implies a disagreement close to 0 (this corresponds to the average maximum distance of opinions on everyone except itself when everyone is assessed at the smallest opinion). Regarding the last step for very large values of σ, the increase of disagreement is simply due to a sensitivity to the smallest values of k slightly higher than for lower values of σ.**Then we don't really confirm our hypothesis H3: there is not really a global negative correlation. Overall, agreement and positivity bias appears quite independently from each other.**

It is now interesting to see how the bias to negativity identified in the Leviathan model evolves compared to the positivity bias. In Figure 8.7, we can observe that the share of underestimated agents (red squares for positivity bias) and the part of negative opinions (purple triangles) tend to evolve closely with the openness. The two indicators decrease as the openness increases until they reach 0.5, indicating there is no bias anymore for an openness equal to 100. This **disconfirms our hypothesis H4.** Indeed, even if as previously observed in Deffuant et al. (2013), **the positivity bias and the bias to negativity are both present for many values of** σ, **they both disappear for very large openness** (σ=100) **even if they are not equally sensitive to the vanity and the influence coefficient, as well as** k. Indeed, Figure 8.7 shows the bias to negativity seems on average little sensitive to k, while this is not the case for the positivity bias.

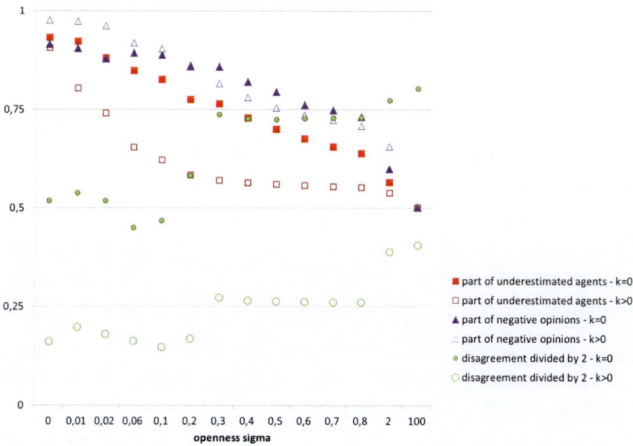

Figure 8.7: Relative share of underestimated agents (squares), relative share of negative opinions in the population (triangles) and disagreement (divided by 2 to be comparable to the two other measures), averaged over the various measures and values of parameters tested in the experimental design, for the case "no gossip" ($k = 0$ in plain squares, circles and triangles), "gossip" ($k > 0$ in empty squares, circles and triangles) and the various tested values for the openness σ (in abscissa). We have added two values for σ to our experimental design (2 and 100) in order to confirm the already observed tendency of evolution of our indicators.

Then, both these biases relate to each other, particularly through the openness σ. The two apparently different positivity biases identified by Deffuant et al. (2013) are finally mainly grounded in the same mechanism. This sounds not so obvious and deserves some explanation.

The degree of openness is very important: the biases are always present for a very low degree of openness, favouring almost only the influence of those held in high esteem. The difference in the consideration of the message about itself of someone held in high esteem compared to someone held in low esteem is responsible for every bias. It is not only that it makes their influence different. **In any case, it also makes the vanity process more determining of what is thought of someone held in low esteem while vanity and influence rule what is thought of someone held in high esteem.** Thus, when openness is slight and vanity is great, the biases are strong. However, even when the vanity is slight, it still makes their influence different and the biases tend to remain until the difference of influence between agents held in low and high esteem becomes insignificant for the dynamics, especially in comparison to the noise δ.

The explanation for the bias to negativity leads to the fact that the punishment of someone held in low esteem has a stronger impact than an equivalent punishment of someone held in high esteem: that is the major explanation for the bias to negativity. This is highly sensitive to one's self-valuation and to the vanity coefficient ω. When the openness is very great, the propagation coefficient tends to be a constant whatever the level of esteem for the speaker and only the influence and the vanity coefficients, as well as the self-valuation, determine the evolution of the speaker (if we consider $k = 0$).

The explanation of the positivity bias leads to the fact that one agent held in low esteem has less chance of becoming held in high esteem than an agent held in high esteem of becoming held in low esteem. Indeed, the strength of the influence of the agent held in high esteem is very high, while that of the agents held in low esteem is close to zero, especially for lower σ and large ρ. In the case of little disagreement between people, it remains possible for people held in high esteem to think themselves lower and influence strongly the listener in this way. In contrast, agents held in low esteem thinking themselves higher have only a little chance of changing the listener's mind. When σ increases, the difference of influence between those held in high and low esteem decreases. This explains why this bias is sensitive to ω and ρ.

When the openness increases, the suppression of the positivity bias occurs quicker than the suppression of the negativity bias. This is due to the complementary sensitivity to the influence coefficient ρ as shown in section 8.4.2. (See Figure 8.5.)

From Figure 8.7, we can also **deduce the link between gossip parameters and emerging structures** (see 2.2.2 for their definitions). The absence of gossip favours disagreement and positive bias. In terms of emerging structure, this means that the structures **elite but mainly equality are favoured by the absence of gossip**. In contrast, as **gossip** strongly decreases disagreement, it **favours dominance, and mainly hierarchy, even more for larger values of openness** favouring positive opinions. We notice that **counterintuitively, the hierarchy is favoured by openness to most people while equality is favoured by a population of very narrow people practically only influenced by people held in higher esteem.** Indeed, an increasing openness increases agreement and positivity, and decreases positivity bias (the relative share of underestimated agents).

The suppression of the positivity bias and the negativity bias for large values of σ indicates that many of the known structures can't be observed for such a value of σ. Indeed, regarding their definition, it seems only hierarchy remains.

8.5   Synthesis and discussion

This section synthesizes our results about the behaviour of the model and the gossip parameters. It considers at first the link between gossip and the emergence and maintenance of a consensus. Secondly, it deals with gossip and bias in favour of oneself or in favour of others. These two properties, consensus and bias, define the observed structures of opinions in the Leviathan model. Overall these three facts are useful to examine how our results can feed the debate of social scientists about the social functions of gossip.

We globally found that **the level of consensus is mainly driven by gossip**, especially by the number of peers discussed during a discussion. On the other hand, **the biases**, i.e. the positivity bias and the bias to negativity, **are grounded in the face-to-face meeting interactions and mainly explained by the openness (depending on the level of esteem in which the speaker is held) to the speaker**. Gossip has only a secondary effect on biases, quite complex and dependent on the coefficients of vanity and influence.

8.5.1   Gossip and consensus

In the Leviathan model, the number of peers discussed during a meeting is the main parameter explaining the emergence of a global consensus about

agents' valuations, especially when the influence coefficient is low. The
main results are: (1) no gossip, almost no consensus about what to say or
what to think about others; (2) increasing gossip decreases the disagreement.
Indeed, a consensus can hardly be obtained in the absence of gossip. It can
emerge from direct meetings only when the influence parameter is high for a
very low vanity coefficient and agents are sufficiently open to each other. In
the absence of gossip under other conditions, disagreement remains. When
gossip is introduced, the disagreement decreases. More generally, the more
people someone talks about, the more conformist is someone's valuation.
This confirms our hypothesis H1: gossip has a strong influence on the level
of consensus within the population.

This is in accordance with the social psychology literature (Wert and Salovey, 2004; Foster, 2004; Emler, 1990). Similarly to what has been outlined
by those authors, gossip in the Leviathan model is a source of reputation. It
maintains the agent's status and thus the group structure. It guarantees a
connection between people and a sufficient level of agreement regarding the
structure.

Following Wert and Salovey (2004), gossiping maintains the structure.
For our model, it is not only responsible for maintenance: the structure
emerges from gossip: the structure differs, depending on the presence and
the strength of gossip (measured as the number of peers discussed during a
meeting).

Indeed, without gossip, people basically structure themselves based on
a very positive bias (very high self-esteem) while they despise others (tendency to negativity). This is the domain of the equality structure mainly
driven by the openness of people to each other, defining the number of
friends of an agent. Regarding "real" groups of individuals, this result tends
to confirm the importance, in the absence of gossip, of the level of openness
for the emergence and maintenance of a structure based on a network of
friendships.

In the model, and always without gossip, for high values of influence,
other structures are able to emerge but it is rare that they are present all over
the parameter space. This tends to confirm the idea proposed by Hobbes
that a strong power is necessary to organize the social space. But in the
Leviathan model, this is only true in the absence of gossip. To sum up, we
can say that norms can't emerge without gossip, unless influence and openness are very large for a low level of vanity but the agreement is just based
on face-to-face interactions and can't be obtained if not everyone meets everyone!

This confirms that gossip is a bonding mechanism as well as a necessity for the maintenance of norms as well as for their emergence. Indeed, even if the Leviathan model does not consider this case, gossip can link agents not talking each other directly.

But it is not only a bonding mechanism, it is also a way to build and transmit social representations. Some authors (Wert and Salovey, 2004) have argued that gossiping maintains the structure by sanctioning people who do not respect the norms. From our results, we can argue in a slightly different way, since the structure is not only maintained by the ability of people to sanction or reward each other through gossiping. It is also because agents reinforce their position in the structure not only talking about themselves but also, from their point of view, relative to the agents they know. This allows the emergence of a consensus since, overall, people not only give one-by-one the opinion they have on a given person but of the structure from their own point of view. The more agents we gossip about, the more completely the description of the structure is discussed and the higher the agreement on the structure. Not only, as said in Foster (2004), does the agent gets a map of his social environment, ... but he can, as outlined by Emler (1990), negotiate its own social identity by influencing its own ranking not only through direct interaction where each other talks about them, but also by directly influencing a large subpart of the ranking trying to transmit its own ranking. Then, finally defending its own reputation through gossip, the agent disseminates its view of the structure and helps the emergence and maintenance of a global structure. This sounds quite in accordance with the literature and confirms the dynamic validity of the theory (Dunbar and Duncan, 1997; Conein, 2011).

In the model, in terms of a structure of the population organized through opinions, while the introduction of gossip mainly favours an elite structure, a larger increase of gossip in terms of the number of discussed peers favours finally a hierarchy. This can be diagnosed since the elite structure is associated to a high positivity bias and a strong disagreement while a low positivity bias and a high level of agreement characterized the hierarchy. In practice, increasing gossip allows passing from a structure based on relationships and led by a network of friends to a structure based on norms and consensus about social representations.

In particular, gossip helps the emergence of recognized leaders. In the Leviathan model, leaders are agents who have a positive reputation: everyone has a positive opinion of them with only small variations from one agent to another. From Deffuant et al. (2013), we know that leaders only appear when reputations are consensual and the propagation coefficient is sufficiently large compared to the vanity. It gives agents held in high esteem the opportunity to impose their point of view on everyone's value since everyone agrees on their higher status. In the social literature, if gossip has been often cited in terms of status maintenance, it has rarely been cited for high status emergence (as far as our knowledge extends, the only exception is Emler, 1990), even if the danger of gossip for reputation has often been discussed (Foster, 2004).

In the Leviathan model, the emergence of leadership is also strongly determined by the asymmetrical influence of people held in low and high esteem. It is probably the most important parameter in the explanation of the emergence and maintenance of status, whatever the form of leadership style or the form of structure. Huet (2013a) has shown in the Leviathan model that since the gossip is introduced into the dynamics for a sufficient level of openness, it is likely that a leader will appear in the population. Also, the number of leaders only depends on the level of openness, since agents practice gossiping.

The question about the characteristics of the leaders and the various associated leadership styles is debated in social psychology (Hogg, 2001; van Knippenberg et al., 2004; Uhl-Bien, 2006; Martin, 2009; Huet 2013a). Investigating further the Leviathan model on this question would be interesting.

8.5.2   Biases and a "conditioned-to-esteem" openness

This section is much more dedicated to our study of the link between gossip and biases. We have investigated the relations between gossip and the positivity bias, positivity bias and the level of agreement about opinions, positivity bias and the bias to negativity of all opinions. Our study shows the great importance of openness to influence by peers. This influence is partly ruled by a sigmoid function parameterized by the openness. This makes more influential an agent held in high esteem: the higher the esteem in which an agent is held, the more influential the agent is. In particular, when the openness is zero, only agents held in high esteem are influential: those considered as lower than oneself are not influential.

The degree of openness explains most of the strength of the biases. Indeed, our main results are: (1) the positivity bias and the bias to negativity decrease when the openness increases; (2) a very large openness makes the biases disappear; (3) while the biases decreases with an increase in openness, disagreement increases; (4) the disagreement level, then the positivity bias, but not the bias to negativity, are sensitive to the absence of gossip (the introduction of gossip decreases the disagreement level and the positivity bias on average).

Openness is very important. Vanity can be seen as a self-enhancement and self-protection mechanism which ensures by itself to an agent a preference for herself. Then the first intuition was to consider vanity as responsible for the bias to negativity as well as the positivity bias. However, following the Leviathan model, it is only a reinforcing mechanism. The main cause of these biases comes from the way people are more or less open to others' influence, depending on their level of esteem for them.

When the openness is small, implying the influence is very conditioned by the level of esteem for the speaker: (1) the punishment of someone held in low esteem has a stronger impact than an equivalent punishment of someone held in high esteem: that is the major explanation for the bias to negativity; (2) an agent held in low esteem has less chance of becoming highly esteemed than an agent held in high esteem has of becoming held in low esteem: that is the explanation for the positivity bias. Both these mechanisms are reinforced by the strength of vanity, which is harsher for people held in low esteem than for people held in high esteem since it is not compensated for by their influence on their own value. Openness makes for the way people define themselves in the dynamics. Especially when it is small, people held in high esteem and people held in low esteem have a very different dynamics. People held in high esteem drive the opinions about themselves through their own influence and neglect what others say about them. They are quite active in the formation and the maintenance of their self-opinion. In contrast, people held in low esteem are very passive regarding their self-opinion. They are much more defined through the vanity process applied by their interlocutor, especially through punishment. This leads them to see themselves more and more negatively through the influence of people they held in high esteem.

When openness becomes larger, everyone becomes equally influential regarding the propagation coefficient. This makes each agent more active in the definition of its self-opinion as well as in the definition of the opinion of others. In such a situation, the biases to oneself (positivity) and to others (negativity) disappear. However, these two biases are not equally sensitive to the level of vanity and the influence coefficient, as well as to k. So it is not possible from our experiments to observe a positivity bias without observing a bias to negativity. But it is possible to observe a bias to negativity without a positivity bias, particularly for a great degree of openness and a low influence coefficient. This remains an utopia compared to the "real" world. Indeed, positivity bias is universal. Its effects can be temporary or locally decreased (Yaniv and Kleinberger, 2000) but remain overall.

The positivity bias and the bias to negativity have never been explained, as far as our knowledge extends, in terms of openness to others or in terms of interaction with others. They take place in the needs of two people to form an opinion of each other to be able to discuss. This opinion is built from a mutual influence which a matter of esteem and from some self-enhancement or self-protection mechanisms. In the literature, gossip is at the same time considered as dangerous for someone's self-suffering from a bad reputation, and explained as a social way to control cheating. That would probably mean that gossip decreases the strength of the positivity bias to discourage cheating. Perhaps it can lead a majority of people undervaluing themselves. On the contrary, the Leviathan model argues that gossip has only a small effect on biases and is not responsible for them overall. It can in fact increase or decrease slightly the biases, depending on the vanity and the influence coefficients, but very poorly compared to the impact of an openness conditioned on the level of esteem. This can explain the difficulty in concluding anything about the impact of gossip. This can also provide some clues to the apparently contradictory results showing that gossip can be either overall negative (Wert and Salovey, 2004) or positive (Ellwardt et al., 2012).

Agreement, then the positivity bias, but not the bias to negativity, are sensitive to the absence of gossip (the introduction of gossip decreases the disagreement level and the positivity bias on average). However, agreement and positivity bias appear quite independent from each other. They are not driven by the same parameters: while consensus is mainly driven by gossip and influence, the biases, particularly the positivity bias, are mainly driven by openness and vanity! Thus this means in practice they can both be present just as well as one present and the other absent. We hypothesized there is a negative correlation between the positivity bias and consensus since it seems counterintuitive that positivity bias and a consensus as to reputation can be present jointly. How can agents, who judge themselves higher than the average, agree on someone's value? Indeed, in the literature, gossip is at the same time considered as dangerous for someone's self-suffering from a bad reputation, and explained as a social way to control cheating (Foster, 2004) to protect norms. This means in practice positivity bias and consensus should be negatively correlated. This can be locally the case in the Leviathan model. But the model does not exclude intermediary situations which have been, even often separately, diagnosed in surveys and experiments. The noise occurring during discussion can also help in reaching an agreement despite the universal preference for oneself.

Overall, the definition of the structures diagnosed in (Deffuant et al., 2013) has to be revised in the light of these results about the effects of the gossip parameters.

8.5.3   Future research

Two types of future research have to be considered: more studies of the Leviathan model, and an investigation using the methods of social psychology of the proposals emerging from the study of the model.

We have seen how gossip participates in the emergence and the maintenance of a structure. However, we have assumed that our population defines only one group of people. In the literature, structures relate not only to one group but mainly to several groups and how they interact with each other, especially through stereotypes (which can be assimilated to the reputation of a given group). This is a particular issue for the Leviathan model which has to be investigated in the future. A possible solution is the modification of the heuristics of talking. At first in the current Leviathan model, everyone talks to everyone on average during one iteration and the interaction takes place in dyads. This is true, according to (Conein, 2011), to the extent that people do gossip more often in dyads. However, that author also has some work showing gossip can occur during a meeting of up to five people. It would probably be very interesting to add such a situation to the Leviathan model. We can also consider differing probabilities of talking or of meeting someone else for agents. This could be through the introduction of a structure of interactions or of communication rules corresponding to various apprehensions to communicate.

Regarding social psychology, one of the main proposals of the model to investigate is probably the link between the strength of the degree of gossip and the type of structure of opinions at the group level. Trying to determine whether such a result can be grounded empirically would be very interesting.

Bibliography

Arrow, H., and K. L. Burns, 2004, Self-Organizing Culture: How Norms Emerge in Small Groups, *The Psychological Foundations of Culture*, M. Schaller and C. S. Crandall, eds., Lawrence Erlbaum. Chapter 8, pp. 171–199.

Beersma, B., and G. A. Van Kleef, 2012, Why People Gossip: An Empirical Analysis of Social Motives, Antecedents, and Consequences, *Journal of Applied Social Psychology* **42**(11), 2640–2670.

Buckley, K. E., R. E. Winkel, and M. R. Leary, 2004 Reactions to acceptance and rejection: Effects of level and sequence of relational evaluation, *Journal of Experimental Social Psychology* **40**(1), 14–28.

Carletti, T., and S. Righi, 2011, Emerging structures in social networks guided by opinion's exchanges, *Advances in Complex Systems* **14**(1), 13–30.

Castellano, C., S. Fortunato, and V. Loreto, 2009, Statistical physics of social dynamics, *Reviews of Modern Physics* **81**(2), 591–646.

Conein, B., 2011, Gossip, conversation and group size: Language as a bonding mechanism, *Irish Journal of Sociology* **19**(1), 116–131.

Deffuant, G., 2006, Comparing Extremism Propagation in Continuous Opinion Models, *Journal of Artificial Societies and Social Simulation* **9**(3).

Deffuant, G., F. Amblard, S. Huet, S. Bernard, N. Ferrand, J. P. Bousset, G. Amon, J. Henriot, N. Gilbert, E. Chattoe, and G. Weisbuch, 2000, Simulation de l'évolution des conversions à l'agriuclture biologique dans le département de l'Allier entre 1994 et 1999 par un modèle multi-agents de la population d'agriculteurs, Cemagref. Rapport intermédiaire du projet Européen IMAGES: 21.

Deffuant, G., T. Carletti, and S. Huet, 2013, The Leviathan model: Absolute dominance, generalised distrust and other patterns emerging from combining vanity with opinion propagation, *Journal of Artificial Societies and Social Simulation* **16**(1/5), 23.

Dunbar, R., and N. Duncan, 1997, Human Conversational Behavior, *Human Nature* **8**(3).

Ellwardt, L., G. J. Labianca, and R. Wittek, 2012, Who are the objects of positive and negative gossip at work? A social network perspective on workplace gossip, *Social Networks* **34**(2), 193–205.

Emler, N., 1990, A social psychology of reputation, *European Review of Social Psychology* **1**(1), 171–193.

Fein, S., and S. J. Spencer, 1997, Prejudice as self-image maintenance: Affirming the self through derogating others, *Journal of personality and social psychology* **73**(1), 31–44.

Fortunato, S., V. Latora, A. Pluchino, and A. Rapisarda, 2005, Vector Opinion Dynamics in a Bounded Confidence Consensus Model, *International Journal of Modern Physics* **16**(10), 1535–1551

Foster, E., 2004, Research on gossip: Taxonomy, methods, and future directions, *Review of General Psychology* **8**(2), 78–99.

Galam, S., 2008, Sociophysics: A review of Galam models, *International Journal of Modern Physics C* **19**, 409–440.

Gargiulo, F., and S. Huet, 2010, Opinion dynamics in a group-based society, *Europhys. Lett.* **91**(5).

Gargiulo, F., and S. Huet, 2012, New discussions challenge the organization of societies, *Advances in Complex Systems* 1250033.

Hobbes, T., 1651, *Léviathan. Traité de la matière, de la forme et du pouvoir ecclésiastique et civil.*

Hogg, M. A., 2001, A Social Identity Theory of Leadership, *Personality and Social Psychology Review* **5**(3), 184–200.

Hogg, M. A., J. C. Turner, and B. Davidson, 1990, Polarized Norms and Social Frames of Reference: A Test of the Self-Categorization Theory of Group Polarization, *Basic and Applied Social Psychology* **11**(1), 77–100.

Hoorens, V., 1993, Self-enhancement and superiority biases in social comparison, *European Review of Social Psychology* **4**, 113–139.

Huet, S., 2013a, Formes de leadership, défense de soi et commérage, *Modèles Formels pour l'Interaction, PFIA 2013*, Lille (France), p. 14.

Huet, S., and G. Deffuant, 2008, Bounded Confidence with Rejection: Clusters or Scattered Opinions? *ESSA, European Social Simulation Association* Brescia, Italy, p. 12.

Huet, S., and G. Deffuant, 2010, Openness leads to opinion stability and narrowness to volatility, *Advances in Complex Systems* **13**(3), 405–423.

Kniffin, K., and D. Wilson, 2005, Utilities of gossip across organizational levels, *Human Nature* **16**(3), 278–292.

Leary, M. R., J. M. Twenge, and E. Quinlivan, 2006, Interpersonal Rejection as a Determinant of Anger and Aggression, *Personality and Social Psychology Review* **10**(2), 111–132.

Lorenz, J., and D. Urbig, 2007, About the Power to Enforce and Prevent Consensus by Manipulating Communication Rules, *Advances in Complex Systems* **10**(2), 251–269.

Martin, J. L., 2009, Formation and stabilization of vertical hierarchies among adolescents: Towards a quantitative ethology of dominance among humans, *Social Psychology Quaterly* **72**(3), 241–264.

Moliner, P., 2001, *La dynamique des représentations sociales*, Grenoble (France).

Moscovici, S., 2000, *Psychologie sociale des relations à autrui*, Paris: Nathan Université.

Myers, D., S. J. Spencer, and C. Jordan, 2009, Social Thinking, *Social Psychology*, F. C. Edition. Canadian, 63.

Srivastava, S., and J. S. Beer, 2005, How Self-Evaluations Relate to Being Liked by Others: Integrating Sociometer and Attachment Perspectives, *Journal of Personality and Social Psychology* **89**(6), 966–977.

Stephan, Y., and C. Maiano, 2007, On the Social Nature of Global Self-Esteem: A Replication Study, *The Journal of Social Psychology* **147**(5), 573–575.

Sznajd-Weron, K., 2005, Sznajd Model and Its Applications, *Acta Physica Polonica B* **36**(8), 1001–1011.

Tajfel, H., 1978, *Differentiation between social groups: Studies in the social psychology of intergroup relations*, London: Academic Press.

Turner, J. C., 1984, Social identification and psychological group formation, In: *The Social Dimension*. H. Tajfel, ed.. Cambridge University Press, pp. 518–538.

Turner, J. C., M. Hogg, J. Oakes, S. D. Reicher, and M. S. Wetherell, 1987, *Rediscovering the Social Groups: A Self-Categorization Theory*, Cambridge, Basil Blackwell.

Uhl-Bien, M., 2006, Relational Leadership Theory: Exploring the social processes of leadership and organizing, *The Leadership Quarterly* **17**(6), 654–676.

Urbig, D., and R. Malitz, 2007, Drifting to more extreme but balanced attiudes: Multidimensional attitudes and selective exposure, In: *4th Conference of the European Social Simulation Association, Toulouse, 10–14 September.*

van Knippenberg, D., B. van Knippenberg, D. De Cremer, and M. A. Hogg, 2004, Leadership, self, and identity: A review and research agenda, *The Leadership Quarterly* **15**(6), 825–856.

Waylen, A. E., M. S. Horswill, J. L. Alexander, and F. P. McKenna, 2004, Do expert drivers have a reduced illusion of superiority? *Transportation Research Part F: Traffic Psychology and Behaviour* **7**(4–5), 323–331.

Wert, S. R., and P. Salovey, 2004, A social comparison account of gossip. *Review of General Psychology* **8**(2): 122–137.

Wood, J. V., and A. L. Forest, 2011, Seeking Pleasure and Avoiding Pain in Interpersonal Relationships. In: *Handbook of Self-Enhancement and Self-Protection*, M. D. Alicke and C. Sedikides, eds. New York: The Guilford Press, pp. 258–278.

Yaniv, I., and E. Kleinberger, 2000, Advice taking in decision making: Egocentric discounting and reputation formation, *Organizational behavior and human decision processes* **83**(2), 260–281.

Chapter 9

Discovering networks of actors behind the dynamics of opinion. A case study

Alexandre Delanoë[1], Serge Galam[2]

9.1 Introduction

A good deal of publications have already been devoted to the theoretical study of public opinion, especially within the framework of sociophysics (Castellano, Fortunato, and Loreto 2009, Galam 2008, 2012). The present paper analyses the diffusion of two competing opinions, whose dynamics leads to the success or the disappearance of one of the opposing views. Dynamics is a selection process seen as repeated local interactions between agents providing a new comprehension of social phenomena (Galam 2002).

But not many real cases have been investigated using real empirical data. The challenge is to compare a descriptive data analysis with a theoretical

[1]Chercheur affilié au Centre d'Analyse et de Mathématique Sociales (CAMS, UMR 8557) et Ecoles des Hautes Etudes en Sciences Sociales (EHESS). Chercheur associé au Centre de Sociologie de l'Innovation (CSI, UMR 9217), Institut Mines Telecom, Mines ParisTech. alexandre@delanoe.org

[2]Centre National de la Recherche Scientifique (CNRS), Sciences Po / CEVIPOF (Centre de recherches politiques de Sciences Po). France (serge.galam@sciencespo.fr)

modelling. This paper presents a case study with a subject dealing with controversial environmental risks. In this case, the treatment of a public problem in the media makes the understanding of the public opinion mechanisms a great challenge. In our context, the application of the precautionary principle is a sensitive issue. The precautionary principle states that if a given policy is suspected of a possible harm to either the public or the environment, in the absence of a scientific consensus about its riskiness, implementing that policy is questionable and should be held up (French Constitution 2004, Galam 2005a).

The empirical data used for this paper deal with the abnormal bees' death, also called colony collapse disorder (CCD) in some countries, including the US. This controversy is emblematic of the question of the risks connected with the burden of making eventual mandatory arbitrage to ban the use of some specific chemical products. The real implementation of innovation leads to public debates, which are inevitably driven by incomplete scientific data. The question arises on how possible risks are translated (Callon 1986) into solid (Tonnies 1922) facts during the ongoing associated public debate (Dewey 1927, Galam 2010)?

Our corpus comprises almost 1500 articles dealing with the topic of the abnormal deaths of bees, published in French newspapers during the period 1998–2010, i.e., 13 years. With a text mining analysis[3] each article is categorized with one of three views about the possible explanation of this abnormal disappearance of bees. The first category gathers the articles asserting that the abnormal deaths are due to only one factor, the use of pesticides. At the opposite pole, other articles argue that many factors combined together produce the abnormal facts, it is thus a multi-factor cause. And a third category of articles states that the abnormal deaths are still an enigma without a clear explanation.

Then, the data are confronted with a model in order to put into question the social meaning of the dynamics. Our hypothesis asserts that the evolution over the years of the importance of support among journalists for each view can be modelled as the result of interactions between journalist agents. Two kinds of agents are used in the model. Inflexible agents, those who never change their mind, and flexible agents, those who can change their mind (Galam and Jacobs 2007, Galam 2005b). The variations exhibited by the data suggest that the number of inflexibles varies from year to

---

[3] An extensive text mining research program has been performed on the same corpus (Delanoe 2004, 2007, 2010) in order to extract the main terms that have a significant weight in the public debate. The text mining method used in this paper benefits from this series of works since it only categorizes articles that do contain some specific terms at least.

year. Those varying numbers of inflexibles are inferred by applying the Galam Unifying Frame (GUF), also denoted the Galam sequential proba- bilistic model of opinion dynamics or the majority model. Applying the model to the empirical data built from a corpus of published articles in news- papers provides a frame with which to go back to the data and question their social meaning with the eventual social mechanisms.

The rest of this paper is structured as follows. The problem is posed in the second section, where the opinion dynamics is quantitatively evaluated with empirical data. The third section highlights the newspaper level to show the profile of inflexible and flexible agents. The GUF is adapted to the problem in section four to determine the proportions of inflexible agents as a function of time in section five. The social meaning behind the data is addressed in section six, illustrating the determining role of lobbying and other externalities. The results are discussed in the last section.

9.2   Behind the dynamics there are many hypo- theses

A Boolean equation dealing with the abnormal death of bees in France dur- ing the period 1998–2010 has led to extracting a corpus of almost 1500 French articles from the Lexis-Nexis and Factiva databases. The collec- tion of papers is taken from the daily, weekly and monthly French speaking press. The annual distribution of the number of articles is shown in Figure 9.1.

A systematic textual analysis of all the collected articles (Delanoe 2004, 2007, 2010) shows that within those dealing with the question of the ab- normal bee deaths, some pointed towards scientific results suggesting a sin- gle cause identified as the use of pesticides (Chateauraynaud 2004, Delanoe 2004), while others emphasized other results suggesting a combination (Ch- iron and Hattenberger 2008) of several different causes (Maxim and van der Sluiis 2010). A combination of words has been established to categorize each view.

1. Articles containing words such as "pesticides" or "insecticides" or "chemicals" and without referencing other factors are categorized in the uni-factor class;

2. Articles containing at least one word in this list are put in the class of the multi-factors cause:

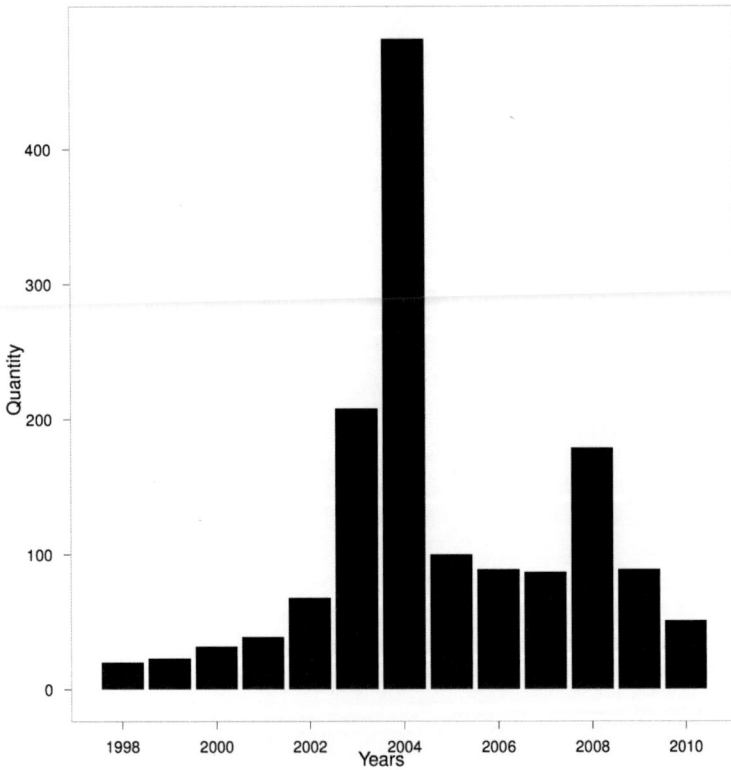

Figure 9.1: Number of articles published each year in the French daily, weekly and monthly press dealing with the bee deaths from 1998 till 2010. Over the 13 years, the total amounts to 1467.

- "Foulbrood" (a bacteria);
- "Nosema" or "Nosemose" (a mushroom);
- "Varroa" (a parasite);
- "Virus" (this represents mainly the Israel acute paralysis virus);
- "Predators" or "galleria mellonella" or "aethina tumida" or "Asian predatory wasp";
- "Monoculture" or "natural toxin of the sunflower" (which refer to agricultural practices);
- "Pollution" or "climate change" or "meteorology" (which represent the external or environmental causes);
- "Multi-factors" or "many factors";

3. Articles containing sentences like those below are assigned to the class claiming there exists no understanding yet (sentences detected with qualitative reading):

  - "While it would be impossible to formally accuse the pesticide of being exclusively responsible for the fall of the hive population";
  - "It is no element of new evidence of anything";
  - "All data analysed does not criminalize formally and exclusively the treatment of sunflower seeds";
  - "The pesticide was evaluated on two occasions over the last three years, and we believe that there is no cause and effect relationship between our product and the problems of orientation of bees".

Semi-automatic textual analysis tools combined with human reading for validation has enabled tagging 84% of the corpus articles, leaving 16% of the articles untagged. Some articles do not infer any cause of the bees' death since the authors were mentioning general facts only. In this specific case, these articles remain untagged. We have restricted the corpus to the tagged articles shown in Figure 9.2.

For the modelling, i.e. the next figures, the years will be numbered from $T = 0$ for 1998 to $T = 12$ for 2010. The corresponding proportions of articles for the uni-factor class, denoted by U_T, are respectively 0.500, 0.60, 0.677, 0.513, 0.627, 0.831, 0.769, 0.517, 0.540, 0.422, 0.544, 0.40, and 0.255 with $T = 0, 1, \ldots, 12$. 'A' indicates the opinion of journalists who

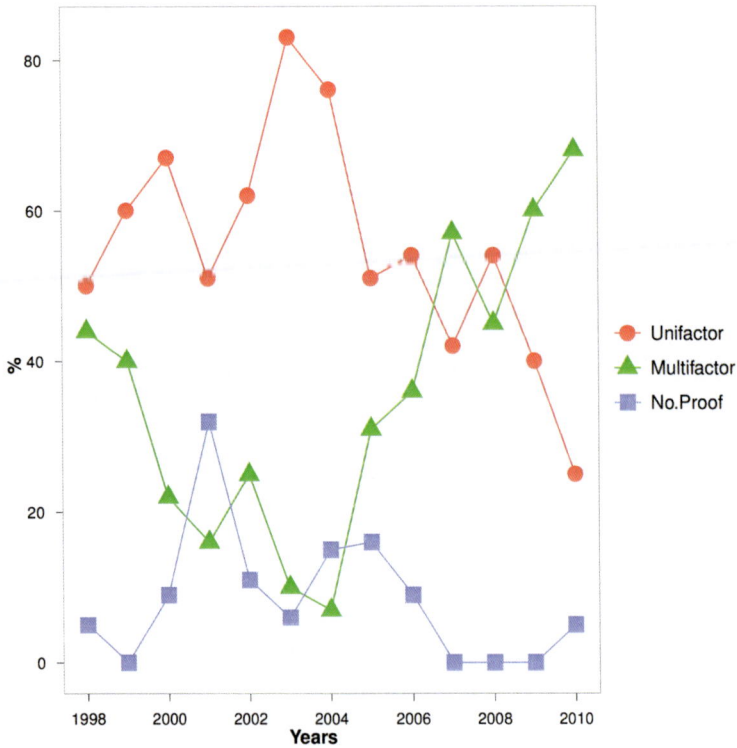

Figure 9.2: Proportions of articles published each year by the daily press dealing with uni-factor, multi-factor, or no-proof categories. 84% of the corpus, i.e. 1233 articles, have been tagged.

belong to the uni-factor class. Simultaneously, the opinion of journalists belonging to either one of both the other two classes, the multi-factor and the no-proof ones, is denoted 'B'.

9.3   The newspaper publications in time highlight the main profiles types

The corpus of selected articles was obtained from almost 60 different newspapers. To model the dynamics of opinion, we need to check if journal contributions exhibit different profiles. Looking at the contributions from *Le Monde* (Figure 9.3), we observe large variations. Indeed, some years present 0% of the articles for either A or B, as in 2001, 2006, 2007, 2008, 2009. Such facts mean that during those years all journalists were either all flexible or flexibles with some inflexibles present only on the opinion side which was advocated at 100%. In other words, a 0% support for one opinion implies the absence of inflexibles on this side. In years for which the dynamics did not reach 100% for one opinion, the inflexibles may have been present on both sides.

While four years, 2001, 2006, 2007, and 2008, are characterized by zero inflexibles for the uni-factor cause, only one year, 2009, feature zero inflexibles for the multi-factor cause. When a year's contributions does not reach 100% and is split over support of A and B, we can infer the possible existence of inflexibles on one or two sides, as for 1998, 1999, 2000, 2002, 2003, and 2004.

The contributions from the newspaper *Sud Ouest* (Figure 9.3) reveal the possible presence of inflexibles on each side every year. Contributions from *Le Figaro* (Figure 9.3) exhibit, as for *Le Monde* several years with 100% polarization, namely 1998, 1999, 2001, 2005, 2007, 2008, 2009, 2010. These years are characterized by zero inflexibles for the uni-factor cause (while for *Le Monde* it occurs for both sides). This position was modified in 2000 and really reversed in 2002–2004, with a slight surge in 2006.

The results of Figure 9.3 hint at the key role played by inflexibles in the making of the data of Figure 9.2. Such a fact would question the social meaning of those results. Accordingly, it is of importance to determine the proportions of inflexible agents present each year. Specifically, the successive brutal changes of trends as exhibited by Figure 9.2 indicate a change of proportions of the inflexibles.

Since the goal is to build the network of actors from the dynamics, implementing the GUF model appears appropriate, as it does not infer the

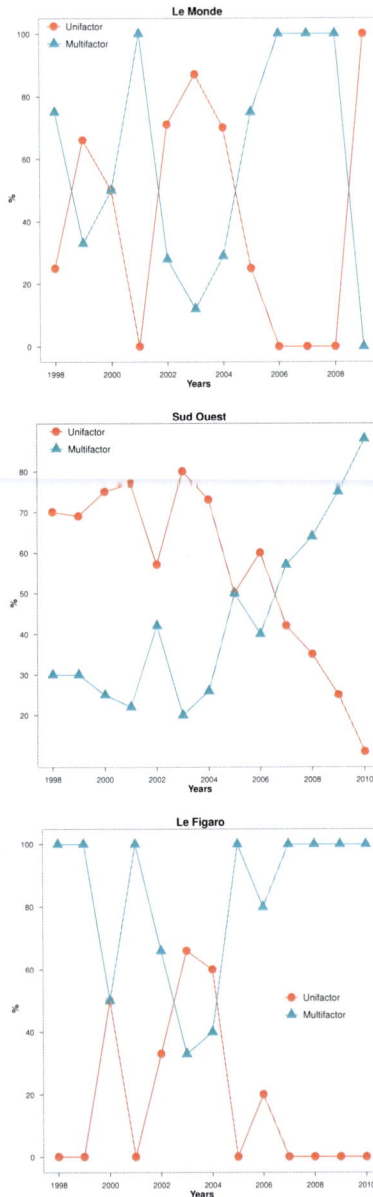

Figure 9.3: Proportions of articles published each year by *Le Monde* (total of 67 tagged articles), *Sud Ouest* (total of 209 tagged articles) and *Figaro* (total of 62 tagged articles) newspapers dealing with uni-factor and not-unifactors which includes both multi-factors and no proof papers.

structure of a network *a priori*. Indeed, this model incorporates only the effect of the inflexibles on the dynamics of opinion among flexible agents (Galam 2010, 2002, Galam and Jacobs 2007, Galam 2005b). Moreover, it has been shown that the size of the local updated groups does not modify the main results, since increasing the group size reduces the number of updates required to reach the attractors (Galam 2002). To keep the equations analytically solvable, group updates of size 3 have been used.

9.4 Using a model to reinterpret the problem

9.4.1 The framework of basic GUF, which depends on the group size distribution

The GUF model investigates the competition between two opposite opinions within a population of inflexible and flexible agents(Galam 2005a, 2002). Each agent has only one opinion. The rules of diffusion assert that flexible agents can change their opinion. Indeed, within a group of agents, a flexible agent takes on the opinion which has the majority. Then the dynamics is implemented via repeated random meetings of agents within small groups of various sizes. For each distribution, the agents' opinions are locally updated according to the local majority in its own group. For groups with an even number of members, in the case of equality, the agents preserve their current opinions.

In real life, people meet and discuss in groups of different sizes. However, these groups are usually small. In the case of the journalists treated here, those meetings occur within the social network of journalists in which they can interact. To take into account this reality, the basic GUF can be extended to include a distribution of sizes, leading to the general update expression

$$u_{t+1} = \sum_{i=1}^{L} s_i \{ \sum_{j=[\frac{i}{2}+1]}^{i} C_j^i u_t^j (1-u_t)^{(i-j)} + \frac{1}{2} k(i) C_{\frac{i}{2}}^i u_t^{\frac{i}{2}} (1-u_t)^{\frac{i}{2}} \}, \quad (9.1)$$

where u_t and u_{t+1} denote the proportion of journalists in favor of A at the times t and $t+1$ and v and w denote the proportions of A and B that are inflexible. L is the size of the largest group, $C_j^i \equiv \frac{i!}{(i-j)!j!}$, $[\frac{i}{2}+1] \equiv$ Integer Part of $(\frac{i}{2}+1)$, and $k(i) \equiv [\frac{i}{2}] - [\frac{i-1}{2}]$ yielding $V(i) = 1$ for i even and $V(i) = 0$ for i odd. The proportion of groups of size i is defined by the probability distribution s_i under the constraint $\sum_{i=1}^{L} s_i = 1$. Including groups of size one accounts for the fact that not all agents have their discussions at the same time in local groups.

Although an infinite number of size distributions $\{v_i\}$ are possible in principle, it happens that the dynamics is qualitatively unchanged: the two attractors $u_A = 1$, $u_B = 0$ and the tipping point $u_t = \frac{1}{2}$ are invariant. The only difference is the number of iterations required to reach either attractor. Larger groups contribute to accelerate the polarization effect. Nevertheless, analytically solving Eq. (9.1) is possible only up to $L = 4$, otherwise, for $L > 4$, numerical solving is required. On this basis, to keep the calculations simple and tractable, we restrict the group sizes to 3 in this paper.

9.4.2   The heterogeneous GUF, mixing inflexible and flexible agents

The uni-factor distributions in Figure 9.2 reveal a series of brutal variations at the years 2000, 2001, 2003, 2005, 2006, 2007, and 2008. According to the model (Eq. 9.1), the opinion reaching the majority is even more dominating. This result is incoherent with the evolution of the empirical data. To counter this systematic increasing trend, external parameters must be included in order to enable a topological modification of the dynamics.

It is worth stressing that in the GUF model, inflexibles do not have more powerful arguments, they still each have one vote, the same as the flexibles. They do obey one person one vote in a group discussion. However once every agent in the local group has written, they do not follow the local majority rule in case they are minority. In the present paper, we consider a population which is a mixture of flexible and inflexible agents. The proportions of inflexibles are external parameters while the respective proportions of flexibles in favor of A or B are internal parameters driven by the dynamics of the local discussions. The possibility of making inflexibility an internal parameter was studied in Martins and Galam (2013) but is not introduced here. Accordingly, Eq. (9.1) becomes

$$u_{t+1} = -2u_t^3 + (3 + v + w)u_t^2 - 2vu_t + v, \qquad (9.2)$$

where v and w denote the proportions of A and B who are inflexible. The associated dynamics has been extensively studied in Galam and Jacobs (2007), Galam (2005b), where the various cases of flow diagrams of opinion have been obtained as a function of all combined ranges of values for v and w.

9.4.3   Fitting the model to the data

The proportions of inflexible agents can be modified every year as a result of the activation of external pressures in favor of either opinion. Then, each year a flexible agent may become an inflexible, and vice versa. During each year, the proportions of the inflexible agents are kept fixed for each successive update[4].

Then the dynamics of opinion is implemented in two steps. First, some fixed proportions of inflexibles are given. And in the second step, n consecutive updates of the flexibles are implemented, keeping unchanged the proportion of inflexible. Then the proportions of inflexibles are modified before n new updates are performed. This two-step dynamics is implemented by modifying Eq. (9.2) to

$$u_{T,t+1} = -2u_{T,t}^3 + (3 + v_T + w_T)u_{T,t}^2 - 2v_T u_{T,t} + v_T, \qquad (9.3)$$

where $u_{T,t}$ and $(1 - u_{T,t})$ denote the proportions of journalists in favor of A and B during year T and intra-time t. The associated proportions of inflexibles v_T and w_T are independent of the intra-time. They depend only on the year T. Given $u_{T,t}$, the GUF determines $u_{T,t+1}$ obtained after one update of opinions for fixed values of v_T and w_T.

To account for the interplay between the two timescales, we note that since $T = 0, ..., 12$ for the years and $t = 1, 2.., n$ for the intermediate intra-time within a year, we have the congruence $(T, n) = (T + 1, 0)$.

In addition, we note that only the fraction $u_{T,t} - v_T$ is flexible, i.e., able to shift opinion under convincing local arguments. We thus have $u_{T,t} \geq v_T$ and $1 - u_{T,t} \geq w_T \iff u_{T,t} \leq 1 - w_T$, which combine to

$$v_T \leq u_{T,t} \leq 1 - w_T, \qquad (9.4)$$

with the constraints $0 \leq v_T \leq 1, 0 \leq w_T \leq 1$ and $0 \leq v_T + w_T \leq 1$. A detailed study of the properties of Eq. (9.3) was performed in Galam (2010), Galam and Jacobs (2007).

9.4.4   Implementing the model to rediscover the data

It is worth emphasizing that we do not aim at reproducing the data exhibited in Figure 9.2. The method aims at evaluating the minimum values of both

---

[4]This constraint is necessary for modelling but questionable as to real journalistic practices. Indeed, why should we slice the social phenomenon by year and fix the behaviour accordingly? In further research, the temporal parameter will be tested with different time slices.

the respective proportions of inflexibles v_T and w_T and the intra-time n, that are compatible with the data for every pair of successive years. Given a pair of values U_T and U_{T+1}, we determine the minimum values v_T, w_T and n, which starting from $u_{T,0} = U_T$ reach $u_{T,n} = U_{T+1}$ with a precision of 10^{-3} after n successive iterations of Eq. (9.3). In the second step, writing $u_{T,n} = u_{T+1,0}$, we evaluate the minimum values of v_{T+1} and w_{T+1} that allow getting $u_{T+1,n} = U_{T+2}$ starting from $u_{T+1,0}$.

More precisely, we start from $u_{0,0} = U_0$ to evaluate v_0 and w_0 such that $u_{0,n} = u_{1,0} = U_1$. Then we evaluate v_1 and w_1 such that $u_{1,n} = u_{2,0} = U_2$. And so on and so forth up to the evaluation v_{11} and w_{11} such that $u_{11,n} = u_{12,0} = U_{12}$. Then, the end of each period of iterations, there begins a new period of iterations with new parameters.

To determine which value n to use, we note that the number of articles for each year period is distributed over three different groups with respectively less than 100 (10), between 100 and 300 (2), and more than 300 (1), as seen in Table (9.1). For each group we determine the minimum value of n which allows implementing $u_{T,n} = U_{T+1}$ starting from $u_{T,0}$ for all cases of each group. We found, respectively, $n = 3, 5, 8$, as reported in Table (9.1).

Table 9.1: Proportions of inflexibles each year to reach the following one, covering 12 annual intervals.

Year	T	v_T^n	w_T^n	$u_{T,0} \to u_{T,n}$	Nb	ΔU_T	n
1998	0	0.091	0	$0.500 \to 0.600$	18	0.118	3
1999	1	0	0.080	$0.600 \to 0.677$	20	0.109	3
2000	2	0	0.285	$0.677 \to 0.513$	31	0.084	3
2001	3	0.081	0	$0.513 \to 0.627$	37	0.082	3
2002	4	0	0.019	$0.627 \to 0.831$	59	0.063	3
2003	5	0	0.169	$0.831 \to 0.769$	178	0.028	5
2004	6	0	0.225	$0.769 \to 0.518$	368	0.022	8
2005	7	0	0.015	$0.518 \to 0.541$	85	0.054	3
2006	8	0	0.169	$0.541 \to 0.422$	61	0.064	3
2007	9	0.209	0	$0.422 \to 0.544$	71	0.059	3
2008	10	0	0.121	$0.544 \to 0.400$	169	0.038	5
2009	11	0.033	0	$0.400 \to 0.251$	85	0.053	3
2010	12			0.251	51	0.061	

From Table (9.1) it can be seen that for any given year n, only one fitting parameter is used, since always either v_T or w_T is equal to zero. The variations of $u_{T,n}$ as a function of successive iterations are shown in Figure

9.4. The error bars are also reported although the GUF values of the series of $u_{T,n}$ recover perfectly the data values U_T for all 13 years. Figure 9.5 exhibits the simultaneous variations of v_T and w_T as functions of T.

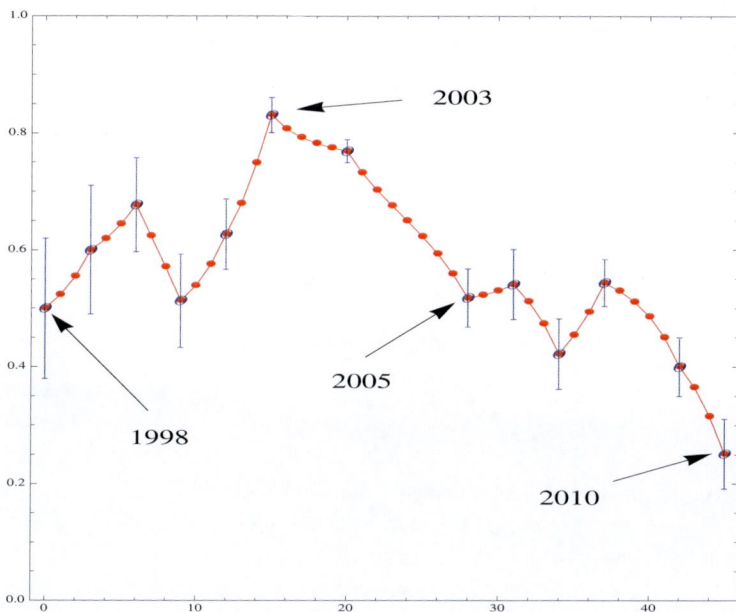

Figure 9.4: Evolution of the proportion of journalists advocating the uni-factor cause as a function of updates using Eq. (9.1) with $9 \times 3 + 2 \times 5 + 1 \times 8 = 45$ updates. Circles show the overlap with the data calculated per year and vertical lines indicate error approximation using the variance of a binomial generator.

9.5 Behind the data another view of the networks

The picture drawn from the modelling leads to a reverse conclusion of what would have been expected. The proportions of categorized articles follow different evolutions than the proportions of agents on each side. From the empirical dynamics, we are led to question how journalists interact in the public debate.

Although in the first year the proportions are equally distributed ($U_0 =$

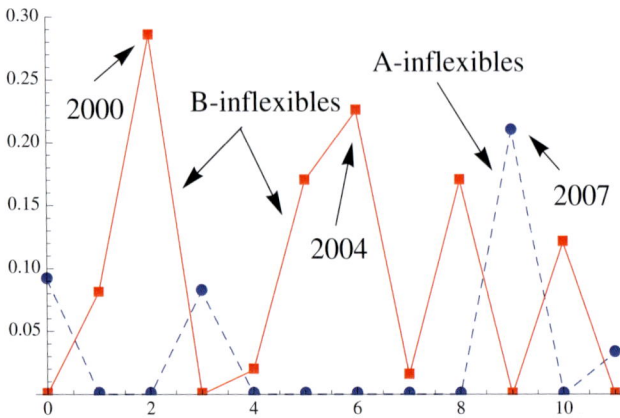

Figure 9.5: Variation of inflexible proportions v_T and w_T as a function of year

0.50), the uni-factor agents appear to have more inflexibles on their side, while none were present on the other side. This result highlights the determining advantage held by the first whistle-blowers about a new controversy.

Once the controversy was launched, the multi-factor inflexible agents emerged, while the uni-factor side turned down its pressure.

From the high level of journalists against the use of pesticides, the industrial side applied significant pressure, getting up to $w_2 \approx 0.285$ inflexibles. The multi-factor expected that the right opinion had then been achieved.

But, with the threshold nature of the dynamics (Galam 2002, Delanoe 2010), the next years brought back the uni-factor side to rather high values in 2002 and 2003 even if these values are lower than expected. Indeed, the beekeepers' view was presented in the public media through political spokesmen during the regional and European elections in 2004. Indeed, pesticides have been banned from sale that year.

With new scientific studies in 2005, the multi-factor side is relaunching the debate in the media.

9.6 Conclusions

This case study shows how empirical data and theoretical modeling can produce a heuristic framework for analysing a controversy. Starting from the empirical data, the use of a model has allowed going back to the data to question their social meaning.

First a text-mining analysis of published articles was performed in order to categorize the articles. Fact reports dealing with the causes of the critical phenomenon have been used to nest the papers. The first category advocates the opinion that the cause is uni-factoral, namely the use of pesticides. The second category asserts a multi-factor cause or the absence of an identified cause.

Second, the evolution of each proportion of categorized articles is assumed to be rebuilt with a dynamics of interactions among journalists. Two types of agents are considered. Some never change their mind with a dispositional view: they are inflexible agents. Some have a positional view since they may shift their opinion. The respective proportions of agents are functions of time and vary only on a year time scale. Between each pair of consecutive years, the fraction of journalists in each class is inferred from the distribution of opinions using the GUF model of opinion diffusion (Galam 2005b). The evolution of the respective proportions of inflexibles is thus obtained for each year.

Third, those proportions of inflexibles extracted from the model are confronted with the empirical data to question the interactions between agents. For example, we can investigate possible pressure on the journalists from the various involved parties.

The GUF model does not suppose any network *a priori* in order to question quantitative text mining evolutions with the results of the simulation. In that context, networks of actors can be questioned with *scenarii*, which include external pressure, social structure, and the frame of the debate.

Bibliography

Borghesi, C., and S. Galam, 2006, Chaotic, staggered, and polarized dynamics in opinion forming: The contrarian effect, *Physical Review E* **73**, 066118.

Callon, M., 1986, *J. Law, Power, Action and Belief: A New Sociology of Knowledge?*, chapter 'Some elements of a sociology of translation: Domestication of the scallops and the fishermen of St Brieuc Bay', pp. 196–223. London: Routledge.

Castellano C., S. Fortunato, and C. V. Loreto, 2009, Statistical physics of social dynamic, *Reviews of Modern Physics* **81**, 591–646.

Chateauraynaud, F., 2004, *La croyance et l'enquête*, volume XV, chapter 'L'épreuve du tangible. Expériences de l'enquête et surgissements de la preuve', Paris: Raisons Pratiques.

Chiron, J., and A.-M. Hattenberger, 2008, Mortalités, effondrements et affaiblissements des colonies d'abeilles, Technical report, AFSSA, November.

Delanoë, A., 2004, Quand les abeilles meurent les articles sont comptés, généalogie et analyse sémantique d'une crise médiatique, *Strategic, Scientific and Technological Watch*.

Delanoë, A., 2007, Analyse des dynamiques managériales face à la contestation sociale : éléments statistiques du cas "ni Gaucho, ni Régent", *Strategic, Scientific and Technological Watch*.

Delanoë, A., 2010, Statistique textuelle et séries chronologiques sur un corpus de presse écrite. Le cas de la mise en application du principe de précaution, *Journées internationales d'Analyses statistiques des données Textuelles*.

Dewey, J., 1927, *The Public and its Problems*, New York: Holt.

French Constitution, 2004, Article 5 - Charte de l'environnement.

Galam, S., 2002, Minority opinion spreading in random geometry, *Eur. Phys. J. B* **25**, 403–406.

Galam, S., 2005a, Heterogeneous beliefs, segregation, and extremism in the making of public opinions, *Phys. Rev. E* **71**, 046123.

Galam, S., 2005b, Local dynamics vs. social mechanisms: A unifying frame, *Euro. Phys. Lett.* **70**, 705–711.

Galam S., and F. Jacobs, 2007, The role of inflexible minorities in the breaking of democratic opinion dynamics, *Physica A* **381**, 366–376.

Galam, S., 2007, From 2000 Bush–Gore to 2006 Italian elections: voting at fifty–fifty and the contrarian effect opinion dynamics, *Quality and Quantity Journal* **411**, 579–589.

Galam, S., 2008, Sociophysics: A review of Galam models, *International Journal of Modern Physics C* **19**, 409–440.

Galam, S., 2010, Public debates driven by incomplete scientific data: The cases of evolution theory, global warming and H1N1 pandemic influenza, *Physica A* **389**, 3619–3631.

Galam, S., 2012, *Sociophysics, A Physicist's Modeling of Psycho-political Phenomena*, Berlin: Springer-Verlag.

Gusfield, J.-R., 1981, *Drinking-driving and the Symbolic Order. The Culture of Public Problems*, The University of Chicago Press.

Katz, E., and P. Lazarsfeld, 1955, *Personal Influence. The Part Played by People in the Flow of Mass Communications*, Transaction Publishers, New Brunswick, New Jersey.

Martins, A., and S. Galam, 2013, Building up of individual inflexibility in opinion dynamics, *Physical Review E* **87**, 042807.

Maxim, L., and J. P. van der Sluijs, 2010, Expert explanations of honeybee losses in areas of extensive agriculture in France: Gaucho ® compared with other supposed causal factors, *Environmental Research Letters* **5**(1), 014006.

Tonnies, F., 1922, *Kritik des öffentlichen Meinung*, Berlin: Springer-Verlag.

Part III

Interactions in social communication

The paper by P-Y. Raccah is a transposition of the mechanical interactions that can appear on a physical device (a 5-ball coupled pendulum, called "the Newton's cradle") and linguistic interactions that occur within a group of human beings. This analogy, because it allows associating deterministic equations to a heuristic model of linguistic interactions is clearly innovative. The article begins with a study of interactions in mechanics and illustrates the response motion to standard stimulations of the 5 coupled pendulums. Then, by analogy with the pendulum behaviour, the author analyzes how the communication can be established inside a group of human beings. The author emphasizes that a large number of schoolbooks and Web pages do not correctly explain the overall behaviour of the pendulum. Such type of error is the result of an ill-posed problem, of untested hypothesis... in short, the search for a shortcut way to explaining the phenomenon. In his article, P-Y. Raccah breaks down the stages of different types of communication and constructs the pendulum/language analogy. He also shows the limits of this model and offers an original approach to communication and its impact on language by explanations providing through the study of educational literature.

The article entitled "Modeling human interactions: A tool for training. A case study with social-sport educators" by F. Glomeron and K. Paret is concerned with the description of the intervention of an educator within the interactions between students. These complex interactions are described through graphs where points represents the aspects of the course situation from interviews with a panel of 10 educators. When a problem is identified, the educator has to choose between resuming, stopping or re-organizing the activity. The authors discuss the effects of these interactions on the evolution of the teaching sessions, the different way educators should identify the origin of instabilities and the complexity of the consequences of educator's reactions.

The teacher' speech is a professional gesture based on a certain design idea of the profession ethic and a certain way of seeing pupils. This is especially true for assessment, firstly because it's a powerful indicator of the teacher's professional identity, and secondly because it has a powerful impact on pupils, the image of themselves and therefore the resources they have to make progress. In his article, Y. Mercier-Brunel analyses the teachers' speech to highlight a certain number of linguistic elements and to suggest a modelling of the interactions in the class.

The paper 'The self-organizing situation, a challenge for physical education teacher training: A complex system of ritualized interaction and proxemic distances', by N. Carminatti-Baeza addresses the problem of learning how to teach and the necessity to build 'teachable', i.e. training situations proposed to students. Teaching can be modeled as a complex system, involving interactions in a proxemic space. This is illustrated by an experiment, in the field of physical education teacher training, in which several students supervise a gymnastics lesson.

The contribution of K. Paret, D. Nourrit and N. Gal-Petifaux is part of a finding based upon research, crossed by a social and economic question: how to improve teaching in University, when this institution has to form professionals with a great challenge to achieve. They have to teach to everyone, everywhere, in an economic crisis context, and in a heated French National education debate. Teachers and educators have to be efficient as soon as possible, need to work in networks and must create new solutions. Their contribution focuses on social sports educators, which takes care of offenders, or people who can't stand school, or young people whose family failed to educate them. To develop efficiency in such a professional context, special workshops simulate the dimensions of these jobs during the initial formation. In these workshops, future educators experience tensions, discomforts and dilemmas and by the fact seem to improve their professional identity construction. They could increase their adaptation abilities when faced with complex situations. Two frameworks provide the opportunity to report this development of adaptation into these workshops: cognitive anthropology (Suchman, 1987)[1] and the course-of-action (Theureau, 2004).[2] The discomfort and dilemmas seem to be parameters of control in the initial formation process.

[1] Suchman, L. A., 1987, *Plans and Situated Actions: The problem of human–machine communication*, Cambridge University Press.

[2] Theureau, J., 2004, *Le cours d'action: méthode él ementaire*, Toulouse, France: Octarès.

Chapter 10

From Newton's cradle to communication using natural language: Are human beings more complex than balls?

Pierre-Yves Raccah[1]

10.1 From ball to balls: Carelessness as a generalized intellectual system?

The simplest and best known models of the interactions between material entities are the ones which describe the exchange of energy, of linear momentum, or of electric charge between small macroscopic objects, such as balls. Let us examine, for instance, the typical scenario of Newton's cradle, when only one of its balls is thrown, as illustrated in Figure 10.1.

The classical description of the observable interactions can be summed up as follows:

[1]CNRS, LLL - UMR CNRS 7270, University of Orléans. Email: pyr@linguistes.fr.

Figure 10.1: A material realization of Newton's cradle.
A material realization of Newton's cradle. [2]

"When the ball at one end is pulled aside and released it collides with the remaining stationary balls and the ball at the other end of the row moves off to reach what appears to be the same height from which the first ball was released. All the other balls are apparently at rest."

[Gauld 2006, p. 597]

From this observation, many school books and didactic web pages (erroneously) conclude that "the effect of the collision simply consisted in the exchange of velocities between both balls"[3]. Considering the situation in which more than one ball is pulled and released allows slightly improving that conclusion, but, as we will see, not to correct it. The cases with more than one ball can be described as follows:

If two balls are pulled aside and released, after the collision two balls move off – again apparently to the same height – with the rest stationary. Releasing three balls results in three balls moving off after the collision.

In those cases, clearly, velocity is not enough and the school books are forced to introduce the notion of momentum conservation and, for some of them, that of energy conservation.

[2]This figure comes from `http://bestgifever.com/?yes=be6a165a2b5efd6d6e5e87362965fb6c`, under free Creative Commons License. Author: Dominique Toussaint; date: 2006/08/08.

[3] Translated from the French Wikipedia page on Newton's Cradle: `https://fr.wikipedia.org/wiki/Pendule_de_Newton` (still visible on February 5 2017).

More precisely, the 'explanation' goes like this:

Ball A (system A) which was separated from the other four (B, C, B', A'), after being released, meets the remaining balls (system [B, C, B', A']), and transmits its momentum to the system [A, B, C, B', A']. In the theoretical conditions of the experiment, it is assumed that there is no loss of momentum or energy. System [A, B, C, B', A'] then transmits energy and momentum to ball A' (opposite of A), which separates in turn; the system is thus in a position symmetrical to the initial situation, in which A' plays the role of A and vice versa.

However, this reasoning does not explain why A' moves off alone (and, more generally, why the number of balls which move off after the collision is equal to the number of balls which provoked the collision). Gauld (2006) reveals an interesting but rather startling fact connected with this lack of explanation: none of the 40 scholarly books he studied considers that question as requiring an explanation...

"About one-third of about 40 tertiary physics textbooks sampled contained some reference to Newton's Cradle [...] However, it is interesting to note that, in spite of the apparent simplicity of the demonstration, the behaviour of Newton's Cradle is not adequately explained in these textbook presentations *and there may not even be a fully adequate explanation available in the physics education literature.*"

[Gauld (2006, p. 598); my emphasis]

Gauld wryly adds, further on:

"One might excuse teaching which is ignorant of little known facts but it is less easy to excuse facts about the apparatus which are more easy to establish such as that the principles of conservation of momentum and kinetic energy are not sufficient, in themselves, to explain the behaviour of Newton's Cradle."

[*ibid.*, p. 615]

And this enormous cheating (if what Gauld pointed out may be called that) was so, in 2006, in spite of the fact that Herrmann and Schmälzle (1981) had shown that the use of the two usual conservation principles (momentum and kinetic energy) is not sufficient to describe the facts observed.

For instance, in the case of an initial impact with one ball (A), energy and momentum conservation would not prevent A's bouncing with a speed equal to one-third of the initial velocity, while A' and B' separate with a speed equal to two-thirds of the initial velocity of A.

Indeed, as Herrmann and Schmälzle (1981, p. 762) develop:

"Imagine now the following final state: ball 1 moves to the left with $v = (-1/3)v_0$, balls 2 and 3 move to the right, both with the same speed $v = (2/3)v_0$. It is easy to confirm that the values of the kinetic energy $[Ek]$ and the momentum $[P]$ of this hypothetical final state are the same as those of the initial state:

$$Ek = (1/2)m\,[(1/3)v_0]^2 + 2(1/2)m\,[(2/3)v_0]^2$$
$$= (1/2)mv_0^2,$$

$$P = m[(-1/3)v_0] + 2m(2/3)v_0$$
$$= mv_0.$$

Thus energy and momentum would be conserved. Nevertheless, the actual experiment always evidences another outcome."

The following fanciful illustration (Figure 10.2) gives an idea of the possible (but not actual, of course) behaviour of the cradle when applying only momentum and kinetic energy conservation principles.

Since that behaviour was never observed, there must be another constraint which the cradle must satisfy.

Herrmann and Schmälzle (1981, pp. 763–764) argue that "In a non-dispersion-free system, energy and momentum are distributed throughout the entire arrangement.", while what we observe is that the total energy and momentum of the incoming balls is transferred to the same number of balls at the other end of the chain. They thus propose a simple explanation involving a third conservation principle, shock wave energy conservation, allowing the kinetic energy transmitted by A to $[A, B, C, B', A']$ to be transferred without loss to A', through the initial shock wave. This constraint allows taking into account the propagation of the double perturbation wave due to the initial shock, perturbations which travel across the system in each of the two directions, and move backwards after reaching the system's ends, *without attenuation*, and therefore, at the same speed. It is at the point where the two wave packets meet that the balls separate; this point must be symmetrical to the point of impact in order for the two wave packets to travel the same distance before they meet. Figure 10.3 illustrates this point:

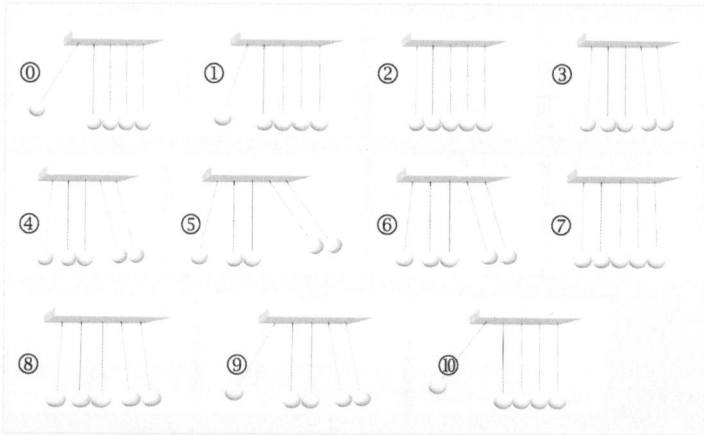

Figure 10.2: Fanciful non-actual behaviour of cradle (compatible with the classical explanation).

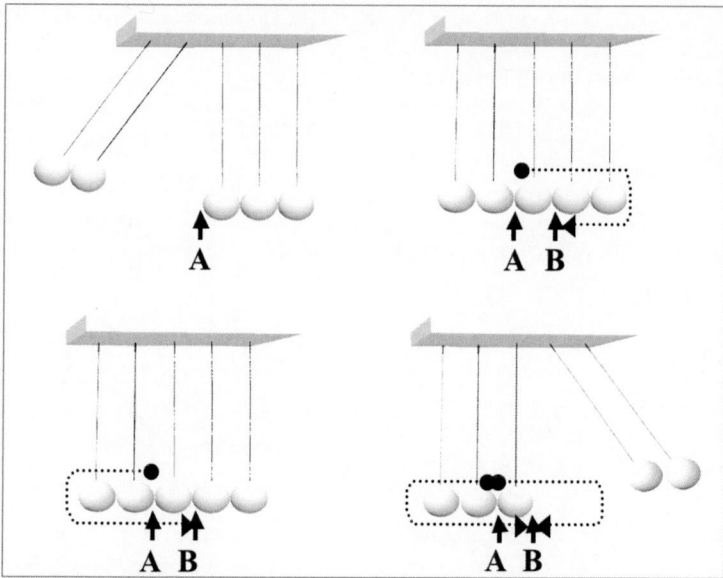

Figure 10.3: Explanation using the (additional) dispersion-free constraint.

I considered the proposal presented in 2010 by a group of secondary school students to the French Physics Olympiad extremely reassuring and fresh, proposal which was awarded the first prize. For they address this specific problem in a way that is both correct and amusing, and propose a solution which is both amusing and ... correct.

In 2010, Nicolas Beliaeff, Romain Heimlich, and Baptiste Jeanin, from the Lycée Jean Eiffel (Dijon, France), formulated their conception of the problem in this way:

"[Nous] nous sommes alors demandés comment la dernière bille, qui n'a pas assisté au cours de physique, a pu savoir qu'elle devait conserver énergie cinétique et quantité de mouvement."

"[We] then wondered how the last ball, which did not follow the physics class, could know that it was supposed to preserve both kinetic energy and momentum"

[Beliaeff, Heimlich, and Jeanin (2010)]

and they immediately propose a hypothetical answer, the validation of which was the object of their exposition:

« Nous avons alors supposé que ces informations devaient être envoyées par la première bille et que le son ou la vibration due au choc devait contenir ces informations. »

"We then made the hypothesis according to which this information was sent by the first ball and that sound or vibration due to the initial shock should contain this information"

[*ibid.*]

I cannot resist the temptation to show the way in which this hypothesis is discussed and illustrated in their work, where *ludic* goes together with *lucid*, and *ludicity* seems to be a real warrant for *lucidity* ... a brief sequence of schemata, shown in Figure 10.4, with their comments, will be enough to give an idea and, I hope, intellectual pleasure.

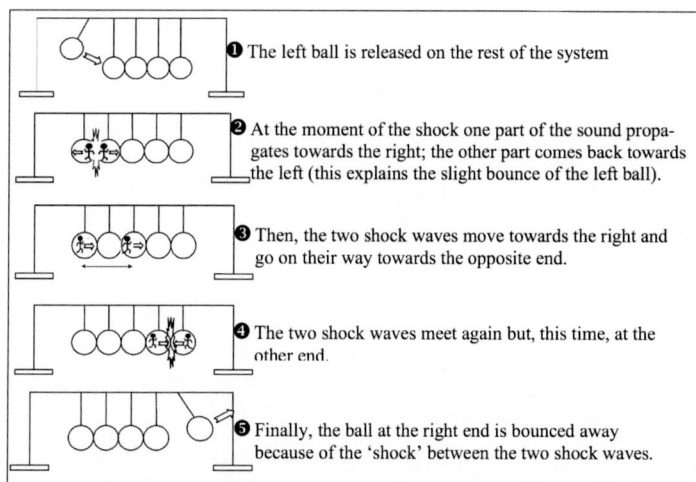

❶ The left ball is released on the rest of the system

❷ At the moment of the shock one part of the sound propagates towards the right; the other part comes back towards the left (this explains the slight bounce of the left ball).

❸ Then, the two shock waves move towards the right and go on their way towards the opposite end.

❹ The two shock waves meet again but, this time, at the other end.

❺ Finally, the ball at the right end is bounced away because of the 'shock' between the two shock waves.

Figure 10.4: Explanation of the process involved in Newton's cradle using the hypothesis 'sound as a messenger' (adapted from Beliaeff, Heimlich, and Jeanin, 2010).

10.2   From balls to Human Being: The ... *impact* of communication

The standard communication model used for signal processing, since Shannon (1948) and Shannon and Weaver (1949), can be summarized in the following proposition:

A *sender* encodes a message and emits the result of the encoding towards a *receiver*, which perceives it in an environment possibly disturbed by other transmissions (noise), and then performs the decoding.

That conception of signal transmission allowed giving an operational definition of *information*: *information*, within signal study, is the inverse of noise[4].

---

[4] Several cynical theorists have argued that that is the reason why the turn of the century, with its ubiquitous glorification of public communication, has drowned information in an ocean of noise; in particular in the realm of scientific research, where the commandment "Publish or perish!" engulfs genuine scientific information in a tsunami of noxious waffle. . .

The communication model underlying that important work on signal was 'sold' to linguistics, though it was intended only for signal[5]. Roman Jakobson (1960) did feel the model was insufficient to account for human verbal communication but believed that, with a few 'patches'[6], it could be used as a first approximation. A version of Shannon and Weaver's signal model, 'patched' by Jakobson and simplified 'for didactic purposes' (hence-forth: STD) is still constantly taught as *the* model of human communication 'discovered' by Jakobson...

Paying attention to the structure of the Signal Treatment Diagram (STD), as it has been adapted for human communication (with 'patches' and 'didac-tic' simplification), one easily realizes that it is even simpler than the one which does not work for Newton's cradle. Figure 10.5 illustrates the model in question.

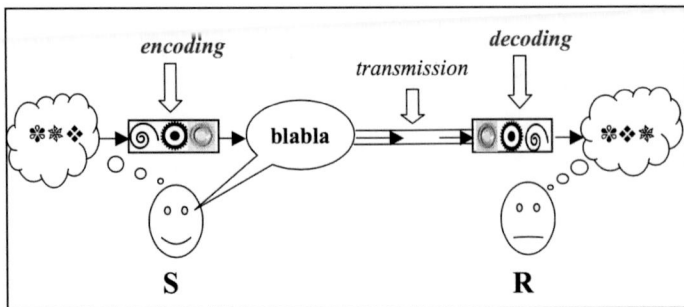

Figure 10.5: STD – An adaptation of the signal model *to human linguistic communication.*

For linguistic studies, the transmission process is not an object of study; on the other hand, neither the encoding and decoding processes, nor the original and 'reconstructed' messages (illustrated here by compositions of ❀,✱,and ❖) are observable.... However, as any linguist is as good a speaker as any other speaker (which is certainly true), most of the linguists (mistak-enly) consider that the non-observability of the processes, of the original message or of the 'reconstructed' message, poses no problem: from the output of the sender, the linguist '*knows*' the message he/she (the sender) encoded; and, at the same time, '*knows*' what the receiver will get when de-

---

[5] It seems that the 'transaction' was done in spite of Shannon's doubts, for whom the signal model was not appropriate for anything other than signal.

[6]*Cf.* the six famous functions of languages he introduced in Jakobson (1960), and which are taught in practically all the classes of linguistics in France.

coding the transmission and 'reconstructing' the message[7]. The discussion of this magic belief is not the subject of this paper[8] and I have sufficiently suggested what I think of it (for some, even too much. . .) I only intend to show, here, that (independently of its magic touch) the model in question, STD, is a caricatural simplification of the one that does not work for Newton's cradle and, as such, is not likely to work for more complex interactions, such as human linguistic communication.

If we want to merge the description of what happens in Newton's cradle and the description of what, according to STD, happens in human linguistic communication, we get something that might look like Figure 10.6:

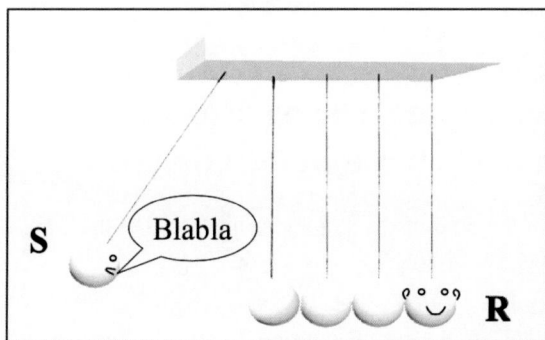

Figure 10.6: A cradle-like approach to the signal-like model of human communication.

where the impact of ball S corresponds to the transmission of the encoded message through the three balls in the middle, and the separation of ball R corresponds to the result of decoding the received encoded message. However, when observing the cradle, the separation of ball R *is* observable (and the properties of that separation – momentum and amplitude – *are* measurable), while, when observing human communication from the point of view of STD, as we saw it in the preceding paragraph, what R understands, which corresponds to the separation of the ball, *is not* observable. In addition, in the cradle, what caused the impact *is* also observable and its properties measurable, while, in the signal approach to human communication, again, what

---

[7] This would seem absurd to any other group of researchers and, probably, to any other human being; but not for most linguists.

[8] For elements of that discussion, see, for instance, Raccah (2005), Raccah (2011a), or Raccah (2011b, pp. 154–161); those who read Russian can find in Raccah (2011c) a few proposals for a linguistic experimental approach to semantics.

precedes the emission of the encoded message *is not* observable (let alone measurable).

To make a short story short, the scheme of human communication based on the signal model (STD) does not predict anything empirically observable: it cannot, thus, account for anything that would not be believed in advance (in particular, it does not say anything about how meaning or *what* meaning is constructed by the recipient); and, of course, it is immunized against scientific refutation.

In the next section, I will discuss several properties a model for human communication should have and, as a conclusion, I will propose a model of linguistic interaction which does have these properties and does not suffer from the weaknesses of STD.

10.3   From human being to human being: The... impact on communication

Regardless of the nonsensical but abundant literature that bad uses of good but limited models produce, there are

  (i) Observable entities which are transmitted from S to R in a human communication between them (excluding meanings, intentions or messages), and

 (ii) Possible observable reactions by R (excluding what R understood), which may allow an observer to produce hypotheses about how R understood S's communicative move.

    Moreover,

(iii) a sufficiently large set of interaction instances *necessarily* gives the cues which allow grasping the semantic rules of a human language; this is so because any dunce can acquire, *and does acquire*, a human language in 18–24 months, being exposed only to speech and human attitudes.

In addition, an acceptable conception of human communication has to take into account other problems that could not even be conceived within STD:

  1. In human linguistic communication, the meaning of what the 'sender' is about to say is accessible to *no one* (not even him(her)self) before (s)he speaks.[9]

---

[9] *Cf.* Raccah (2008), pp. 72–75.

2. If *sound* is transmitted indeed (as in the school students' conception of the cradle), meaning is not: it is constructed by the 'receiver' (in particular, on the basis of the stimulations that the sound occasioned).

3. Understanding a discourse (or any utterance of a human language sequence) does not involve encoding or decoding[10]

4. Understanding a discourse (or any utterance of a human language sequence) is an irrepressible and unconscious activity[11].

If we go on exploring what is beyond the sole signal, it is clear that what S transmits to R (sound), (a) S too perceives it, (b) R (sometimes) reacts to it, (c) S too (sometimes) reacts to it, (d) S reacts to what she/he perceived of R's reaction, etc. The picture, thus, looks much more like the playful approach to the cradle, as presented by the high school students (with something running in both directions) than like the static simplification of the 'linguistic' model based on the *signal model*. Figure 10.7 is an attempt to merge these considerations into Figure 10.6:

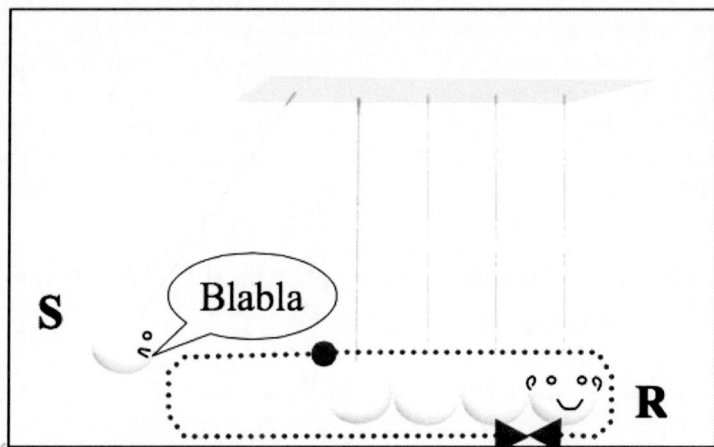

Figure 10.7: Figure 10.6 with feed back loop.

The difficulty we underlined in the last section, the fact that *reactions* are not necessarily directly observable (in particular, *what is understood* by R is not directly observable), is not solved but, here, it is not hidden:

[10]See Rastier (1995), Raccah (1998), and/or Grillo (2003).
[11]See Bruxelles and Raccah (1992).

R's reaction is present in the schema, and the experimentalist semanticist must design experiments which will allow indirectly observing these reactions. The inclusion of the sort of feed-back loop in the presentation of what happens in communication allows avoiding two incredible mistakes, due to STD, which are quite often present in the implicit background of semantic work:

1. Discourse meaning, that the semanticist indirectly observes, and that will help him/her describe sentence meaning, is not the one (s)he him(her)self built, but one (s)he can show that this is what the receiver constructed: the last one requires arguments, while the first is based solely on intuition.

2. In order to be in the position to make (and argue for) the hypothesis that the 'receiver' built an utterance meaning S for discourse D, in the situation i, it is necessary to take into account the reactions of said 'receiver', and make abductive testable hypotheses about the discourse meaning, which could explain these reactions.

Verbal reactions of R are not to be favoured: they require interpretation by the semanticist, and thus introduce subjectivity into his/her observation. In the lack of better empirical material, one can be happy with them for a first approximation. However, this subjectivity regards the same area of research that we want to scientifically explore: there might soon be biases and conflicts of interest...

10.4   As a conclusion: A proposal for a model of human communication

We saw that the conception of human communication, STD, implicitly or explicitly used by most linguists all over the world, is a simplified version of the model used for the treatment of signal. We also saw that that model is a model of interactions that is even simpler than the one mistakenly used to account for what happens in Newton's cradle: in addition to conservation of momentum and kinetic energy, another principle is needed, namely what I called the *shock wave energy* conservation principle. I showed that the communication model STD, derived from the signal model, could make no empirical prediction, and I pointed out that several of its implicit assumptions are erroneous. I suggested that modifying those assumptions implies a radical change in the conception of human communication, and adding an analogue of the shock wave conservation principle (which allows taking feedback into account) could yield a much more interesting conception of human communication.

If I have convinced the reader, I could be satisfied, even though one might argue that criticizing is easier than proposing.... Luckily enough, I do have something to propose, which stems from all that critical work: a conception of *human communication using language*, which proceeds from a more general Communication Activity Representation conception (CAR). One of the advantages of this conception (from the point of view of the present paper) is that it does use the reflection developed here about Newton's cradle and the contrast between what is observable (for instance, the separation of the last ball) and what we use in order to account for it (principles), which is not observable. Another advantage (which might appear, at first, as an inconvenience) is that, once the proposal is understood, it seems to be so evident that it becomes difficult to envisage it as a genuinely original conception of communication. This characteristic is an advantage because any move that is made to defend the idea that the said conception is too obvious to be a genuinely original proposal is, *ipso facto*, evidence in favour of that conception...

The proposal, called the *Manipulatory Conception of Human Communication Using Language* (in short, MCC) is presented in Figure 10.8:

When S communicates with R, S manipulates R to get R to construct the meaning S wants R to construct

Figure 10.8: Illustration of the manipulatory conception of human communication using language.

The sounds emitted by S do not carry meaning, but act on R to make R construct a specific meaning for S's utterance. The meaning R will create is, as we said, influenced by S's utterance, but also depends on R's knowledge and beliefs: what S produces (S's utterance) functions as a set of instructions which R carries out or implements, and which have the effect of making him/her build this specific sense. R's internal process is not conscious: obviously, R may build a conscious *a posteriori* picture of 'what happened' in his/her mind, but the understanding process is not accessible while it takes place. The process itself is not directly observable: no one can access such internal processes (and, as we just saw, not even the person in which it takes place), but, sometimes, the outcome of the process provokes observable facts, which can be used to formulate abductive hypotheses about some aspects of the process. The process of meaning construction is also *irresistible*: the state of being a speaker of some natural language *forces* one to understand any understandable utterance using that language. For instance, about one-half of the Anglophone American inhabitants might feel uncomfortable when hearing the following utterance:

John Doe is a Republican but he is honest

This is so because the hearers cannot understand the utterance if they do not – at least provisionally, during the time of the process – accept that there is some kind of opposition between being a Republican and being honest. Interestingly enough, the other half (if we except the few semanticists that might belong to it) usually do not even notice that they have to accept such a point of view about Republicans: the semantic instruction acts unconsciously and irresistibly until it eventually leads to a point of view which is inconsistent with the hearer's. When this contradiction does not occur, there is no reason for the hearer to be conscious of all the points of view (s)he had to accept in order to understand. When the contradiction does occur, the hearer often feels (s)he is the victim of an aggression: for good reasons, since (s)he has been forced to accept (even provisionally) a point of view which (s)he does not share.

What has just been said about the oral modality is relevant, with very little changes, for written material. This might be surprising, because oral and written material differ in nature: each time one reads a written segment, (s)he produces a new utterance, different from one produced by another reading, and which is usually treated as new oral material (a sort of *inner voice*). However biologically different might be the initial effect of a real sound input from the more abstract effect of a reading, one of the very interesting properties of human language processing is that, at the level of complexity of semantic treatment, they should not, and *cannot*, be differentiated: as far as meaning assignment for natural language expressions is concerned, no distinction is to be made between an *inner* voice and an *outer* voice. This astonishing property, which is characteristic of a separate scientific field for the study of language, is certainly related to the level of abstraction at which the complex processes of understanding take place; it does deserve more investigation, but this aspect of the question is both out-side the domain of research on language as such, and outside the scope of the present paper: we will have to limit ourselves to indicating the question.

Although MCC is too general to be a model, it may yield sound models of human communication, grasped through its interactive aspect (as we saw in section 2, this is not true about the conception based on STD): in MCC, there is an observable input (sound or written material) and there may be observable outputs (reactions), in such a way that the non-directly observ-able intermediates (meanings, representations, points of view, beliefs, etc.) can be indirectly observed, in a way analogous to how momentum, for in-stance, can be indirectly observed. The relation between the added value of MCC and the correct account of Newton's cradle is the fact that, in MCC, S knows whether his/her action on R succeeded by interpreting R's reactions: the global effect of the interaction lies thus at the meeting point between the two packets of action.

Clearly STD and MCC are two incompatible, competitive CARs. Now why should you buy the new CAR I propose? The first reason is, obviously, because your old CAR is out of order.... More seriously, STD blocks the evolution of accounts of human communication and of human language se-mantics because the oversimplified concepts it uses presuppose, as we have just seen, facts about communication and about language that we now know are not the case. It may have been useful when the tracks leading to find-ings about human languages were narrow; but the map of knowledge on this subject has evolved since that time and you should not use your old CAR on the expressways that are available for scientifically exploring this territory.

Bibliography

Barthes, J., 2010, La pétanque de Newton. *Bulletin de l'Union des professeurs de physique et de chimie*, **926**, 877–884.

Beliaeff, N., R. Heimlich, and B. Jeanin, 2010, *La pétanque de Newton. Mémoire de concours : 1er prix de Olympiades 2010 ; 2ème prix – au concours CASTIC à Canton (Chine) en août 2010.* Lycée Eiffel (Dijon).

Bruxelles, S., and P.-Y. Raccah, 1992, Argumentation et sémantique : le parti-pris du lexique. *In* W. de Mulder, F. Schuerewegen, and L. Tasmowski (eds.) *Enonciation et parti-pris*. Amsterdam: Rodopi.

Gauld, C. F., 2006, Newton's Cradle in Physics Education. *Science & Education* **15**, 597–617.

Grillo, É., 2003, Parler la même langue. *MAG Philo*, revue électronique, 9 : *Langages*.

Herrmann, F., and P. Schmäälzle, 1981, Simple Explanation of a Wellknown Collision Experiment, *American Journal of Physics* **49**(8), 761–764.

Jakobson, R., 1960, Closing Statement: Linguistics and Poetics, *in* Thomas Sebeok (ed.) *Style in Language*.

Raccah, P.-Y., 1998, L'argumentation sans la preuve : prendre son biais dans la langue. *Interaction et Cognition*, **2**(1–2)

Raccah, P-Y., 2005, What is an empirical theory of linguistic meaning a theory of? *In* Zygmunt Frajzyngier et al. (eds.) *Diversity and Language Theory* Studies in Language Companion Series, John Benjamins.

Raccah, P.-Y., 2008, Contraintes linguistiques et compréhension des énoncés : la langue comme outil de manipulation. In *Entretiens d'orthophonie*. Paris: Expansion Formation et Éditions, pp. 61–90.

Raccah, P.-Y., 2011a, Linguistique critique : une exploration cognitive. *Intellectica* **56**, 305–314. Special issue: *Linguistique cognitive : une exploration critique*, ed. Jean-Baptiste Guignard.

Raccah, P.-Y., 2011b, Les questions de rhétorique sont-elles des questions sémantiques ? Réflexions sur une théorie de la signification, informées par des études de sémantique contrastive. *RSP* **29–30**, 151–173.

Raccah, P.-Y., 2011c, *A semantic description of lexicon: A means to efficiently assist intuition?* Presses de l'Université d'État de Novossibirsk. Série: Linguistique et Communication Interculturelle, **9**(2), 96–120 (in Russian).

Rastier, F., 1995, Communication ou transmission ? *Césure* **8**, 151–195.

Shannon, C. E., 1948, A Mathematical Theory of Communication. *Bell System Technical Journal* **27**(3), 379–423.

Shannon, C. E., and W. Weaver, 1949, *The Mathematical Theory of Communication*, Urbana, Illinois: University of Illinois Press.

Chapter 11

Modelling human interactions: A tool for training. A case study with social-sport educators

Frédéric Glomeron[1], Karine Paret[2]

11.1 Introduction and context

Intervening with problematic young people in the area of training and education currently appears difficult and unforeseeable (Johannès, 2013). For professionals, uncertainty and a lack of motivation can generate suffering on the job, which results in educators' quitting the job and a high turnover rate. Interaction with a public with special needs (young school drop-outs, with social and/or family difficulties, deprived inner-cities sometimes under judicial warrant) cannot be envisaged without a rigorous professionalism of the social-sport educators (SSE). Our exploratory study attempts, through modeling, to clarify the complexity of the various interactions during sport

---

[1]GREF, Université d'Orléans. Email: frederic.glomeron@univ-orleans.fr.

[2]MAPMO – UMR -CNRS 7349, Université d'Orléans. Email: karine.paret@univ-orleans.fr.

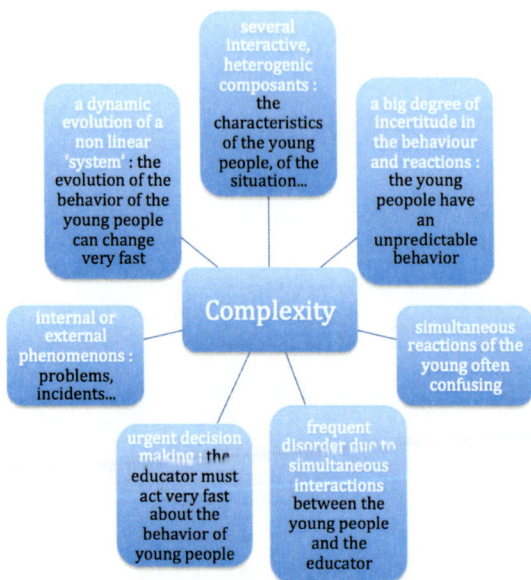

Figure 11.1: Representation of a complex system.

interventions. We advance the hypothesis that the environment composed of the situations where the educators intervene can be considered a complex system.

11.2   Sport social interactions, a complex system

We assume that the interventional situation, which is the subject of our study, is a complex system (Bonami et al., 1996; Clergue, 1997; Le Moigne, 2002; Morin, 2005). In each situation, we have the actors, the stakes, the obstacles, the resources, the tools, and the temporality which gives a dramatic dimension during a crisis and for the results. In (Lave, 1988, p. 150), two dimensions were proposed to describe the environment: the 'Arena', which is the objective aspect, something permanent in the situation, and the "Setting', which is constituted by the singular, subjective dimensions of the situation, the point of view of an actor. In this paper, we focus on the 'arena'.

Figure 11.1 illustrates the complex nature of interventional situation.

Our study takes into account the role of the SSE exposed to a complex environment. During the decision making process, the SSE combine with the contextual constraints, available resources (their level of expertise) and their conception of the appropriate reaction (moral and educative values). The educator must also employ self-preservation when confronted with the pressures in this complex environment (Granon and Changeux, 2011). The educator has a balancing role, maintaining the link that will enable the situation to remain viable.

11.3   Question for the training course

How to train and prepare for action in a difficult context? The training of educators can only rely on the predictive model: even a very thorough and rigorous planning of the activity cannot eliminate the hazards and uncertainties. The development of routines, even numerous ones, would be insufficient to encompass the daily issues (Barrère, 2002; Tardif et Lessard, 2000). A professional training that prescribes predefined solutions will not allow the educator to adapt ideally to incidents and hazards. In order to deal efficiently with disruptions, an active strategy must be adapted and transformed (Gelignier, 1979).

Figure 11.2 shows that the intention of the intervention is an arrow pointing towards the planned activity. Some 'noise', i.e. disturbances, disorganizes the activity. The actor must simply change their activity (adapted activity) or completely redefine the activity (transformed activity).

'The action depends on the complexity, that is to say, hazard, chance, initiative, decision... ' (Morin, 1990; 2005) Training within a complex environment and including complex tools might be a solution. How can a training module provide a genuinely complex educational situation? How can the student educator be confronted with and understand the risks, dysfunctions, and surprises linked to the context? How and with what limitations can this complexity be used for professional development?

11.4   Typical environments which generate tensions and complexity

The law of the freedom and autonomy of the French universities ("LRU in French) of August 2007 redefines the public service missions for higher-level teaching (or university level teaching) by including the orientation and

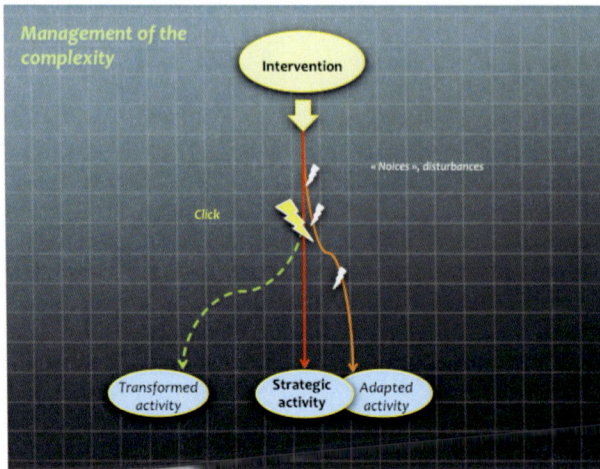

Figure 11.2: Transformed activity.

professional insertion of the students. Diplomas must now be in phase with the needs (Gayraud et al., 2011).

Clergue (1997) suggests educating 'with and by complexity'. The "mobilization of complex educational situations: that is, situations where the causes tangle and accumulate as in reality, and so reinforce the motivation of the student by transforming the immediate failure into reasons for inventing new solutions" (...). "Every time new needs emerge from his or her environment, the student progresses" (Clergue, 1997 p. 14). In the present article, we propose to adapt this view to the training of SSE, which means taking into account the interactions between the different actors.

The analysis of these interactions is at the centre of our study. The definition of the word *interaction* is the action of entering a situation with the intention of influencing how it plays out, or to engage in an action, an authority, to play a role in the situation, to act deliberately. The French etymology of the verb *intervenir* dates to 1363 (*intervener*, *interagir* in French) and explains that it means to *entrevenir*, i.e. 'to go out and meet other people' (Bloch and Von Wartburg, 1986). Also, 'intervention' (dated 1322), a word borrowed from the Latin *entevenire*, is considered in the legal area to be a 'mediation'. De Montmollin (1977, pp. 20–21) defines social interactions as the sum of the effects that result from the use of words or the actions of others on the individual answers to one's environment. According to this definition, reciprocity is discussed within the answer of the individual to the social environment.

11.5   The theoretical framework: A systematic approach; complex and situated actions

Professional didactic theories (Samurçay and Pastré, 2004) and the situated action (Suchman, 1987) will be mobilized in order to chart the exercise of the SSE profession. It is necessary to examine the activity in a given situation, in a familiar environment, in order to diagnose what is needed for the training. The activity and technical registries (Martinand, 1994; Glomeron, 2001) provide us with an analytical activity grid with its limitations and effects.

An activity is not detachable from the context in which it has occurred, and is the proof of the adherence of the actor to his environment. The complex approach sheds light on the actor's decision making process. The precise moment that marks the arrival of a disturbance signals the start of the actor's capacity to adapt (Bénaïoun-Ramirez, 2013).

In order to clarify our postulate, it is possible to introduce into the training the complexity of the real-life situations. This apprenticeship in complexity (Clergue, 1997) seems possible if the reality of the interactions is correctly and systematically interpreted. This is why it is necessary to fully understand the interactions in real-life situations and in their evolutions. We seek the appearance of the release mechanisms and analyse them. We identify the breaks, intermittences, and changes which arise from disorders in the situation. In other words, we ask the question: what is the state of the system?

Our hypothesis is that the analysis of the situation, its dynamics, the disturbances, and the reorganizations that occur in response, can allow a better understanding of the adaptation and the regulation of the competencies of the SSE. The transformations are proof of the professional apprenticeship of the future educators.

11.5.1   Methodology

The activity in reference is the socio-sports intervention. The physical activity is organized by the SSE in such a way that the level of performance is not the only objective. Sport here is a tool, an object, a social practice, a way to allow these young people, who lack social and relationship skills,

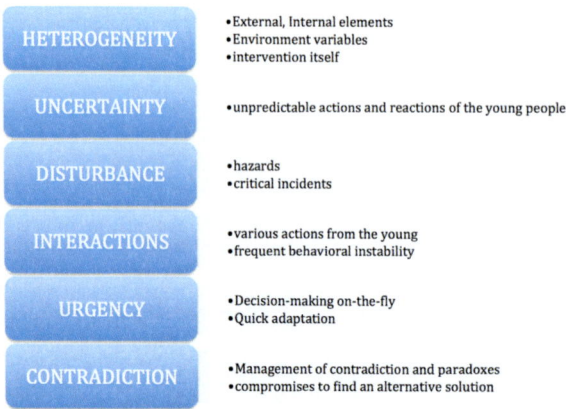

Figure 11.3: Social-sport intervention as a complex system.

to progress. The SSE are constantly aware of the need to deal with these vulnerabilities during sport activities. The educational choices of the SSE are guided by this end point. In the complex, educational interaction situation, priorities are thus defined. For the actors (the SSE), the social-sport intervention is characterized as a complex system, as represented in Figure 11.3.

Our real-life study led us to make observations in those social areas targeted by the training (Martinand, 1994). These areas are legal youth protection centrs, detention centres, children centres, local associations, and regional government bodies. The professional activity of 5 educators during 10 sport sessions is the basis for this study. The observation grid used for the data collection includes the following guide points:

- The elements doing the interacting;

- Types of interactions;

- Luck, incident;

- Marks of destabilization of the actors (attitude, visible emotive expression);

- Alternative decisions: paradox, contradictions to be resolved;

- Decision made on the action;

- Effect on the action.

The educators being observed filled in their own worksheets. Their notes related to the ongoing intervention or were filled in a while afterwards. They were used in the modeling of the complex systems: the components, the links identified, the interactions and events of the actions, etc. Interviews with the 10 educators provided further notes for the worksheets. During each interview, two or three significant situations were discussed. Critical moments, such as destabilization, incidents, or hazards (Glomeron, 2001) were chosen to observe the professional ability of the SSE in action. From the actor's point of view, these critical moments reveal a specific part of the system that should be modeled, and so we obtained an instantaneous modeling at the desired time. This is valid for the analysis of interactions, evolutions, and the parts of the targeted system.

The analysis was based on the data collected. The components were identified. The interaction network was constructed. The disorders were characterized. The alternatives and decisions of the actor were presented. The evolution of the system was identified.

11.5.2   Results

We were able to analyse the modeled interventional situations and to identify the moments, triggers, that led to destabilization. The analysis of the inter-actions and the evolution of this complex system supplies elements about the abilities used in the form of an 'organized improvisation' (Perrenoud, 1999). This provides professionals with the capacity to make decisions rapidly:

- Type of interactions and the 'trigger' events;

- Contradictions and alternatives that have to be managed;

- Characterization of the interactions;

- Technical actions types in interactions.

Our study illustrates the different phases of decision-making during a volley-ball session. The components are depicted in Figure 11.4 by points, while their active interactions are shown as lines. The components are important elements for the actor (observed or evoked). The system is an open one, and certain components are linked to outside elements. For example, the equipment can be shared with another actor. The local project can be linked with other projects. The 'citizens' agenda' is at the heart of the project's objectives. It is linked to other aspects for the young, e.g. good health. Figure 11.4 includes the components that are related to and which

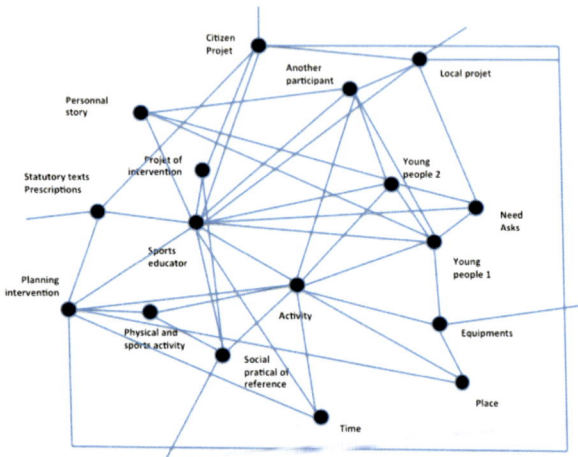

Figure 11.4: Intervention, a complex system.

are important for the actor. The actor appears in the heart of the action. We can observe interactions with the various components of the situation.

Let us apply this model to sports through the example of volley-ball.

Figure 11.5 represents the different phases of decision making during a volley-ball session:

1. In red the main interactions. SSE focus their attention on the activity of two young people. There is a risk of a sudden altercation between them;

2. The central elements of the interaction represented with the large circles;

3. SSE increase their attention on this interaction: focus;

4. The public becomes tense. The initial situation is now perturbed. The relationship installed up to this point is broken. This is a key moment for the evolution of the system.

It thus becomes necessary to reorganize the structure of the situation in order to keep moving on.

We identify this as a decisive moment in the decision making process. The SSE have to rely on educative tools systems. A choice must be made between:

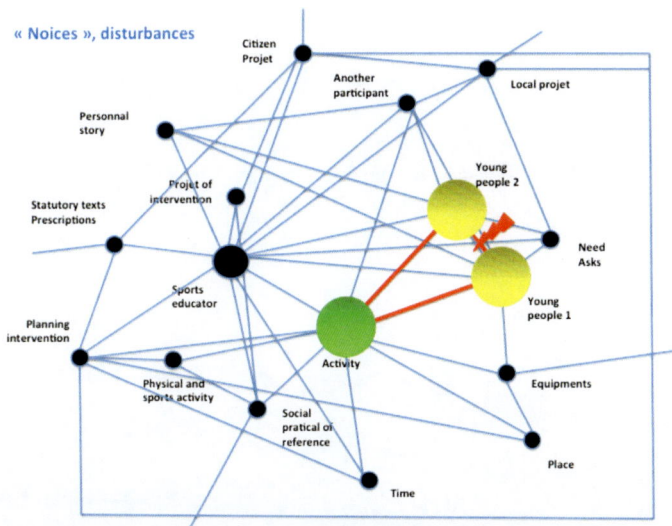

Figure 11.5: Altercations in sports: Volley-ball example.

1. Maintaining the progress of the session and staying in conformity with what was planned. Risk: could be contagious and increasingly disturbing.

2. Stopping the progress of the entire group in order to deal with the two perturbing elements.

3. Reorganizing, then resuming the activity.

The observational answer is the following one: the situation is reorganized in its normal context and the incident replayed. This incident becomes the basis for a new activity. It can be said that some elements of the system are inhibited, others liberated. The strategy consists in changing the direction of a given situation without stopping it. A secondary complexity emerges, shown in Figure 11.6.

The objective of this first study was to describe and explain this complex system with its interactions, issues, and schedule reorganizations. While observing a social sport exercise oriented towards a fragile public, several analytical levels of the SSE activity can be perceived, which can combine and influence each others, as shown in Figure 11.7. However, while this analytic division is easy to understand, it does not function so well when

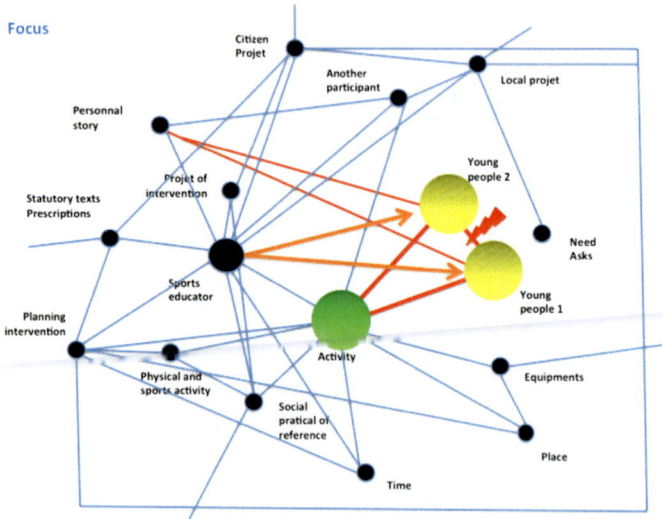

Figure 11.6: Conflict regulation and inhibition.

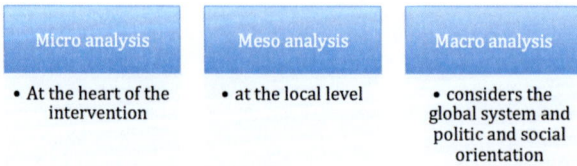

Figure 11.7: Different analytical levels of the SSE activity.

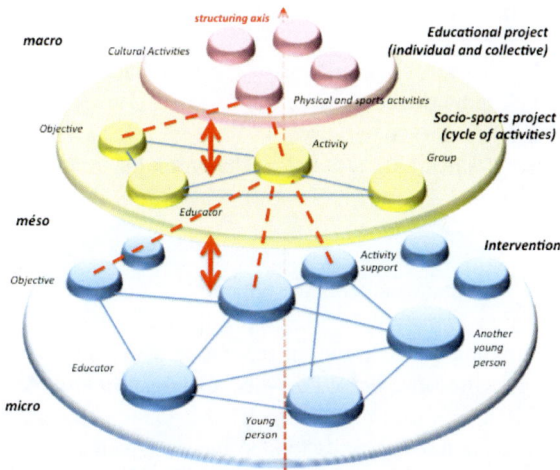

Figure 11.8: Intervention analysis layers.

applied to the decision making process. Even for a local level decision, the SSE must take into account the interactions between the different levels. Appropriate decisions must be made within a systematic and multi-layered network. In order to illustrate this, we have attempted a multi-layered approach, as presented in Figure 11.8. Even if the links between the levels are taken into consideration, the analytical approach does not adequately reflect reality. These different levels are constantly and simultaneously interacting. We comment on the repetitive structure at the different levels. 'We can characterize the complexity such as the emergence of different levels within the same system' (Clergue, 1997, p. 18). Thinking should be organized systematically from micro to macro and vice versa.

One method could be to focus on one level. However, when concentrating on the macro level, we would ignore elements from the micro level that might affect the macro level. Thus in our comprehensive method we must analyse the system in its entirety. A structuring axis exists that organizes all levels so that they perform coherently. This axis represents the endpoint of the profession of the SSE, i.e. the ability to educate and reintegrate the young minors. We perceive the different levels to be coherent. The micro level is the situation proposed by the educator. The meso level is the group of activities with an objective leading to transformation for the young (a

Table 11.1: Opposite feelings SSE have to face.

Commitment – Pleasure	vs	compliance – constraint
Opposition – individualism	vs	cooperation – sharing
Urgency – immediate – pulse	vs	long term
Contingency – uncertainty	vs	adaptation planning – anticipation
take care of one person	vs	assume collective project

sport project). The macro level is the educative project for the young (individual and group). In the previous example, the incident observed at a micro
level (the altercation between two of the young people) is the beginning of
the rebuilding of the activity by the SSE, keeping in mind the educational
objectives while concentrating on some of the young people. The ability of
the SSE is reflected in the capacity to face and manage these incidents. Their
decision takes into account several factors as discussed below. They might
intervene and maintain an equilibrium while protecting themselves, with no
renouncement of values and bearing in mind the feasibility and compatibility of the educative target. In this work, we focus on the 'trigger' moments.
These are the moments when a change of direction, disorder, destabilization, but also creative adaption appear. 'we are always tempted to control
through order but disorder allows for creative activity' (Le Moigne, 1999).

11.6   Typical events

We have identified temporal and functional markers of the decision-making
process. These 'triggers' generate a need for adjustment, adaptation, decision making, and action in context. The triggers identified in this study
are the following: altercation – conflict; resistance to the action – inertia –
refusal to share – fold; Non-compliance – inappropriate action; risk taking
(for self and / or others); questioning of authority; form of 'anomie' (foiling
rules – social disorder). We also found different types and modes of interactions: verbal and non-verbal modes (gestures, eye contact), close to the
point of conflict or at a distance (inter-individual distance).

Human interactions in the sports situation comply with the nature of
the actions described in the life of a modeled complex system, shown in
Figure 11.9.

In fact, the actor manages the contradictions and paradoxes depending
on how he experiences the opposite feelings presented in Table 11.1.

In making a decision, the SSE are exposed to contradictions and antag-

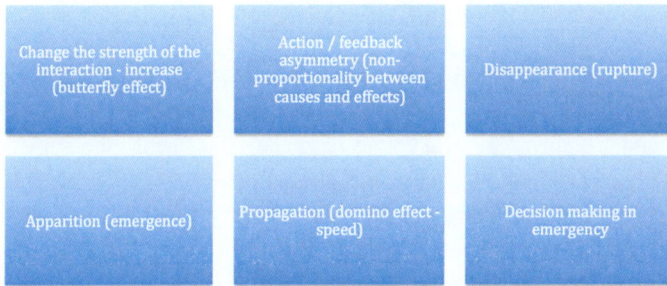

Figure 11.9: Modeled complex system.

onisms. This is another element to consider in training. Clergue (1997, p.131) wonders whether 'learning, which is essentially a nonlinear dynamic phenomenon, could not also find these state changes involving profound changes?' We believe that these bifurcations are elements favouring change and adaptation.

11.7   Action registers, registers of technicality

The action in these situations requires some technical interactions. They are technical (Combarnous, 1984) when three interacting elements are present: a technical rationality in process, the actor taking a role, a specialization, and the use of machinery (equipment, appliances). This technicality is implemented in different registers of action (Martinand, 1994). More precisely, four registers of technical actions can be used:

The register of control is represented by the professionals. The register of participation is the ability to play a guided role; for example a beginner, novice. The register of reading and interpretation is the ability to 'read' the practice, to identify indicators. The register of modification implies innovation, transformation, invention.

This should be kept in mind while building the training session. However, the results of the characterization of the registers (Glomeron, 2001) show that a training that insists on control of the action is not the most relevant: this register being insufficient. An action of the SSE is often composed of several registers. To this end, some extracts from interviews speak for themselves: The register of control: 'However, it is a sport that I have practiced for a long time at the club and I felt at ease, probably too much so.

Finally, I was not able to handle the situation.' That is a good example of inefficient control and planning.

The register of participation: 'Young people do not want to participate. However, afterwards with the speaker, I gained confidence. I saw that I could regularly intervene with young people, even to change their positions.'

The register of reading and interpretation. 'Even when I'm not near the action I constantly monitor and listen. As soon as I feel it, I can intervene very quickly. That is what is most important to me.'

The register of modification: 'We have changed roles and I made concessions regarding compliance to the rules. If I don't act as is, the whole group will disagree with me. On the contrary, they were ok with my decisions. As they have really made an effort, I think next time I will be able to come back to the real rules.'

In teaching physical practice, a continuous and strict control of the rules and techniques is not relevant, because of the uncertainty and difficulty of the settings. The tool used, i.e. the sport, even when perfectly mastered by the teacher, does not absorb the hazards generated by an audience of young offenders, within the socio-sporting interventions. The same study shows that technocratic rigour can block the overall project. A too rigid strategy can inhibit planned action and stop the operation. The efficiency of the SSE in maintaining the viability of the system depends on its ability to manage different registers of technicality.

11.8   Discussion

Some principles of action appear repeatedly in these cases and contrasting situations, and are summarized in Figure 11.10. This is the reason why a professional's actions are often 'do-it-yourself actions', and appear to come from 'controlled improvisation' (Perrenoud, 1983). This 'controlled improvisation' is a contextualization, a possibility of action in a personal adaptation to the situation. It is based on constructed, organized, but flexible, principles.

11.8.1   Modes of action and decision making

The management of uncertainty in professional situations requires the ability to think at several technical levels. The SSE must be able to adapt themselves to the environment and be flexible. It also seems that this kind of flexibility helps to keep some 'clarity' and ability to 'think' adaptation. This

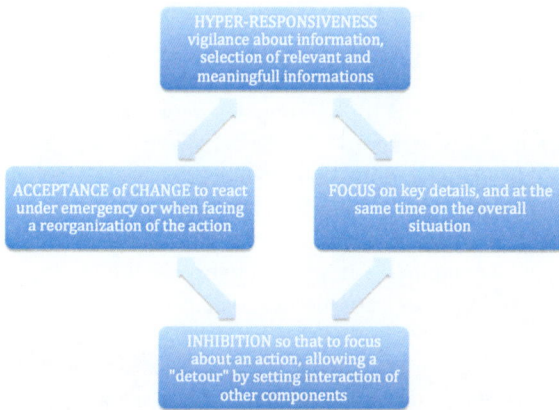

Figure 11.10: Common principles of action.

is the reason why the analysis of decision making is made at different levels of the system (micro, meso, and macro) and in various time frames.

11.8.2 How to train in the complexity of the interactions

We were able to identify the criteria of 'authenticity' and relevance of the situations that will expose students to the complexity of reality. Some examples are listed below:

- Confront the students with professional problems;

- Allow triggers, bifurcations, emergence;

- Enforce the treatment of contradictions and decision making;

- Need the commitment of the actor;

- Provoke reactions, emotions to be managed;

- Put the player in a dynamic adaptation;

- Avoid blocking or permanent withdrawal;

- Seek collective commitment (support, resources, complementarity);

- Solicit qualities of 'hyper-responsiveness' and high level of vigilance;

- Develop the acceptance of tests and error.

11.9   Conclusions

Through this study, we have verified our hypothesis: the complex approach provides valuable elements for training. Therefore we can say that the SSE can be confronted in workshops with the complexity of reality. It had major benefits for their training. Situation modeling, or interaction modeling in a multi-level approach, provide the keys for training. In training, the system of interactions in sports clearly shows the importance of work in an environment made of complexity. It is useful to expose future educators to these environments so they can learn the many factors influencing the sequences of actions.

Multilevel modeling reveals the need for joint actions that incorporate the concerns of other professionals. Once shared among professionals, the structuring axis gives coherence to all the actions of the various actors, and fits into decision making at this level.

11.9.1   To open up new horizons

Beyond this study, this is a new challenge for training the SSE: to accept and integrate uncertainty into the initial training. Thus, the professional specializing in difficult publics will be able to adapt. This uncertainty, which is desired and mobilized by the trainers, can contribute to building important skills to overcome difficult environments.

By analysing the activity of professionals, action logics can be characterized with some degree of 'authenticity'. These modeling actions can contribute to the situations of training and help to become aware of the complexity of the reality. To intensify this point of view, Rogalski and Leplat (2011, p. 24) show 'the importance of the variability of situations and disruptions in (the acquisition of) basic skills.' For the future, we hope to validate the findings of this study in two different manners: by testing them with more cases, and with different contexts. We may want to transpose to other work environments facing the same issues, that is to say, intervention in schools. We can also think about the existence of the typical profiles of the actors and/or the existence of the preferential registers of action which show a trend to a more important flexibility and adaptability in emergencies in a complex system.

Bibliography

Barrère, A., 2002, *Les enseignants au travail. Routines incertaines*, Paris: L'Harmattan.

Benaioun-Ramirez, N., 2013, *Traitement d'imprévus et incidents par un enseignant expérimenté, extrait d'une étude de cas*, Congrès AREF, Montpellier, France.

Bloch, O., and W. Von Wartburg, 1986, *Dictionnaire étymologique de la langue française*, Paris: Presses Universitaires de France.

Bonami, M., B. De Hennin, J.-M. Boque, and J.-J. Legrand, 1996, *Le management de la complexité*, Brussels, Belgium: De Boeck.

Clergue, G., 1997, *L'apprentissage de la complexité*, Paris: Hermès.

Combarnous, M., 1984, *Les techniques et la technicité*, Paris: Messidor / Editions Sociales.

De Montmollin, G., 1977, *L'influence sociale. Phénomènes, facteurs et théories*, Paris: Presses Universitaires de France.

Gayraud, L., G. Simon-Zarca, and G. Soldano, 2011, "Université : les défis de la professionnalisation", NEF, 46.

Gelinier, O., 1979, *Nouvelle direction de l'entreprise*, Editions Hommes et Techniques.

Glomeron, F., 2001, *Unité et cohérence de la formation des professeurs de technologie – Contribution à la définition des registres de technicité*, Thèse de doctorat de l'ENS Cachan.

Granon, S., and J.-P. Changeux, 2011, "Deciding between conflicting motivations: What mice make of their prefrontal cortex", *Behavioural Brain Research* **229**, 419–426.

Johannès, F., 2013, "Le lent naufrage de la Protection judiciaire de la jeunesse", *Le Monde* (Paris), 09.02.2013.

Lave, J., 1988, *Cognition, Practice*, Hillsdale, New Jersey, USA: Lawrence Erlbaum.

Le Moigne, J.-L., 1990) *La modélisation des systèmes complexes*, Paris: Dunod.

Martinand, J.-L., 1994, "La didactique des sciences et de la technologie et la formation des enseignants", *Aster* **19**, 61–75.

Morin, E., 1988, "Le défi de la complexité", *Chimères* **5–6**, 79–94.

Morin, E. 1990, *Science avec conscience*, Paris: Seuil.

Morin, E., 2005, *Introduction à la pensée complexe*, Paris: ESF Edition.

Perrenoud, P. 1983, "La pratique pédagogique entre l'improvisation réglée et le bricolage", *Éducation & Recherche* **2**, 198–212.

Perrenoud, P., 1994, *La formation des enseignants entre théorie et pratique*. Paris: L'Harmattan.

Perrenoud, P., 1999, *Enseigner : agir dans l'urgence, décider dans l'incertitude. Savoirs et compétences dans un métier complexe*. Paris: ESF.

Ria, L., 2009, "De l'analyse de l'activité des enseignants débutants en milieu difficile à la conception de dispositifs de formation", in M. Durand, L. Filliettaz (eds.), *Travail et Formation des Adultes*. Paris: Presses Universitaires de France, pp. 217–243

Rogalski, J., and C. Leplat, 2011, "L'expérience professionnelle : expériences sédimentées et expériences épisodiques", *Activités* **8**(2), 4–31.

Samurçay, R., and P. Pastré, 2004, *Recherches en didactique professionnelle*, Toulouse, France: Octarès.

Suchman, L., 1987, *Plans and Situated Actions: The Problem of Human–Machine Interaction*. Cambridge University Press.

Tardif, M., and C. Lessard, 2000, *Le travail enseignant au quotidien*. Brussels, Belgium: De Boek-Westmael.

Chapter 12

Modalizing a teacher's evaluative speech in order to study the student's position[1]

Yann Mercier-Brunel[2]

12.1 Introduction

In the vast amount of research focusing on the movement of speech at school, all agree on one point: the teacher speaks a lot, and the student mostly listens. The teacher's speech is estimated at from 60% to 80% in an oral classroom context (Dolz & Schneuwly, 1998; Bucheton, 2009; Mercier-Brunel & Jorro, 2009). The remaining time is then shared among the pupils, depending on their abilities and on their status in the class (Sirota, 1988). These linguistic dimensions, as well as anchoring identities and the values and plans that are affiliated to them, are too often left implicit and related to the teacher's skills and charisma (Bucheton, 2008).

Our aim is to analyse the language used by teachers when talking to their pupils, in order to highlight a certain number of linguistic elements (De Nuchèze, 2001) and to suggest a modelization of the interactions in a class,

---

[1]Translation by L. Mirzoyan, S. Sadrin, and H. Manas
[2]Laboratoire Ligérien de Linguistique - UMR CNRS 7270, Orléans, France. Email: yann.mercier-brunel@univ-orleans.fr.

knowing that these interactions have repercussions on learning, as regards a linguistic but also a cognitive aspect (Guerin, 2011). School-related issues seem to be considerable.

12.2   To think speech in class

12.2.1   From the institution's point of view

Since the early 20th century, curriculums have significantly evolved in terms of the role of oral communication in class, expressing the fact (or not) that school used to consider that it was part of its duty to teach pupils how to communicate orally. Studying the last hundred years alone, it appears that the teaching of French in the 1923 and 1938 curriculums aimed at "a beautiful language and a beautiful literature", as well as a strengthening of national unity. The 1971–1972 curriculums appear to be exceptions, for they define the oral according to oral exchanges deriving from "real experience situations, or game situations', highlighting a kind of speech authenticity. The 1985 curriculum backtracks on this progress and lapses into a concern for form to the detriment of substance. In 2002, 2007, and 2008, official instructions gradually placed the oral at the service of disciplinary fields, within the framework of the triptych speak/read/write, where the speech of the pupils had to be useful and obey precise school models. The common core of competences leads teachers to split off an oral taught in a syntaxic aim (competence 1) from an oral instructed in a discursive purpose (competences 5 and 6), "as if, from all eternity there had been language first, and then the ways of using it" (François, 1998).

12.2.2   About research

Research contributions to the oral in class have also significantly evolved. In the 1970s, researchers like Romian and the *Groupe Langue Orale* tried to develop the problematic of the oral in school. In the 1980s, several linguists worked to renew the approach to verbal exchange by no longer considering the oral as a degraded form of the written (Peytard, 1970), but as a specific object (Culioli, 1983), inducing another way to mean (Blanche-Benveniste & Jeanjean, 1987; Gadet, 1989). Analytical tools were created (Kerbrat-Orecchioni, 1986; 1990), and corpus analyses were made (Bange, 1987; Cosnier & Kerbrat-Orecchioni, 1987).

However, the educational sciences really started focusing on pupils' speech in the 1990s. For example, the analysis contained in Nonnon's research has continually evolved, from verbal interaction (1986; 1990) to discursive behaviour (2001). She closely studied the existing links between speech and cognition. Halté and the CRDF team from the University of Metz have also analysed classroom interactions based on the hypothesis of a language activity constituent of learning, and observed that the "oral was less taken into account" (Halté, 2006). The Geneva Schneuwly team built a paradigmatic classification of didactic situations relating to the oral, and suggested a series of pedagogic devices (Dolz & Schneuwly, 1998), such as interview and radio programs. These are based on the language development theories of Piaget and Vygotski, as well as on Bakhtin's secondary speech genres. There are also the publications of the INRP, with its two main research teams: one of them based on the corpus analyses collected in the classroom, and suggesting various conclusions (Grandaty & Turco, 2001), the other addressed to teachers in order to show them the importance of working on oral skills in the classroom, with the final aim of offering them sessions, as well as offering them some thoughts on discovery learning, assessment and remediation (Garcia-Debanc & Plane, 2004).

All of this research has enabled, among other things, showing that it is in the course of actual exchanges, i.e. with actual issues, other than those implied by answering a teacher's questions about the current lesson or by mechanically repeating a sentence, that the development of language know-how really happens, and that this has real consequences for the development of knowledge in general, for language mastery, and also for the way of comprehending the world.

12.3   The teacher's speech

12.3.1   Evaluative speech

We are interested in the analysis of teachers' speech regarding its relatively stable aspects, and because it influences the pupils' positioning.

The first point is to find the element of this speech which calls out to the listener's singularity regarding his referential, emotive, and conative functions at the same time. Indeed, we decided to focus on this subject since "the individual is recognizable by its elocutions which can be qualified as evaluative" (Ricoeur, 1990), and evaluative utterances are ubiquitous in classroom situations.

There is no doubt that this evaluative speech has a strong influence, whether because academic functioning implies a dialectical relationship between the assessment by the teachers and the learner's self-regulating process (Allal, 2007), or because academic judgment has undeniable psycho-logical consequences for the pupils (Barlow, 1992; Bressoux & Pansu, 2005).

Actually, evaluative utterances seem to be heterogeneous in the class-room's daily context, with rather immediate consequences but varied forms, from the closure or verdict, to an indicative, evocative, or inchoate appear-ance (Jorro, 2003). The question of the teacher's efficiency seems to hinge on this kind of feedback about the pupils' actions, depending on their con-tingency, targeting, and a corrective dimension in the case of a student's mistake (Carhay, 2007).

12.3.2   Definition of the selected linguistic indicators

Based on what was stated above, we took an interest in the nature of the contingency of evaluative feedbacks and the way of approaching this. The first studied indicators within the teacher's speech are the types of state-ments corresponding, according to Benveniste (1966), to "three fundamen-tal behaviours of speaking and acting through speech": the assertive, the interrogative and the imperative (injunctive).

These types then raise the question of certainty (Ducrot & Schaeffer, 1995), introducing three modalities: the possibility/capacity or apodictic truth, an alethic modality; the epistemic modality, i.e. what the teacher faces during the exchange or what he or she is sure of because it falls within his knowledge in the field; and deontic modality, i.e. what he or she allows and what he or she imposes.

But other than these aspects, which could be defined as illocutionary, there are indicators referring to perlocutionary values present in the utter-ances, that is to say, the perlocutionary effect intended and/or attained by the teacher's speech, in terms of demands, respectively, persuasion, induc-tion, or guidance, according to which effect is observed in the behaviour, respectively, beliefs, feelings (Vermersch, 2007), or cognitive processes of the pupils.

From that point, we then established a first series of elements that aims at a pre-categorization of the teachers' evaluative utterances, studied here and presented in Table 12.1.

12.3.3   Complementary enunciative indicators

An interesting aspect of evaluative utterances lies in the strong enuncia-tive dimension that cannot be perceived by the previous pre-categorization, which is more directed towards other functions of language. We thus sug-

Table 12.1: Categorization of the teacher's evaluative utterances.

Indicators concerning the teacher's utterances	Types	Assertion
		Question
		Injunction
	Modalities	Alethic (*apodictic vs hypothetical*)
		Epistemic (*knowledge vs supposition*)
		Deontic (*right vs duty*)
	Perlocutionary values	Demand (*order, request*)
		Persuasion (*argumentation, demonstration*)
		Induction (*manipulation, rhetoric*)
		Guidance (*support, inquiry*)

gest returning to look at some of Kerbrat-Orecchioni's works, in which she tries to explore the traces of the utterer's subjectivity in his or her words, using what she calls *subjectivèmes* (subjective markers) (Kerbrat-Orecchioni, 2006). To do so, she relies on the *affective* and *evaluative* dimensions – in qualitative and quantitative terms and she notices, among other things, three types of indicators.

Deictic markers make it possible to place the action on a self-centered spatial-temporal level. We will study *personal pronouns*, subjects, and complements that make it possible to place the process in terms of its author and its object. Here we will focus on a pragmatic dimension, especially in the context of exchange situations in the classroom, that is to say the type of address and the way the teacher creates a link with the group. The differences are noticeable between a general and a targeted *you*: e.g. is "You didn't learn that lesson" addressed to the whole class or just some of the pupils? Similarly, is the use of *we* in "Do we agree?" purely rhetorical or does it really speak to the pupils? Amongst the deictic markers, there are also *spatial demonstratives*, which are often responsible for the difficulties pupils have decoding: in "This isn't correct" or "You made a mistake here", the reference is often about a sum of elements that pupils have difficulties distinguishing. *Temporal demonstratives* also raise a problem: "What did I tell you last week?" does not help the pupils who know precisely what they will learn and who establish links with the previous lessons, which is rarely the case with the pupils in difficulty to whom these questions are often addressed.

Affective markers are another kind of *subjectivèmes* that mark an emotional commitment from the speaker to what he or she is referring to. It should be noted that if certain types of speech claim they proscribe these kinds of terms, notably because they aspire to a form of neutrality, they remain in fact very present in evaluative synthesis – "Disappointing results" – and in dialogues – "I really like your text", "Your test turned out terrible". The matter is less about the evaluated object strictly speaking than it is about what it provoked in the assessor, while a competent assessor goes unnoticed (Jorro, 2000).

Non-axiological evaluatives have a qualifying aspect in respect to a certain scale. They rely on a double norm inside the object and specific to the speaker. In other words, the use of an evaluative adjective is related to the idea that the speaker has of the evaluative norm. For instance, "There are too many mistakes" typically refers to the acceptable amount of mistakes that can be made while doing a given task, according to the assessor.

Finally, *axiological markers* bring a value judgment to the evaluative utterance and pupils rarely put enough space between performance and capacity. It is thus common that pupils go from "Your maths test was bad" to "I am bad at maths." One can wonder if an "It is very poor", thrown out while giving back a test, is in fact addressed to the performance or to the pupil? (Barlow, 1992). Field observation shows the importance of the axiologicals in the teachers' evaluative utterances, often supported by non-verbal actions: a hand on the shoulder versus arms crossed, an open and smiling face versus eyes lifted towards the ceiling. "Congratulations" and "well done" are not as clear as they seem either: if they refer to a *good* action, the question is then *good for whom*? For a teacher that has set himself or herself objectives, "good" might refer to everything that helps him or her reach them (Mercier-Brunel & Jorro, 2009). Now, not all these objectives are oriented towards the quality of the learning.

12.4   Application to a corpus

12.4.1   Research site

The data was collected in 2008 in four CE2 classes (third grade) from the same school in order to minimize the *institution effect* (consequences of the sociological context). The pupils were between 7 and 10 years old – most of them were 8. It was decided to work at this level because the Official Instructions are more precise regarding oral practices within various frameworks, and they impose a precise definition of the school subjects (for instance, History, Geography and Science belong to a field called "world discovery").

The school is located in a rather privileged neighbourhood in the centre of Nice. Half of the pupils have a rather wealthy background (children

Table 12.2: Teachers' profiles.

	Christine	Geneviève	Léa	Mathieu	Viviane
Age	45	52	29	35	48
Experience as a teacher	17 years	10 years	5 years	12 years	17 years
Experience teaching CE2 (third grade)	13 years	4 years	4 years	4 years	11 years
Number of pupils	26	27	26	27	26

of middle management and liberal professions). The other half come from more modest or even precarious environments: social housing and emergency shelters can be found in the same area where the school is located. Two-thirds of the more modest families belong to the working class and are of foreign origin.

These four classes have 26 to 28 pupils, which is a lot for a primary school. On average, the school has classes of 25.5 pupils. As the school is not in a deprived area, this is a standard rate for a city as important as Nice.

12.4.2 Data collection process

The five filmed sessions all followed identical protocols: a 20-minute correcting session after lunch. The corrected exercises had to correspond to the classes' daily work. However, the duration of each session differed, for practical reasons, breaks, the teachers' wishes, and for consistency, longer or shorter activities, faster-working or slower-working pupils. The five sessions below were recorded between the end of October and the end of November of 2008.

The profiles of the teachers are quite different. They are summarized in Table 12.2.

Christine was filmed during a 34-minute group correcting session. For a grammar exercise, the pupils had to find individually "what the pronoun referred to".

Mathieu was filmed for 44 minutes, during which time he and his class corrected several exercises related to types of sentences and the conjugation in the present of the verbs "have", 'be", and "go".

Geneviève corrected exercises on the values of numbers and the various figures composing numbers following the decimal system. Her session lasted 31 minutes.

Viviane corrected mathematics for 36 minutes: in the various exercises, pupils had to find a specific number in a list following given instructions, such as parity or the figures composing it.

Léa corrected three geometry exercises for 22 minutes. The pupils had to find right angles in various geometric shapes, using a set square.

12.5 Categorization of the teachers' discourses

Although none of the teachers can be defined as using a single type, modal construction, or *subjectivème*, we noticed some unexpected combinations, sometimes with alternations between two combinations.

During the first part of her session, **Christine** is characterized by her use of several injunctions, a few highly inductive questions, and more and more axiological markers. This seems to put a correcting routine into place, each questioned pupil being firmly guided to give answers in a specific order. This discourse has been identified as *routine creating*.

Teacher (T): (to Anouar) So, have we to read the *one*?
Anouar: No
T: No we haven't / it will be for doing exercice you're right / good / we?

A little while after that, Christine starts to speak less but gets into longer dialogues with the pupils she is questioning (reaching 4 to 6 Q&A with a single pupil). Her questions become more about guiding than inducing an answer (i.e. the pupils can no longer guess the answer because of the question, but they know where to look), the affective subjective markers are more numerous and the personal pronouns are in the singular.

T: She / the teacher / it's feminine / and so? / it was good what you've just tell before
Angelo: She is alone
T: So you've chosen? /
Angelo: The / the / singular
T: Right / very good / thanks

She also seems to focus on a selection of pupils, guiding them both to find the answer and to secure them emotionally using a *supervising* discourse.

Geneviève alternates two types of combinations during the entire session. The first one is based on injunctions and a few highly inductive assertions, a lot of axiological markers and deontic modal constructions. Her discourse is about "imposing a routine". Unlike Christine, this routine is previously created instead of being co-built during the sessions.

T: So / how many one hundred euro bills would you give / for nine hundred and twenty-nine?
Emie: Nine
T: Nine / how many ten euro bills / Steven / for nine hundred and twenty-nine?
Steven : Two
T: Two / and how many one euro coins will you give?

The second combination is essentially made of highly inductive questions (the pupils guess the answer half the time), numerous deictic markers indicating the key elements to solve the problem, and alternations between personal pronouns in the singular (the student being questioned) and in the plural (the class being questioned).

T: So / who doesn't agree? / (to interrogated pupil) how do you read that? / the number you wrote? / where do you have to place the decade?
Margot: Six hundred and one
T: (to Solène) Wait / she realizes she was wrong

Her discourse is about "taking charge of the procedure".
Léa, the youngest teacher, seems to be using a single combination made of injunctions (personal pronouns in the singular and deontic modal constructions) and interrogations (personal pronouns in the plural and epistemic modal constructions). Her speech is only slightly inductive and contains just a few affective and axiological markers.

T: Elyes / how do you put your template?
Elyes: Well like this
T: You put the corner / on the angle's corner / you put a side / on the figure's side

It seems like she wants to firmly regulate the pupil's position individually and their procedures collectively using a *normalizing* discourse.
Mathieu is the one who uses the highest number of assertions, with many alethic and epistemic modal construction aimed to guide. His feedbacks are not very explicit, which means there are few subjective assessment

markers and many inductions.

T: Yes rather it's an order / it could be an advice / depending on the context
[...]
T: Yes it's right / it's right they resound / how could I explain / there are
verbs / which are / euh / which you didn't use / with every person / so / it's
tricky

He seems to present himself as owning the knowledge, using an *apodictic* discourse. Then, as he is pressed for time by the end of the session, he
uses more evaluative axiological markers, inductive injunctions, and a lot
less guidance, therefore opting for a *session leading* discourse.

Viviane started her session with many highly inductive questions, a few
injunctions, and very short assessment markers. She focuses on the pupils'
theoretical knowledge, about which she asks for explanations through an
epistemic discourse. Later on, her speech becomes shorter, the pronouns in
the singular are replaced by plurals, and she focuses more on the rest of the
class than on the pupil whom she is correcting to validate or invalidate the
answers. Her questions are very slightly inductive and can sometimes guide
but are mostly neutral.

T: I'm odd / so it's 63 / do you agree with him / others? [...]
T: Kevin doesn't / who is wrong? / could you explain to me where it come
from? one of the two who are wrong?

It becomes difficult for the pupils to know what she wants and if their
answers are right, and they have to build their approach with the class rather
than with the teacher. Her discourse is based on *postponed assessment*.

12.6   Synthesis and perspectives

This categorization of the teachers' evaluative speeches, which we chose
not to develop any further, enables us to rebuild a certain pedagogical intentionality and find a way to involve the pupils through the words of the
teachers. Consequently, we can see positions developing, with a withdrawal
in supervision encouraging the pupils' investment with Viviane, securing
and targeting for Christine, turned towards leadership with Mathieu, and
towards correcting with Léa and Geneviève.

These various discourses allow the pupils to be part of the exchange in different ways. For instance, we have noticed that pupils tend to simply answer the questions with Mathieu or Léa, while they try to verbalize their intellectual approach with Geneviève or Viviane. In the first case, the method is pre-established, but in the second, the method is created on the spot. Furthermore, we have noticed that the secure feeling felt with Christine allows inferences that are not always relevant to the exercise but interesting for the understanding of the thought process.

The data presented here is part of a wider set of results and shows why a linguistic categorization of the teachers' evaluative discourse is necessary to reveal its effects on the implication of pupils in the class exchanges, which we know to be at the basis of school education. In relation to this subject, we only refer to the socio-constructivist works in didactics done on mathematics and French. This data is therefore an interesting tool for research as well as training: teachers can understand the impact of their evaluative comments on their pupils' education.

Bibliography

Allal, L., 2007, "Régulation des apprentissages: orientations conceptuelles pour la recherche et la pratique en éducation". In L. Allal (ed.) *Régulation des apprentissages en situation scolaire et en formation.* Brussels, Belgium: De Boeck Université, pp. 7–23.

Bange, P., 1987, *L'analyse des interactions verbales. La dame de Caluire : une consultation.* Bern: Peter Lang.

Barlow, M., 1992, *L'évaluation scolaire, décoder son langage.* Lyon: Chronique Sociale.

Benveniste, E., 1966, *Problème de linguistique générale.* Paris: Gallimard

Blanche-Benveniste, C., and C. Jeanjean, 1987, *Le français parlé: transcription et édition.* Paris: Didier.

Bressoux, P., and P. Pansu, 2005, *Quand les enseignants jugent leurs élèves.* Paris: Presses Universitaires de France.

Bucheton, D., 2008, "Professionnaliser ? Vers une ergonomie du travail des enseignants dans la classe de français". In D. Bucheton and O. Dezutter (Eds.), *Développement des gestes professionnels dans l'enseignement du français.* Brussels, Belgium: De Boeck Université, pp. 15–27.

Bucheton, D., 2009, "Le modèle de « l'agir enseignant et ses ajustement »". In D. Bucheton (Ed.), *L'agir enseignant: des gestes professionnels ajustés .* Toulouse, France: Octarès, pp. 25–68.

Cosnier, J., and C. Kerbrat-Orecchioni 1987, *Décrire la conversation.* Lyon: PUL.

Crahay, M. 2007, "Feedback de l'enseignant et apprentissage des élèves: revue critique de la littérature de recherche". In L. Allal (Ed.), *Régulation des apprentissages en situation scolaire et en formation.* Brussels, Belgium: De Boeck Université, pp. 45–70.

Culioli, A., 1983, "Pourquoi le français parlé est-il si peu étudié ?" *Recherches sur le français parlé* 5, 291–300.

De Nuchèze, V., 2001, *Sémiologie des dialogues didactiques.* Paris: L'Harmattan.

Dolz, J., and B. Schneuwly 1998, *Pour un enseignement de l'oral*. Paris: ESF.

Ducrot, O., and J.-M. Schaeffer, 1995, *Nouveau dictionnaire ency-clopédique des sciences du langage*. Paris: Editions du Seuil.

François, F., 1998, *Le discours et ses entours*. Paris: L'Harmattan.

Gadet, F., 1989, *Le français ordinaire*. Paris: Armand Colin.

Garcia-Debanc, C., and S. Plane, 2004, *Comment enseigner l'oral à l'école primaire ?* Paris: Hatier.

Grandaty, M., and G. Turco, 2001, *L'Oral dans la classe. Discours, métadis-cours, interactions verbales et construction de savoirs à l'école primaire*. INRP.

Guérin, E., 2011, "Sociolinguistique et didactique du français: une interac-tion nécessaire", *Le français aujourd'hui* **3**, 139–144.

Halté, J.-F. 2006, "Entre enseignement et acquisition: problèmes didac-tiques en apprentissage du langage". *Mélange CRAPEL* **29**, 13–28.

Jorro, A., 2000, *L'enseignant et l'évaluation. Des gestes évaluatifs en ques-tion*. Brussels, Belgium: De Boeck Université.

Jorro, A., 2003, "L'évaluateur est un autre !" In J.-P. Astolfi (Ed.), *Education et formation: nouvelles questions nouveaux métiers*. Paris: ESF.

Kerbrat-Orecchioni, C., 1986, *L'implicite*. Paris: Armand Colin.

Kerbrat-Orecchioni, C., 1990, *Les interactions verbales*. Paris: Armand Colin.

Kerbrat-Orecchioni, C., 2006, *L'Énonciation*. Paris: Armand Colin.

Mercier-Brunel, Y., and A. Jorro, 2009, "La parole évaluative de l'enseignant". *Les dossiers des Sciences de l'Education*, **22**, 9–24.

Ministère de l'Education Nationale, 2002, *Qu'apprend-on à l'école pri-maire ?* Paris: CNDP.

Ministère de l'Education Nationale, 2005, "Loi No 2005-380 du 23 avril 2005 d'orientation et de programme pour l'avenir de l'école". *Bulletin Officiel No 18*, 5 May 2005.

Ministère de l'Education Nationale, 2008), *Bulletin Officiel No 3, Hors-série du 19 juin.*

Nonnon, E., 1986, "Interactions verbales et développement cognitif chez l'enfant: aperçu des recherches psycholinguistiques récentes en langue française". *Revue Française de Pédagogie* **74**, 53–86.

Nonnon, E., 1990, "Est-ce qu'on apprend en discutant ? Un exemple d'interaction maître-élèves en Section d'Education Spécialisée". In F. François (Ed.), *La communication inégale* Neuchâtel, Switzerland: Delachaux et Niestlé, pp. 147–212.

Nonnon, E., 2001, "La construction d'objets communs d'attention et de champs notionnels à travers l'activité partagée de description". In P. Grandaty and G. Turco (Eds.), *L'Oral dans la classe, Discours, métadiscours, interactions verbales et construction de savoirs à l'école primaire* Paris: INRP, pp. 65–102.

Peytard, J., 1970, "Oral et scriptural. Deux ordres de situations et de descriptions linguistiques". *Langue française* **6**, 35–48.

Ricœur, P., 1990, *Soi-même comme un autre.* Paris: Editions du Seuil.

Sirota, R., 1988, *L'école primaire au quotidien.* Paris: Presses Universitaires de France.

Vermersch, P., 2007, "Approche des effets perlocutoires : 1/ Différentes causalités perlocutoires : demander, convaincre, induire", *Expliciter* **71**, 1–23.

Chapter 13

The self-organized situation, a challenge for physical education teacher training: A complex system of ritualized interactions and proxemic distances

Nathalie Carminatti-Baeza[1]

13.1   Introduction

In his letter of October 5, 2010 (see Appendix), the Minister in charge of Higher Education and Research wished that the initial training teacher education would be further developed, more particularly by insisting on how this reform could be implemented in higher education institutions. The importance of teacher training for any nation does not need to be reiterated.

---

[1]IRES. UFR STAPS. Université d'Orléans. France 2, Allée du Château, BP 6237, 45062 Orléans Cedex 2. Email: nathalie.carminatti@univ-orleans.fr

Similarly, it would be an illusion to believe that this training could follow an immutable pattern with which we alone could define the contents and modalities. Society is evolving and changing, so it is essential that teacher training integrates these changes (Lessard, 2000). Moreover, international experience clearly shows that all nations are also involved in the recurring process of reforming the initial training, without an ideal solution having achieved a consensus among all the stakeholders (Jolion, 2011). Therefore, reforms are not to be questioned.

Moreover, it seems a pity that although the competitive exam (to become a teacher) is a necessity, no one has imagined a system which would allow measuring the number of students, the relative flux of the various sectors, monitoring the student profiles whether in degrees preparing for the exam or without preparing.

This feeling of disconnection between the competitive exam and the teaching is unanimously felt and criticized by students when they prepare themselves to become primary school teachers or secondary school teachers (Jolion, 2011). This point will be widely commented upon as it is the major pitfall of this reform. Contest requirements sometimes seem distant from what is required as a teacher. Teaching is interacting with students. Teaching is transmitting knowledge and making sure that students learn something. The competitive examination does not prepare for this interaction.

The aim of the reform is to improve the training of future teachers and enable them to acquire a greater vocational qualification.

Teaching is not an art but a profession (Mérieu, 1989) that can be learned and its learning can and must be an accompanied one. Who can doubt that today's pupils, living in a culture of immediacy, who are both at the centre of the world and more isolated than ever, require a renewed pedagogy that cannot be reduced only to individual charisma?

The 'masterisation' involves changes in the implementation of teaching and in the reflective skills to be acquired. A new analytical framework is born around interactions as the basis for the training of future teachers. It is all about innovation and helping future physical education teachers use problem-solving skills to figure out learning situations on their own. The question of complexity (Clergue, 1997; Morin, 2005) is an essential issue, and we are confronted with a real problem of didactics. The framework of complexity highlights the fact that it is not only a sufficient conceptual idea, but it calls for major developments towards the analysis of the activity in a training situation: what we call here the construction of the 'Teachable'. 'Building the Teachable' raises the question of proposing training situations to students (spatial and temporal organization, interaction with pupils, choice of teaching contents, providing feed-back to students), allowing them to encounter a part of the reality of their future occupation (Blanchart-Laville, 1998). Figure 13.1 represents this analytical framework.

Figure 13.1: Analysis model of physical education teacher training.

The complexity lies in the interaction between the implementation of the training situations, the didactic obstacles that arise, the institutional framework, and the construction of professional skills.

In their face-to-face relations in the public arena, human beings are engaged in scanning or reading each other and, in turn, presenting themselves through externalization so that they are read in appropriate ways by others who are scanning them. Frame analysis reveals the complexity of social activities and brings out the arbitrary nature of any fixed dichotomy based on social domains or activities, between what is 'staged' and what is 'real'. The teaching profession manifests the presence of a complex system. This system is built from interactions through a proxemic space that is ritualized and temporal (Baeza, 2012).

'Symbolic interactionism' is a micro-level perspective which focuses on how daily interactions create and maintain social order. Social interaction generates an order that changes adaptively to internal and external constraints, resources, and power. Complex adaptive transformation focuses on the self-organizational behaviour of complex, dynamic systems, and the interaction of such systems with each other. "Complex systems have a self-organizational process that emerges out of the nature of the properties of their component parts" (Siegel, 2003).

The aim of the present article is to demonstrate that the establishment of self-organized situations generates a real dynamic of change among students, the future Physical Education teachers. These self-organized situation allow building the "teachable".

13.2   Framework

The disciplinary orientation used is that of 'situated action' (Suchman, 1987), which is based on the idea that any action is highly dependent on the material and social circumstances in which it takes place. The actions are still socially and physically situated and the situation is essential to the interpretation of the action. Suchman (1987) insists on the changing complexity of situations and on the 'opportunistic' reactivity of the actors to environmental contingencies.

This approach will be complemented by the theoretical framework of symbolic interactionism. Interactionism, and especially the paradigm developed by Goffman (1974), focuses on the rituals of interaction and face to face interaction. From the different types of rituals proposed by Goffman, it is possible to transpose into a constraining school context the following:

- **"Presentation Ritual"**: Our behaviour enables our pupils to perceive how we consider them and how we will deal with them in the next classes. The distance between the teacher and his pupils reveals the presence of this ritual. In order to belong to the space of the group, the teacher must come in.

- **"Confirmation Ritual"**: Conveying an image of ourselves to others as well as emphasizing the attention we pay to others. The teacher has some obligations (pedagogical and institutional). For that purpose, he uses pedagogical and didactic strategies and tools.

- **"Reparation Ritual"**: This takes place when some incidents may damage the pedagogical relation. It is all the more important when the teacher wants to build a dynamic by meeting the expectations of the pupils and making the group work. When people are together, unexpected events may occur that can break the initial balance within that group. Under those circumstances, the teacher will have to restore the balance in order to maintain his authority.

Goffman (1974) has assumed that this maintenance activity constantly occurs in ordinary interactions since it is the foundation of encounters. Thus the word 'ritual' is used because this activity, although very simple, represents the effort we make in order to present a symbolic image to others, viz., the pupils.

Communication is an activity regulated by rules and set standards for collective interaction. Finally, we will complete the theoretical framework by a proxemic analysis. Proxemics (Hall, 1966), or the study of the perception and use of space by humans, is defined as the physical distance established between the people caught in an interaction, an exchange of communication. Hall discusses the structure of conditioning experience by studying the culture and the cultural dimension in a vast communication network with multiple levels. Four interpersonal distances are highlighted and presented in Figure 13.2: the intimate distance (7–45 cm), personal distance (45–125 cm), social distance (1.25–3.60 m) and public distance (3.60–7.50 m).

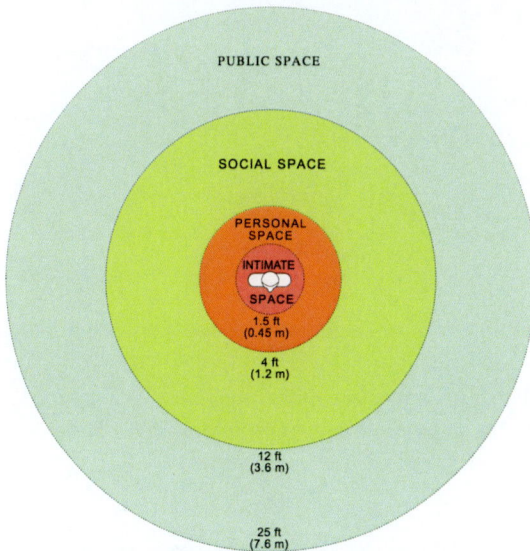

Figure 13.2: Proxemic distance (Hall, 1966).

13.3   Hypothesis

The construct of the "teachable" implies a new perspective on the part of the trainers. The interaction between the trainer, the student, knowledge, and context, is a process to study. We need to emphasize what escapes the constraints of the interaction (environmental constraints, communicational constraints) rather than the constraints themselves (the interaction order) (Goffman, 1974). We make the hypothesis that the observation of the proxemic phenomena (Hall, 1966) in interaction rituals is likely to enhance a didactic approach in the training of future teachers. The proxemic phenomena result in the measurement of the distances between the actors. These distances reflect the nature of the rituals. They are shown in Figure 13.3.

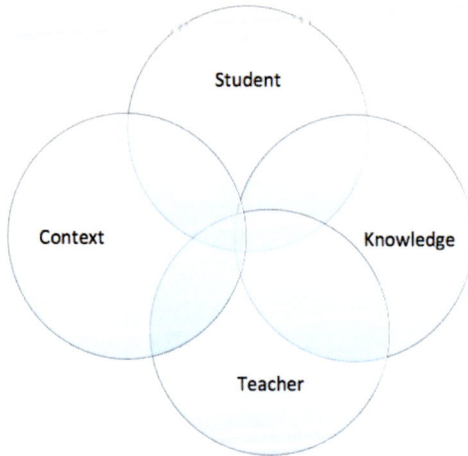

Figure 13.3: New model of physical education training teacher.

This model shows a new form of complexity. The students, knowledge, context of the training, and trainers are interwoven. These four categories operate together and are in direct interaction.

13.4   The situation of intervention

The trainee teacher is subject to fixed constraints in a self-organized situation of training. The environment will be used to situate the professional practice (see Figure 13.4).

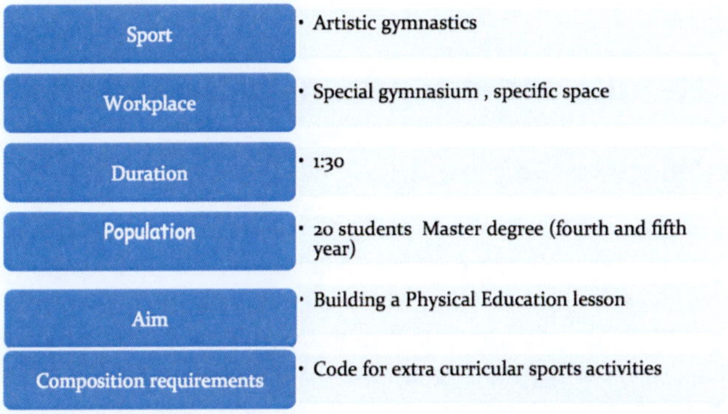

Sport	• Artistic gymnastics
Workplace	• Special gymnasium , specific space
Duration	• 1:30
Population	• 20 students Master degree (fourth and fifth year)
Aim	• Building a Physical Education lesson
Composition requirements	• Code for extra curricular sports activities

Figure 13.4: Self-organized situation of teaching.

The aim is to present a lesson from a theme imposed by the trainer. It is assumed that these requirements operate as didactic variables (Robin, 2012). The context plays the role of a cognitive artifact: envisaging the action as a self-organized coupling involves considering the context, and its properties, such as offering the 'resources for action' available to the actor according to his intentions (Norman, 1993).

13.5   Population

Seven fifth-year students were supervising a gymnastics lesson. This lesson was completely filmed and provides support for analysing the activity of the trainee teacher.

13.6   Methodology

Three registers of analysis are used for this exploratory study: (1) planning lessons for understanding how the trainee teacher structures self-organized situations, (2) the video to read the activity of the trainee teacher in situ. The collection of field data will be analysed from a proxemic analysis grid, (3) the collective post activity discussions to understand the analysis made by the trainee teacher on the action in the situation.

Here we only consider the 'space', which will be analysed from the point of view of proxemics, studying the signifying structure of human space. We observe the proxemic distances established between the actors during the interactions in gymnastics lessons in physical education.

13.7 Results and discussion

13.7.1 Space in the planning

Planning allows seeing the workspaces as they are located and organized according to professional prerogatives (spatial organization, temporal organization, methods of grouping students, implementation of several learning situations). The space is controlled by the student. The use of media, the materialization in workspaces before the beginning of learning, is also reflected. Each student group is confronted with a specific learning situation with media to consolidate the instructions given by the teacher-student. The space is arranged to promote specific motor skills.

13.7.2 Space in the lesson

The results are presented in Figure 13.5. Regarding the ritual of presentation, the distance is 2.50 meter for 100% of the population. This social distance permits delivering general instructions for the lesson.

Relating to the ritual of confirmation, we can observe that 29% use private distance, 57% personal distance, and 14% social distance. These distances differ, depending on the interaction between the pupils and the teachers.

Concerning the ritual of reparation, two types of distance appear: 86% use personal distance and 14% social distance. When there is a contract breaking-off, the teacher prefers to keep their distance in interactions with the students.

13.7.3 Space in collective debriefing

The feedback highlights three key points:

- The workshops are located in the workspace;

- Students are divided between the workshops;

- The workshops are organized according to the activity and needs of the pupils.

The results show three major points:

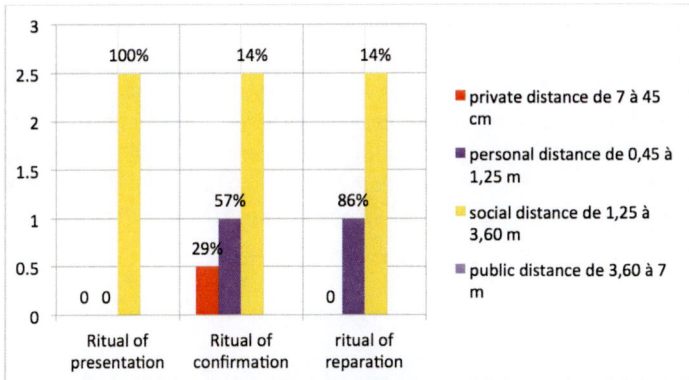

Figure 13.5: Space in the gymnastic lesson (video results).

1. Trainee teachers have built a planned design of the "teachable", taking into account the institutional, pedagogical, and didactic requirements.

2. The teachable is an unfinished construction because the institutional, pedagogical and didactic requirements are realized in interaction with the students. The implementation of the requirements raises a problem for the trainee teacher, who must learn to manage the complexity of the act of teaching and to think of the properties of the transmission process. It is the implementation of the "teachable" that remains to be built.

3. Trainee teachers build strategies to control some variables of the 'education' system. What seems to raise a problem is the instability that could result from the interaction between the variables (institutional, teaching, and learning). The "teachable" is accompanied by a change in posture and acceptance of some kind of disorder.

13.8   Conclusion and perspectives

Changing the posture of the student in a self-organized situation generates a dynamic engagement. The interaction between the trainee teacher, the pupils, knowledge, and the trainer, generates the construction of a balance, a 'middle ground' between adaptation to the constraint of the task and the stability built in the self-organized situation.

The workshop of professional practice is a tool to build the 'teachable' in a training framework that includes complexity. By adopting a pedagogy of simulation, the workshop turns into a self-organized situation where we will put forward the necessary cooperation of the actors; the goal is to build and bring out a changing reality through interconnected variables (institutional, pedagogical, and didactic) that create or oppose change. This cooperative game is indicative of the process of interaction (Figure 13.6). This new reference is a true interaction model in a complex system.

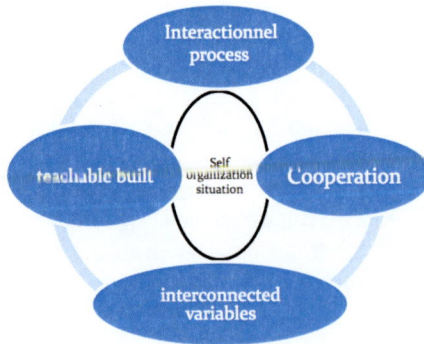

Figure 13.6: Professional situation, a real organized-situation.

Bibliography

Baeza N., 2012, *Interaction ritual, proxemic: Two paradigms to teach formation*. Seminar AMISC.

Blanchard-Laville, C., and D. Fablet 1998, *Analyser les pratiques professionnelles*, Paris: L'Harmattan.

Clergue, G., 1997, *L'apprentissage de la complexité*, Paris: Hermès.

Goffman, E., 1974, *Interaction Ritual*, Paris: Minuit.

Hall, T. E., 1966, *The Hidden Dimension*, New York: Doubleday.

Lessard, C., 2000, 'Evolution du métier d'enseignant et nouvelle régulation de l'Education', *Recherche et Formation*, **35**.

Mérieu, P., 1989, *Enseigner, scenario pour un autre métier*, Paris: ESF Editeur.

Morin, E., 2005, *Introduction à la pensée complexe*, Paris: Points.

Norman, D.A., 1993, *Things That Make Us Smart*, New York: Addison-Wesley.

Rapport, Jolion, 2011, *Masterisation de la formation initiale des enseignants*.

Robin, J. F., 2012, La gymnastique : un jeu de règles, *e-JRIEPS*, **25**, 27–42.

Siegel, A., 2003, 'Représentation des Systèmes Dynamiques substitutifs non unimodulaires', *Ergodic Theory and Dynamical Systems*, **23**, 1247–1273.

Suchman, L.A., 1987, *Plans and Situated Actions*, Cambridge University Press.

Appendix A Letter from the minister in charge of higher education and research

Monsieur le Président,

La réforme des conditions de recrutement et de la formation des maîtres doit permettre d'améliorer la qualification des personnels, en vue de faciliter la mobilité au sein de l'Union européenne et de renforcer la réussite des élèves.

L'enseignement supérieur s'est largement engagé, dès cette année universitaire, dans cette réforme de la formation initiale des enseignants (professeurs des premier et second degrés, professeur documentaliste et conseiller principal d'éducation) dans le cadre de cursus universitaires débouchant sur le diplôme national de master.

En votre qualité de Président du Comité de suivi du cursus master, je souhaite que vous puissiez approfondir la question de la mastérisation de la formation initiale des enseignants que j'évoquais dans ma lettre de mission du 30 mai 2008, en étudiant les modalités de mise en œuvre, par les établissements d'enseignement supérieur, de la circulaire n° 2009-1037 du 23 décembre 2009 relative à la mise en place des diplômes nationaux de master ouverts aux étudiants se destinant aux métiers de l'enseignement à compter de la rentrée universitaire 2010.

Plus particulièrement, je vous invite à étudier les réponses apportées par les établissements, dans leurs masters, aux grands enjeux de la formation initiale des enseignants en termes de contenu des formations (spécialisation et professionnalisation progressive, initiation à la recherche, apport en pédagogie, connaissance concrète du système éducatif, analyse de situations professionnelles, stages, préparation aux concours, ouverture sur d'autres métiers notamment en cas d'échec aux concours), cohérence académique et nationale de l'offre.

En effet, la lisibilité et la qualité de l'offre de formation sont essentielles afin de permettre à l'étudiant, aux différentes étapes de son cursus, de faire des choix d'orientation positifs et de faciliter dans tous les cas son insertion professionnelle, qu'elle intervienne dans les métiers de l'enseignement ou dans d'autres secteurs professionnels. De même, un maillage harmonieux de nos territoires est indispensable pour garantir le renouvellement du corps enseignant et l'égalité de traitement des candidats.

Monsieur Jean-Michel JOLION
Président du Comité de suivi du cursus master
Université de Lyon
Quartier Sergent Blandan
37 rue du repos
69361 LYON cedex 07

Chapter 14

Hazard and discomfort in complex situations: Resources for the training of social sports educators

Karine Paret[1], Déborah Nourrit[2], Nathalie Gal-Petitfaux[3]

14.1   Introduction

This contribution is related to a university training course for a job we named 'social sport educators' (SSE). These professionals work in a complex environment, primarily because the public which they take care of has accumulated many problems: social, psychological and medical needs, with intricate dimensions. To overcome these needs, socio-sports educators support these people through sports and arts practices. The pedagogy has to be very well adapted to this unstable public. Many factors come into play

[1]MAPMO - UMR 7349, Université d'Orléans. Email: karine.paret@gmail.com.

[2] MAPMO - UMR 7349, Université d'Orléans. Email: deborah.nourrit@univ-orleans.fr.

[3]Laboratoire Acté -EA 4281 - UFR STAPS, Université Blaise Pascal Clermont Ferrand II. Email: nathalie.gal-petitfaux@univ-bpclermont.fr.

in the daily decisions of these professionals, so they are exposed to extreme stress, associated with the particular areas in which they are involved. These special 'schools' accommodate young people who suffer from different levels of social and physical deprivation. Consequently, these educators are often acting in an emergency, at the crossroads of various influences. Actually, they must constantly evaluate the priorities, according to their own sensibilities, and to their own experiences too. They have to admit that the only certainty they might have is that nothing can be entirely planned in advance. They must admit they will have to create solutions that nobody could teach to them. To manage such complex situations, newcomers in education have to develop competences. For example, 'the capacities of problem-solving, risk taking, trust in collaborative process, ability to cope with change and commitment to continuous improvement as organization" (Hargreaves, 2003, p. 3) are necessary qualities to be educating in the society of the 21st century. In fact, they have to take into account a major parameter of the situation they are involved in: themselves. Here we speak about their own feelings, their own sensations that will have a major impact on the decision process, especially at the start of their career. The complexity of the situation, all the interactions between the many parameters, the unpredictability of these contexts, are some of the characteristics of what some well-known scientists, such as Morin (1999), Delignières (2009), and Le Moigne (1999), have called 'complex systems'. In this contribution, we focus on the impact of discomfort and hazard in the simulated teaching process, and we examine how these troubling sensations seem to participate in the development of qualities of adaptation, which could be a kind of 'elegance in action' (Berthoz, 2009).

The University of Orléans trains professional educators who intervene through sports with 'fragile' populations in institutions that are professionally harsh (homes for people with social problems, homes for young offenders, prisons). These audiences are known to practice the concept of 'teaching misunderstanding' (Garnier, 2003), inducing the speaker, the teacher, to adopt 'survival strategies' (Woods, 1990, p. 96). In the professional context, the socio-sports educator adopts a state of alertness that allows for the correct and competent reaction when confronted with these moments of rupture. All day long, they are under pressure to make decisions, and must stay efficient to manage tensions: it's a matter of a sustainable career, often at the risk to their own health.

This paper is part of an 'action research' within the scope of a bachelor's degree called the 'Professional Licence in Development and Mediation by Sport' (LP DSMS). To do this, in their training courses, students develop professional practices with organized, interventional working groups included in the training unit. Under agreed partnerships between the University and the institutions, these working groups, or workshops, expose students 'in a controlled manner' to a public having singular characteristics (social disruption, school leavers, under court order). These working groups are authentic moments of training as they preserve the uncertainty and the randomness in the professional domain and emotionally involve the student. These workshops are 'complex training situations' (Clergue, 1997), or 'encouraged action spaces' (Durand, 2008), and therefore require the adaptive development of human processes.

What is involved is being able to adapt one's professional acts according to the possibilities and the constraints of the situation, and to reasonable and possible objectives. We then talk about moments of expertise and consider the possibilities of simplifying complex situations. More specifically about our terminology, we will speak of 'Simplexity', a concept introduced by the neuroscientist and physiologist Berthoz in 2009. The use of Simplexity does not imply reducing the dimensions of a given situation and thus hiding its complexity, but rather insists on the importance of prioritizing, taking into account the situation's current capabilities. The trainer should not over simplify for the sake of preservation, but rather 'confront and consolidate' and so trigger qualitative changes to their professional expertise, to achieve a kind of 'elegance in action' (Berthoz, 2009)

'Simplexity is a way of living with the world. It favors elegance over dullness, intelligence over formal logic, subtlety over rigidness, diplomacy over authority. Simplexity is Florentine; it anticipates rather than reacts, suggests laws and interpretive grids, is forbearing. It is adaptative rather than normative or prescriptive, probabilistic rather than deterministic. It takes into account the perceiving body as well as consciousness, and it considers context. Simplexity is intentional.' (Berthoz, 2012, p. 209)

14.2   Frameworks, object, model

The work of the social-sports instructor is conceived of as a 'situated action'. In this sense, we employ the framework of Cognitive Anthropology (Suchman, 1987). We also choose the theoretical and methodological framework of the course-of-action (Theureau, 2004; 2006) because we will look at the work of the student training, engaged in a teaching situation that is 'significant for him, i.e. presentable, relatable and which can be commented on by him at any time, under favourable conditions' (Theureau, 2004, p.48).

This self-confrontation, a kind of re-enactment through material traces (for this study, video traces) of the teaching activity, gives access to what Theureau calls 'pre-reflective consciousness' and testifies to the presence of a 'structural coupling'. The dynamical 'structural-coupling', according to Durand's study in 2008, is the asymmetric rapport established between the actor (he calls a system) and his environment. 'This coupling both builds and changes the structure of the system', we could say the SSE structure in our study, in a relationship we could characterize as asymmetric.

Moreover, 'this coupling is asymmetrical insofar as it is the system that defines what, in its environment disturbs him, that is to say, in fact, what is relevant and meaningful to him. This structural coupling creates a perspective, that of the system (the SSE), which changes every moment according to its (his) dynamic and disturbances coming from the environment.' (Durand, 2008, p. 107).

This theoretical and methodological framework is based on the assumption of the *autopoiesis* of living systems (Maturana and Varela, 1980).

The course also provides recent discoveries in neurophysiology on action, perception and decision (including Berthoz, 1997; 2002; Damasio, 2010).

14.2.1   Assumptions

For the moment, two hypotheses have been established in this part of our research project. These two hypotheses need to be supported by more extensive data.

Hypothesis 1: We assume that the hazard and discomfort experienced by the intervener in the case of a local socio-sporting action in complex situations generates emotions and sensations with 'magnifying effects' (Ria et al., 2001).

Hypothesis 2: We assume that these emotions mark changes in quality levels in the process of development and learning of the student in training (Gal-Petitfaux et al., 2010)

14.2.2   Participants and methods of data collection

Students animate and manage sports workshops with a sensitive public, at 4 sequences per student in the six-month course, each one for a duration of 40 minutes. All the sequences are filmed. Each student then participates in a self-confrontation interview recorded from the video of the day's session, helped by the trainer, according to Theureau's method.

Two training assessments are prepared: one at mid-training and the other one at the end of the course. During these assessments, prepared in advance with a guide provided, the sequences and self-confrontation interviews are discussed from elements that make sense to the student, who is able to self-assess his own competency. Audio recordings are made on this occasion for verbatim transcription.

Two classes of students have participated on this research protocol.

In this study, we will mobilize the self-confrontation interviews of two students (Justine and Hélène) and their two individualized assessments conducted on December 21, 2012 and April 17, 2013. At this point of our discussion, some information about the experience of these students could enable us to better appreciate the collected data.

When Justine begain the LP DSMS program, she had already completed, with difficulty, a degree as a sports intervention educator for standard schools. Due to her lack of confidence and her deep recurring difficulties in face-to-face teaching, she was a candidate to enter the LP DSMS with the objective of extending her expertise. She is a basketball coach, an activity for which she feels legitimately qualified.

Hélène has a degree in STAPS, which can be explained as 'Sciences and Technological training course in Sports and Physical Education'; she is a gym coach, an activity at which she thinks she is very competent. The qualities she claims for herself are: methodical, organized, and perfectionist. In interviews and appraisals, Hélène always appears as very willing and engaged in the analysis of her performance and its evolution.

These data, before training, during training, and at different stages of the training process, allow us to speak here of a 'course of life' (Theureau, 2004).

We have to notice that these two future SSEs were particularly comfortable with this kind of qualitive method, even though they had been inclined to speak about discomfort during their teaching time.

14.3   Data

A. Justine - extract from individualized assessment, April 17, 2013, and re-evocation of the 4th workshop in intervention practice.

Justine (J): The trigger that I had, it was with the video analysis (...) really evoking what worked, didn't work (...) why I was like that (...) it gave me confidence for the following sessions (...)

Trainer (T): 'It didn't destabilize you, it gave you confidence?'...

J: Yes, yes ... The first video session before going there, I was not great, huh ... because I was stressed, I was apprehensive, well like first practice session actually... But I told myself alright... need to take advice...

T: Had you finally started systematically taking into account your positive points?

J: Oh yes, well, yeah right ... I'm very pessimistic about myself ... now I said, I am pleased with myself, I made some progress... I'm out of there... there are positive things... (...) The third year was a total collapse [she is evoking here the previous training]... I said: Am I made for this?

T: Now what?

J: Now, that's fine. (...) I know what I'm like; it's not just a matter of pedagogy.

Then, referring to her fourth intervention this year in basketball,

J: Yes, I enjoyed myself. (...) In all these years of college, this is the session which I enjoyed the most (...) it happens late maybe (...) I was with them (...). I felt useful...

B. Hélène - extract from individualized assessment, April 17, 2013 and re-evocation of the first workshop on October 24, 2012.

Hélène: I'm not the same person in a club and on Wednesday [at workshops intervention]... The stress of there being a public makes a change for me [young minors under mandate of justice] (...) I am less afraid... I manage myself... I do not like being out of my routines

Trainer: Why?

H: ... it replays my position, it hurts me... discord, it is like an attack.

T: You are a perfectionist?

H: I'm too demanding with others and myself ... I did not realize ... All of life is like that, it is a facade.

I must go out of gym... be shaken... Being reactive, adapt, that's the problem I have in my internship ... Even in everyday life, by the way... in my intervention I do it, but it does not please me (...)

I do not remember the positives

14.4 Results and discussions

14.4.1 Justine

Justine freely evokes a 'course-of-life' (Theureau, 2004), that is to say, the conjunction of several originally separate 'courses of action'. At this point in her explanations, Justine speaks about singular moments of her activity during the training as a whole, of those of the elements which became significant and conscious. She remembers the difficulties felt; she analyses a chaotic and painful journey. She evokes, by naming it, the video as the source of a click under feelings of stress, like a transition to another type of behaviour, or feelings, self-awareness ('I was with them', 'I felt useful'), relating to a stage of a dynamic identity.

'Professional identity can be considered as a complex and multidimensional phenomenon incorporating a historic process of building, filled of questions , of losses and re-appropriations, that lead to forms of identity reconstructions which help people building a coherent world, and to give it sense. Feelings of membership, involvement, recognition are parts of a dynamic that integrates strategies and adaptations to the context, to the discipline, to the others, and this in relation with the potential of the actor (many facets related to the diversity of the actor), sometimes questioned by the events.'

'At this point, we enter in a "logic of subjectivation" the actor is committed to maintain the feeling of being the author of his own life and begins to put at a distance, if necessary, his role or previous positions. His identity is formed (trans- formed) by its tension with the surrounding world, causing it to evolve constantly.' (Roux-Perez, 2011, p. 146, translated by the authors)

Considering Justine's point of view, here we could be at a tipping point between the feeling of being just a follower student, and the feeling of becoming useful as a professional SSE: becoming useful, that major aim she mentioned when she joined course. She seems to switch identities. Before that clearly identified tipping point, she did not consider herself competent, and she doubted her choice of career. This workshop and maintenance are elements that seem to trigger an awareness, a more benevolent view of the self. After discomfort, doubt, and pessimism comes pleasure, a more positive self-image, a gain in confidence, and a feeling of being useful.

14.4.2   Hélène

Immediately, Hélène evokes the fear of losing face experienced by newcomers. She feels threatened. Thus she re-evokes a workshop that mobilized her in an extreme way emotionally, and did not sleep well the night before. She identifies the stress generated by the public as a constraint, which leads her to deviate from her routine as a gymnastics coach. Planner and anticipatory, she veers towards a more open attitude to the environment, even if it remains unpleasant for her. She also reviews that as a perfectionist, she does

not focus enough on the positive aspects of her intervention. Hélène transforms her power of action, even if the process has only just been engaged
and needs to be confirmed.

For these two students, intervention workshops, true 'authentic learning
situations' (Clergue, 1997) are an opportunity for 'revelations'.

'Learning is what helps to make sense and the meaning emerges from
the interaction of subjects with a world of objects and subjects according
to their basic needs and desires. This implies recognition of complexity
of things that carries themselves, like a paradox, this doomed to remain
unsatisfied need to have control of everything.' (Clergue, 1997, p. 143,
translation by the authors).

According to the sense of this quotation, these two 'SSE' are emotionally involved in the workshops that they re-live with the help of the video
sessions. They doubt of their abilities to face the context. But step by step,
they use precise and personal words to characterize what they do, what they
think, and what they feel during these experiences of teaching, according to
their own perception of the environment ('structural coupling'). Because of
these workshops, they take advantage of the nature of the situation ('authentic learning situations') to switch to a new self-knowledge in action, toward
a new professional identity, a kind of solution ('revelation') for overcoming
the complexity they were afraid of before.

Clergue then cites the thought of Barel 'Is complex at the same time
what threatens or destabilizes, and what can cope with this situation. Complexity is the problem and the solution of the problem.' (Amiot, Billiard,
and Brams, 1993, translation by the authors).

This increasing knowledge could be a solution to re-engage one's confidence. To overcome complex professional situations, SSE's self-confidence
could generate a dynamic that inspires new compliant solutions, far from
stereotyped routines. In reality, these routines work as self-benchmarks
(when Helene speaks about her job in association) from past experience,
but not as solutions adapted to the present, which are much more sensitive.
These routines could operate as a sort of barrier when faced with a specific
context, and the feeling of discomfort might operate as a emotional alert that
suggests to the 'SSE' giving up the planned action and taking into account
other dimensions of the present context. To a certain extent, we could speak
about a kind of emotional invitation in favour of the re-examination of the
'structural coupling'. We evoke here the sensitive question of the perceived
discomfort level, which seems to work as an adaptation variable in order to
adapt to the teaching context.

Both SSEs come to the training with accrued routines, but also with questions about their competency. Paradoxically, it is in discomfort and stress, facing the reactions of a public in a state of social or family exclusion, that they gain relevance and knowledge of themselves. They clearly identify a tipping point in their training courses, because the training situation keeps all the complexity required for the emergence of random triggers: unpredictable reactions, defiant attitudes, disobedience. These hazards cause emotions that are 'complex and ongoing process of adaptation to the environment. In this respect, the action of teachers is guided by these sensitive signals allowing to perceive things and to act in the classroom' (Ria and Visioli, 2010, p. 14, translation by the authors), linked with their capacity to adapt, their sensitivity to context in order to keep in balance a didactic and extremely tenuous contract.

Thus, emotions when training are essential to enable the development of skills, and can hardly arise outside of authentic learning situations:

'The relation feeling–knowledge is at the heart of professional development: emotions are purveyors of knowledge and the relevance given to them depends on sensitive elements conferring a peculiar emotionally marked meaning (e.g. waiting for a pleasant situation or rather unpleasant related to past experiences.' (Ria and Visioli, 2010, p. 14, translation by the authors).

The cases of Hélène and Justine fit perfectly in the announced hypotheses, and are consistent with the results of the cited research. These intervention workshops and associated training tools seem to encourage the development of qualitative adaptation with the involvement of the emotions due to the complexity of the environment rebuilt at the University, as part of a professional training.

In this study, the dynamic adaptation process observed in the workshops, in other words, the 'complex training situations' (Clergue, 1997, translation by the authors), or the 'encouraging action spaces' (Durand, 2008, translation by the authors) looks like a process described in the framework of the dynamic systems approach to motor coordination and training, and considering motor skill acquisition as a dynamical process with transition (Haken, 1985, translation by the authors).

In our research, we focus on the hazard and discomfort that play a role as contraints – resources for acquiring gains in development and adaptation and which could be conceived as parameters of control (Nourrit et al., 2003). With this effect 'contraints – resources', the SSE puts aside the stereotyped routines developed in other contexts (basketball clubs, gymnastic clubs), as a phase transition. This may be the means of accelerating the development of an adaptation to sensitive professional contexts in the initial training course.

Bibliography

Amade-Escot, C., 2003, 'La gestion interactive du contrat didactique en volley-ball: agencement des milieux et régulations du professeur', in C. Amade-Escot (Ed.), *Didactique de l'éducation physique: Etat des recherches*. Paris: Revue EPS, pp. 255–278.

Amiot, M., I. Billiard, and L. Brams, 1993, *Système et paradoxe: autour de la pensée d'Yves Barel*. Paris: Ed. du Seuil.

Berthoz, A., 1997, *Le sens du mouvement*. Odile Jacob.

Berthoz, A., 2003, *La décision*. Odile Jacob.

Berthoz, A., 2009, *La simplexité*. Odile Jacob

Berthoz, A., 2012, *Simplexity: Simplifying principles for a complex world*. New Haven, CT: Yale University Press.

Clergue, G., 1997, *L'apprentissage de la complexité* Paris: Hermès.

Damasio, A. R., 2010, *Self comes to mind: Constructing the conscious brain*. New York: Pantheon Books.

Durand, M., and N. Gal-Petitfaux, 2001, 'L'enseignement de l'éducation physique comme action située: proposition pour une approche d'antropologie cognitive'. *STAPS* **55**, 79–100.

Durand, M., 2008, 'Un programme de recherche technologique en formation des adultes'. *Education & Didactique* **2**(3), 97–121.

Delignières, D., 2009, *Complexité et compétences : Un itinéraire théorique en éducation physique*. Editions Revue EPS.

Gal-Petitfaux, N., C. Sève, M. Cizeron, and D. Adé, 2010, 'Activité et expérience des acteurs en situation : les apports de l'anthropologie cognitive'. *In* M. Musard, G. Carlier, and M. Loquet (Eds.), *Sciences de l'intervention en EPS et en sport*. Paris: Editions Revue EP.S., pp. 67–85.

Hargreaves, A., 2003, *Teaching in the Knowledge Society: Education in the Age of Insecurity*. New York: Teachers College Press.

Haken, H., J. S. Kelso, and H. Bunz, 1985, 'A theoretical model of phase transitions in human hand movements'. *Biological Cybernetics* **51**(5), 347–356.

Le Moigne, J.-L., 1999, *La modélisation des systèmes complexes*. Paris: Dunod.

Maturana, H. R., and F. J. Varela, 1980, *Autopoiesis and Cognition: The Realization of the Living*. Berlin: Springer-Verlag.

Morin, E., and J. L. Le Moigne, 1999, *L'intelligence de la complexité*. Paris: L'Harmattan.

Nourrit, D., D. Delignières, N. Caillou, T. Deschamps, and B. Lauriot, 2003, 'On Discontinuities in Motor Learning: A Longitudinal Study of Complex Skill Acquisition on a Ski-Simulator'. *Journal of Motor Behavior* **35**(2), 151–170.

Perez-Roux, T., 2011, *Identité(s) professionnelle(s) des enseignants: les professeurs d'EPS entre appartenance et singularité*. Paris:: éd. EP&S.

Ria, L., J. Saury, C. Sève, and M. Durand, 2001, 'Les dilemmes des enseignants débutants : Etudes lors des premières expériences de classe en Education Physique'. *Science et Motricité* **42**, 47–58.

Ria, L., and J. Visioli, 2010, 'L'expertise des enseignants d'EPS'. *Science & Motricité* **71**, 3–19.

Suchman, L. A., 1987, *Plans and Situated Actions: The problem of human–machine communication*. Cambridge University Press.

Theureau, J., 2004, *Le cours d'action: méthode élémentaire*. Toulouse, France: Octarès.

Theureau, J., 2006, *Le cours d'action: méthode développée*. Toulouse, France: Octares.

Woods, P., 1990, *Teacher Skills and Strategies*. London: Taylor & Francis.

Part IV

Interactions between economic agents

The paper proposed by Ch. Garrouste and E. Courtial discusses the prevision of employability in the domain of Earth Sciences. As energy issues, including economy policies and sustainable development, are increasingly present, employment in the fields of Earth Sciences is steadily growing. Within this frame, universities need to adapt their training offer, their programs and have to manage their flow of graduate students in this domain. It would be interesting to have dedicated tools for that purpose in order to anticipate and bring an efficient strategy in terms of training policy. Thus, the authors propose the CIPEGE tool as a combination of econometric models (macro and micro) and a predictive control strategy. This name comes from the CIPEGE Center (Centre International de Prospective pour l'Emploi en Géosciences et Environnement), in charge of studying the strategic foresight of employment in Earth and environmental Sciences. This study then aims to ensure that the forecasted employability tracks the reference trajectory for the French graduates. The targeted employability is estimated using a labour market matching approach. The tracking problem is treated with a Model Predictive Control (MPC) approach, suitable for dealing with tracking problems. Simulations show a drop in the demand for higher education skills in the coal industry and in the oil and gas sector at the 2020 horizon. In the case of the metal industry they anticipate a smooth increase in the demand for higher education skills. Thanks to the CIPEGE tool, the authors indirectly address a real challenge to society, which is among other things the management of energies and mineral resources.

The article by F. Nemo aims to show how the modelization of market interactions could be greatly improved by taking into account some of the parameters which have proved to be crucial to the understanding of the relation between language, communication and information, namely the performative, attentional and sequential dimensions of interactions. The author first insists on the performative nature of the price of goods, which refers to the fact that suppliers are not price-takers but can influence the price of the good they sell. The author secondly focuses on the attentional constraint inherent to human interactions. This attentional constraints contrast with the perfect information hypothesis made in economic theory. According to the author, any modelization of interactions which postulates that everyone is related to everyone and is aware of everything describes an impossible world and an absolute chaos. Accounting for attentional constraints is thus important because they quantitatively limit the global flow of information and at the individual level considerably constraint his/her economic and non-economic behavior. The author finally studies the importance of sequentiality in modeling market's interactions. Indeed, not only everybody's actions are also reactions to previous actions, but also that these actions are orders chronologically and that such a scheduling is crucial to the analysis of market interactions and their outcome.

In "Kinetic collision models and interacting agents in complex socio-economic systems", S. Cordier illustrates how collisional kinetic theory can be used to model interacting economic agents. He considers a simple model of an open market economy, modelling exchanges between individuals and speculative trading. The aim is to study the repartition of wealth after a large number of transactions, i.e., when reaching stationary states. This requires to simplify the model and under conditions the stationary distribution exhibits a Pareto power law tail.

This book is concluded by an article of C. Piatecki, who analyses the choice of studies' length in an evolutionary game. Since Becker's (1964) contribution,[1] the accumulation of human capital has been developed only within the framework of individual choices based upon wage expectations. For over forty years, this approach showed itself fruitful. However, with the rarefaction of jobs during periods of crisis, it does not seem anymore able to explain most of the decisions regarding human capital accumulation. Indeed, because anybody can be sure to obtain an employment which will yield a positive return to education, the decision to acquire further education can only be apprehended as a strategic decision in a situation of imperfect information. This individual decision depends on the answer to the following simple question: *With the level of studies which I intend to acquire, will I have bigger chances to obtain a job than my classmates who stopped their studies at a lower level?* It's the reason why the author models studies' length as a two strategies evolutionary game to show under which condition the population of players splits in two class of strategies in equilibrium.

---

[1] Becker G. (1964), *Human capital*, The University of Chicago Press, Chicago.

Chapter 15

Forecasting Employability in the Earth Sciences: The CIPEGE tool

Christelle Garrouste[1],
Estelle Courtial[2]

15.1   Introduction

In the past two decades, energy consumption, sustainable development, and environmental protection, have become priorities of the energy policies in most countries (the Kyoto Protocol, Grenelle environment in France, Carbon plan in the United Kingdom, 20-20-20 targets in the European Union). The transition to renewable energies is leading to many changes in the economy as a whole, in the labour markets' structure and dynamics, in societal behaviour, and in research and education. Among the study and research disciplines the most affected by these changes is Earth Sciences (ES). ES includes the study of the atmosphere, hydrosphere, oceans, biosphere, as

[1]Laboratory of Economics of Orléans, University of Orléans, CNRS, UMR7322, Orléans, France (e-mail: christelle.garrouste@univ-orleans.fr).

[2]Corresponding author: Laboratory PRISME, University of Orléans, EA4229, Orléans, France (e-mail: estelle.courtial@univ-orleans.fr).

well as the solid Earth. ES can help meet the major challenges faced by society and the environment, such as the management of mineral resources (green mining) with its direct impact on consumption and labour. For instance, rare-earth minerals (REM) are increasingly used for the production of high-tech items, such as smart phones and laptops, but also in magnets for wind turbines, hybrid-car batteries, etc. After a misleading forecast of its own consumption needs, the U.S. lost its position as leading producer in favour of China, and is now constrained to import, at a very high price, the goods it used to produce domestically. As a consequence of the closure of its mining operations in the 1990s, the U.S. also decreased its investment in the training of solid ES scientists. While it now faces the need to re-open its REM mining sites to satisfy an increasing demand for high-tech goods, it suffers from a deficit in qualified Earth System scientists. A similar deficit affects Australia and Canada.

With its technical and scientific competences, its historical assets, and its first-class actors in the field of ES, France aims at becoming a worldwide leader in ES training.

In 2011, the French Ministry of Higher Education nominated the so-called project VOLTAIRE as a LABEX (Labouratory of excellence). Among the tasks of this project is to construct an anticipation tool to ensure the employability of Earth System scientists trained in French universities. The CIPEGE center (Centre International de Prospective pour l'Emploi en Géosciences et Environnement – International Centre for the Strategic Foresight of Employment in the Earth and Environmental Sciences) was created to handle this task. In the sequel, the anticipation tool will be called the CIPEGE tool. This paper presents the innovative forecasting strategy adopted to develop the CIPEGE tool.

The CIPEGE tool builds upon several methods to handle the complexity of the system it aims at predicting. As will be explained in this paper, the system is composed of several elements measured at different levels of units of analysis (individual, institutional, national, international), each with a non-negligible internal structure. The elements are linked by non-linear interactions. The system is exposed to external factors at different scales and there is a reaction of the collective behaviours upon the individual behaviours. This means that the elements will collectively modify their environment, which in turn will constrain them and modify their possible behaviour. Hence, the behaviour of the global system can only be predicted by knowing the properties and behaviour of each element and the properties of the interactions that link the elements.

Therefore, the CIPEGE tool combines economics (macro- and microeconometric modeling) and control process strategies (model predictive control). The targeted employability of French ES graduates is estimated using a labour market matching approach, controlling for European macroeconomic trends. The measured and forecast employability are then derived from a microeconometric multinomial-conditional logit approach. The task of this study is then to ensure that the forecast employability tracks the reference trajectory of employability for the French graduates in ES. From a control theory perspective, this objective can be viewed as a tracking problem. Among the existing advanced control laws, Model Predictive Control (MPC) is a control strategy well-adapted to deal with tracking problems (Alessio and Bemporad 2009, Camacho and Bordons 2007). The success of MPC in several industrial sectors is due to the ease of formulating the control objective in the time domain and also to its ability to handle constraints (Qin and Badgwell 2003). A wide variety of applications has been reported in the literature but no application in economics exists to our knowledge. MPC is based on the direct use of an explicit model to predict the future behaviour of a process. This model plays a crucial role in the MPC strategy. In our case, the econometric model is used to forecast the behaviour of the process (the flow of ES graduates in France) over a finite prediction horizon. In the context of this study, the term 'prediction' actually refers to an 'anticipation' or 'forecast'.

The main advantage of this strategy lies in the structure of the MPC approach, which enables us to systematically and continuously link all the elements composing the employability of young ES graduates. This structure allows the definition of different time horizons (in contrast to classic econometrics) and allows continuously correcting the trajectory, thanks to the feedback mechanism of the MPC approach. Another advantage is the fact that this strategy can take into account both estimation errors and modeling errors to correct the reference value at each run. This two-step error correction procedure constitutes a valuable tool for obtaining more robust estimates. Moreover, the MPC approach determines the control inputs exogenously in such a way that it is applied simultaneously to the process and the model. In an econometric model, the effect of the exogenous control input is usually estimated inside the model and is considered as endogenous to the process. For the first time, the control inputs are defined exogenously and policy makers can test an unlimited range of interventions.

The remainder of this paper is organized as follows. Section 2 presents the definition of the employability retained for the CIPEGE tool and the model applied to estimate the employability of reference for French ES graduates. Section 3 deals with the principle of Model Predictive Control and details the control structure used. Section 4 addresses the way of combining the econometric model used to measure the process of employability and the predictive control approach. Then, in Section 5, different simulations serve to test the feasibility of the control strategy. In the last section, we synthesize our preliminary results and outline the perspectives of the proposed approach.

15.2   Employability

Measuring the employability of graduates is a controversial issue due to the difficulty in applying a straightforward definition (Gazier 1998, McQuaid and Lindsay 2005, Arjona Perez, Garrouste, and Kozovska 2010). Employability is a complex and multi-faceted concept. Therefore, either because of a lack of compatibility between dimensions or a lack of data, a holistic measure of employability has so far been recognized to be impossible. Employability measures are instead reduced to the most pertinent dimensions for the study at hand.

15.2.1   Definitions

In McQuaid and Lindsay (2005), the authors highlight the existence of two alternative perspectives in the employability debate. One focuses only on the individual's characteristics and skills, referring to the individual's potential to obtain a job. The other perspective takes into account external factors (e.g. labour market institutions, socio-economic status) that influence a person's probability of getting a job, of moving between jobs, or of improving his or her job. In De Grip, van Loo, and Sanders (2004), these factors are called 'effectuation conditions', i.e. the conditions under which workers can effectuate their employability. In addition, the literature also considers the aspects of the time lag between leaving education and employment (Boateng, Garrouste, and Jouhette 2011), the degree of matching of skills between one's educational background and his or her occupation, and the type of contractual arrangement (full-time vs part-time; permanent vs temporary) (Arjona Perez, Garrouste, and Kozovska 2010).

Employability is about having the capability to gain initial employment, maintain employment, and obtain new employment if required (Cedefop 2008). In other words, the employability of a graduate is the predisposition of the graduate to exhibit the attributes that employers consider necessary for the effective functioning of their organization (Harvey, Geall, and Moon 1998). Hence, employability is a combination of capacity and willingness

to be and to remain attractive for the labour market, for instance, by antic-
ipating changes in tasks and work environment and reacting to them (De
Grip, van Loo, and Sanders 2004). For a given person, employability depends on the knowledge, skills and attitudes he/she possesses, as well as the
way he/she uses those assets and presents them to employers (Hillage and
Pollard 1998).

In the context of the CIPEGE tool, we define employability as the capacity of a French Earth Sciences graduate to be employed at a fulfilling job
that enables him/her to make use of the skills acquired during the training,
given the evolution of the demand of the relevant sectors of activity at the
European level.

15.2.2 Energy prices, environmental policies and employ-
ment trends

In a recent report for the European Commission's Employment Directorate
General, Cambridge Econometrics (2011) conducted an in-depth analysis of
the employment consequences of the implementation of policies to achieve
the key European environmental targets of a 20% cut in emissions of green-
house gases by 2020 (compared to 1990 levels), an increase in the share of
renewable energy to 20%, and the objective of a 20% cut in energy consumption (i.e. the so-called "20-20-20 targets"). The findings from that
project suggest that after an initial cost to the European Union (EU), the
implementation of the EU 20-20-20 targets will lead to a modest positive
outcome for GDP growth and employment over the longer term, increasing
by around 1%–1.15% (in net terms) by 2020.

However, when looking at specific industries, these impacts prove to
be much differentiated, with some sectors, such as iron, steel, cement and
petroleum, experiencing a decrease in employment; and other sectors, such
as renewables, construction and transport, experiencing a growth in jobs by
2020. The occupations with potential benefits from a low-carbon transition were identified to be Research and Development, manufacturing and
installation or engineering, operations and maintenance, management, administration, and sales.

The shift towards a greater reliance upon renewable energies will create
a demand for the engineering and technical skills related to generating elec-
tricity from wind, marine, and solar sources. The forecasts of employment
from the Energy-Environment-Economy model of Europe (E3ME), developed by Cambridge Econometrics, suggest a greater demand for professional, associate professional, and (to a slightly lesser extent) skilled trade

workers. Still, these analyses do not go into detail on the implications for specific areas such as science, technology, engineering, and mathematics (for further details on that topic see, for instance Wilson 2010).

Further evidence suggests that the shift towards renewable energies may result in an increasing demand for specific types of engineers and technicians who are not only highly qualified and skilled in their general disciplines, such as electronics engineering, but can also apply their skills within a renewable or green policy environment. Hence, there is likely to be an increasing demand for a form of hybrid skill (the general engineering or technical discipline, plus specific knowledge or experience of renewables) (Cambridge Econometrics 2011, p. 172).

Further, there is also an increasing demand for more renewable-specific skills related to hydrology, hydraulics, aerodynamics, ornithology, environmental impact assessment, etc. Both the hybrid and renewable specific skills need to be deployed in a number of functions, including RnD, design, operations, and maintenance. Although the number of workers required to fill these jobs might be relatively small, they are critical to the success of the renewable sector.

The problem that currently arises in Europe is that there will be potentially more and more markets that will display insufficient capacity to fulfill the demand for these new technologies because of a deficiency of human capital. For example, if Europe is not able to establish an economically viable wind turbine or solar panel manufacturing capacity prior to the maturity of the technology and its uptake by users, there is a high risk that non-EU producers will take over the market, once established. This would clearly have a negative impact on the overall employment potentials within Europe. Another similar example could be given of innovation in the environmental and eco-technology sector.

Among all European Member States, the United Kingdom (UK) is the only one currently investing in upskilling at the tertiary level to meet the new "green" occupation needs. The main focus of the UK is on the upskilling of tertiary engineering qualifications in energy, especially in the installation and maintenance of low-carbon technologies and customer service skills. Another UK training focus is on tertiary qualifications in commodity trader and broker, with the development of practical skills in the functioning of the carbon market and the understanding of trading tools. Other EU Member States are instead focusing on the green upskilling of the workforce at the vocational level (e.g., Denmark, Estonia, France and Germany) (GHK Consulting 2010).

15.2.3 Modeling

Let y_j^F denote the number of employed individuals in France who graduated in Earth Sciences at degree level j, with $j \in \{3, 5, 8\}$, such that $j = 3$ for a 3-year degree (i.e. the bachelor's degree); $j = 5$ for a 5-year degree (i.e. the master's degree) and $j = 8$ for an 8-year degree (i.e. the PhD).

For a given degree level j, the number of ES graduates employed at time t is

$$y_j^F(t) = S_j^F(t) - un_j^F(t) \qquad (15.1)$$

where S_j^F is the stock of ES skills on the French market and un_j^F is the stock of unemployed ES graduates in France.

We model the employment of French ES graduates as a matching function, as suggested by Mortensen and Pissarides (1994), to describe the formation of new relationships ('matches') from unmatched individuals of the appropriate types. In our case, we are interested in the formation of matches, at the European level, between the number of unemployed ES j-level graduates and the number of job vacancies in ES domains at the j-level. In other terms, we are interested in the European demand for workers with ES skills at each j-level. We assume our matching function to have the following Cobb–Douglas form:

$$m_j^{EU}(t) = M(un_j(t), v_j(t)) = \mu(un_j(t))^a(v_j(t))^b \qquad (15.2)$$

where $m_j^{EU}(t)$ is the number of new matches created at current time t on the European market, and μ, a and b are positive constants. While un_j is now the stock of unemployed ES graduates in Europe, $v_j(t)$ is the number of job vacancies in ES field at degree level j. The matching function is increasing, concave, and homogeneous of degree 1. As reviewed by Petrongolo and Pissarides (2001), the Cobb–Douglas form of the matching function can be justified by empirical evidence of constant returns to scale, i.e. $a + b \approx 1$.

If the fraction of workers separating from a firm per period of time (due to firing, quitting, and so forth) is δ, then the change in employment from one period to the next is calculated by adding the formation of new matches and subtracting the dissolution of old matches. Combining equations (15.1) and (15.2) yields the following representation of the dynamics of employment over time:

$$y_j^F(t+1) = m_j^{EU}(t) + (1-\delta)y_j^F(t) = \mu(un_j(t))^a(v_j(t))^b + (1-\delta)y_j^F(t) \quad (15.3)$$

The evolution of unemployment is given by

$$u\dot{n}_j = \delta(1 - un_j) - m_j^{EU}(un_j, v_j) \qquad (15.4)$$

Under the assumption that the matching technology[3] exhibits constant returns, this equation has a unique stable steady solution for every vacancy
rate v:

$$un_j = \delta/(\delta + m_j^{EU}(v_j/un_j, 1)) = \delta/(\delta + \lambda(\theta)) \qquad (15.5)$$

where $\theta = v_j/un_j$ indicates the tightness of the market while the unemployment spell hazard is represented by $\lambda(\theta) = m_j^{EU}(v_j/un_j, 1)$. Drawing
equation (15.5) in the vacancy–unemployment space generates a Beveridge
curve, i.e. a negative relation between vacancies and unemployment that
is convex to the origin, due to the properties of the matching function (for
details see Pissarides and Mortensen 1999).

Empirical evidence shows that the job destruction flow, δ, is not constant, especially at business cycle frequencies (see Davis, Haltiwanger, and
Schuh 1996). Following Pissarides and Mortensen (1999), we therefore allow future job productivity p to vary according to the relative value (in terms
of required competences) x of the product or service. Because $x \in [0, 1]$,
px can take more than two values[4].

When an unemployed worker and an employer with a vacancy meet,
wage bargaining takes place. The outcome is a wage $w(x)$ that divides the
quasi-rents associated with a match between worker and employer, according to the value of x. The value of a filled job is a function of the future
job productivity, the bargained wage, and the probability of destruction of
the job. Job creation takes place if all rents from new vacancy creation are
exhausted, i.e. if $v_j = 0$. A job is destroyed only if its idiosyncratic productivity falls below a critical level rs (equilibrium reservation productivity),
i.e. if $x < rs$.

15.2.4   Empirical specifications

Except for the matching function, all parameters of the targeted employability are estimated, for the years 1990 to 2011, using the annual Employment Survey microdata collected by INSEE (the French National Institute
of Statistics and Economic Studies).

The matching function m_j^{EU} is estimated using an extended version of

---

[3]A matching technology, like a production technology, is a description of the relation between inputs, search and recruiting activity, and the output of the matching process, the flow
rate at which unemployed worker and vacant jobs form new job–worker matches (Pissarides
and Mortensen 1999).

[4]According to Pissarides and Mortensen (1999), Mortensen (1994), Cole and Rogerson
(1996), the fact of regarding p as a stochastic process characterizing an aggregate shock is
consistent with the time series characteristics of job creation and job destruction series reported
by Davis, Haltiwanger, and Schuh (1996).

the E3ME model to forecast skills supply and demand in Europe (Wilson 2010). Thanks to the structure of the E3ME, m_j^{EU} captures the influence of international environmental and economic shocks on the demand for skills in Earth Sciences.

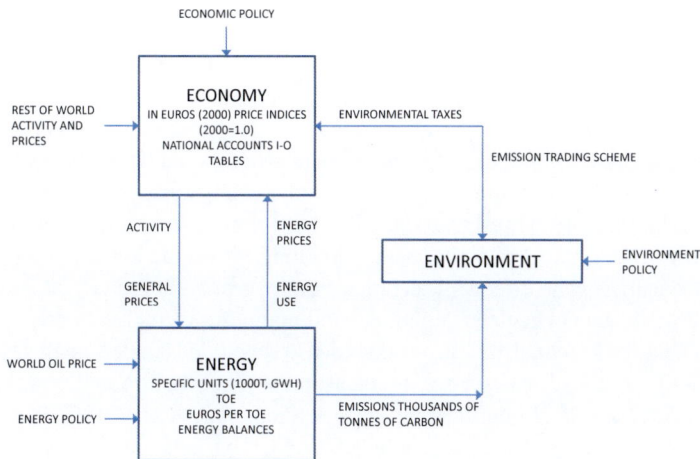

Figure 15.1: The three dimensions of the E3ME: energy, environment and economy (Source: Cambridge Econometrics 2011, p. 212).

Figure 15.1 shows the linkages between the three modules (Energy, Environment and Economy) of the E3ME. Exogenous factors are shown on the outside edge of the chart as inputs into each component. For the European Union (EU) economy, these factors are economic activity and prices in non-EU world areas and economic policy (including tax rates, growth in government expenditures, interest rates and exchange rates). For the energy system, the outside factors are the world oil prices and energy policy (including regulation of energy industries). For the environment module, exogenous factors include policies such as a reduction in SO_2 emissions by means of end-of-pipe filters from large combustion plants. The linkages between the modules are shown explicitly by the arrows that indicate which values are transmitted between components.

The economic module is solved as an integrated EU regional model, dis-aggregated at the industry and country levels. The labour market is treated with sets of equations for employment demand, labour supply, average earnings, and hours worked. The equations for labour demand, wages, and hours worked are estimated and solved for 42 economic sectors (industries), defined at the 2-digit level of the European statistical classification of economic activities (NACE). Labour participation rates are disaggregated by gender and five-year age bands, and multiplied by Eurostat population data to obtain the labour supply.

Employment is modeled using national accounts data, as a total headcount number for each industry and region. This stock is a function of the evolution of unemployment and job vacancies (depending on industry output, wages, hours worked, and technological progress) and of global (worldwide) changes in environmental policies and in energy prices and consumption. The industry output is assumed to have a positive effect on employment, while the effect of higher wages and longer working hours is assumed to be negative. The effects of technical progress are ambiguous, as investment may create or replace labour, depending on the sector of activity.

The E3ME model has been extended to include detailed analyses of the skills' demand and supply, as measured by occupation and qualification. Three levels of qualification were defined, namely low (ISCED 0–2), medium (ISCED 3–4) and high (ISCED 5–6). This extended version computes the labour demand by sector, occupation, and educational level.

We ran the extended E3ME to generate, for the overall European market, the labour demand at the highest level of qualifications (i.e. ISCED 5–6), by sector and level of occupation. Using the annual INSEE Employment Survey (1990–2011), we identify matrices of sectors and occupations that employ workers with a j-level ES degree in France. We merge the E3ME output with the individual data from INSEE on the basis of these matrices and derive the number of demanded ES workers with j-level skills, m_j^{EU}. The current employment of ES graduates in France, $y_j^F(t)$, is estimated by Ordinary Least Squares (OLS) regressions, for each level of degree, using the INSEE microdata, controlling for age, gender, and sector of activity. The job destruction flow, δ, is also estimated from the INSEE data. The estimated current matching on the European market is then integrated into equation (15.3), alongside the current employment level in France $((1 - \delta)y_j^F(t))$, to estimate the number of employed j-level ES graduates in France in the next period.

15.3   Model Predictive Control (MPC)

This section addresses the principle of MPC and the control structure considered in this study is detailed.

15.3.1   Principle

MPC is a mature control strategy. Initially developed for linear systems in the 1970s, MPC has been extensively studied for nonlinear systems with constraints and successfully been applied in numerous industrial domains (Alessio and Bemporad 2009, Qin and Badgwell 2003). The MPC strategy is based on the receding horizon principle, and is formulated as solving on-line a nonlinear optimization problem (Camacho and Bordons 2007). The basic concepts of MPC are the explicit use of a model to predict the process behaviour over a finite prediction horizon N_p and the minimization of a cost function with respect to a sequence of N_c controls, where N_c is the control horizon. At the current instant t (see Figure 15.2), the process output is measured and the MPC algorithm computes a sequence of N_c control inputs by minimizing the tracking error (the difference between the reference trajectory and the predicted model output) over N_p. Only the first element of the obtained optimal control sequence is really applied to the process. At the next sampling time (see Figure 15.3), the finite prediction horizon moves a step forward, the measurements are updated, and the whole procedure is repeated.

15.3.2   Internal Model Control (IMC) structure

Predictions based on data are inevitably subject to disturbances and modeling errors. To gain in robustness, the well-known Internal Model Control (IMC) structure (see Figure 15.4) is used in this approach.

    The process (the physical system) is described by its mathematical model. The control inputs u are simultaneously applied to the process and the model. The difference between the process output y_p and the predicted model output y_m provides an error signal e. This signal embeds the disturbances and modeling errors and constitutes the feedback information affecting the reference trajectory y_{ref}. The feedback information is taken into account in an original way, rarely used in economics. Due to the acquisition of the sampled data, a discrete-time formulation is used, where t is the current time iteration.

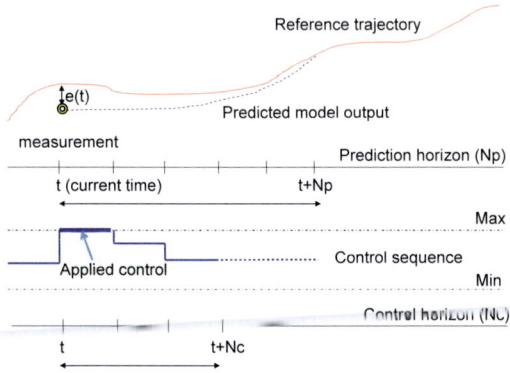

Figure 15.2: Principle of MPC at the current time t.

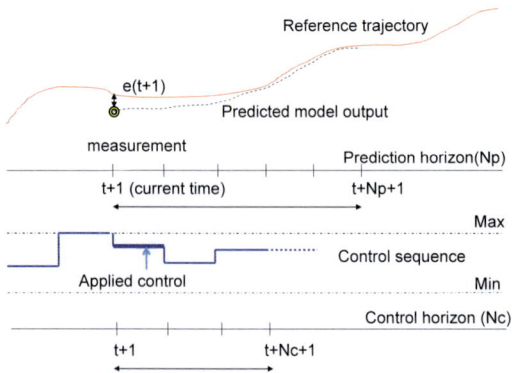

Figure 15.3: Principle of MPC at the current time $t + 1$.

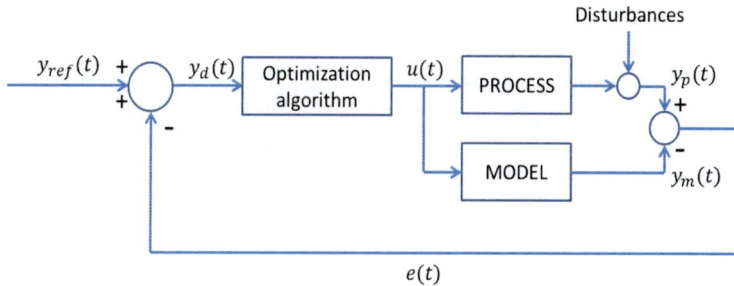

Figure 15.4: Internal model control structure.

According to Figure 15.4, we can write

$$\begin{aligned} y_d(t) &= y_{ref}(t) - e(t) \\ y_d(t) &= y_{ref}(t) - (y_p(t) - y_m(t)) \\ y_d(t) - y_m(t) &= y_{ref}(t) - y_p(t). \end{aligned} \tag{15.6}$$

The tracking of the reference trajectory y_{ref} by the process output y_p is equivalent to the tracking of the desired trajectory y_d by the model output y_m.

Remark: the feedback signal e can be filtered to avoid sudden changes due to measurement noise. In practice, a low-pass filter is well suited to this.

15.4 Econometric model-based predictive control

This section addresses the way of combining the econometric model and the MPC approach. The common points to all predictive strategies are discussed according to the control objective: the improvement of French students' employability in the field of ES.

15.4.1 The reference trajectory

The reference trajectory corresponds to the expected behaviour of the process. In our case, the reference trajectory y_{ref} to be tracked corresponds to the employability of French ES graduates. This reference has been determined off-line by estimating equations (15.1), (15.2) and (15.3) using E3ME outputs and French microdata (INSEE Employment Survey). The following figure (see Figure 15.5) presents the reference trajectory for the employability in France of ES graduates at the master's degree level, i.e. at level $j = 5$, for 2003–2025, taking into account labour market shifts at the European level. While the goodness of fit of the reference trajectory for 2003–2012 could be tested with the observed values of the output variable, the trajectory beyond 2012 was drawn as a linear extrapolation of the previous period. In economics, because of the high degree of unpredictability of individual behaviour, fitted values are likely to vary within a 90% confidence interval.

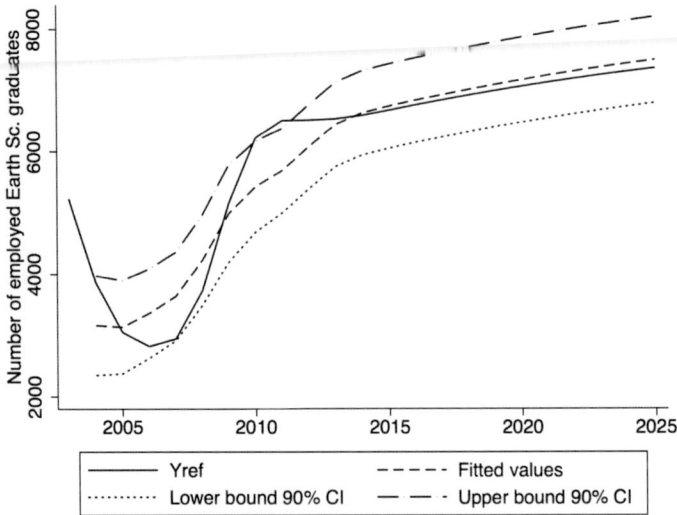

Figure 15.5: Reference trajectory for the employability in ES at master's degree level.

The significant drop in employability observed between 2003 and 2007, followed by a significant increase between 2007 and 2012, was mainly due to the changes in the graduation system imposed by the European Bologna Process. In 2004–2005, most European Member States, including France, launched a harmonization process of their tertiary degrees to converge towards a three-level system composed of the bachelor's degree, the master's degree, and the PhD. All intermediary degrees were progressively cancelled from the tertiary systems. In France, this new structure has replaced the previous five-level structure (DEUG-Licence-Maîtrise–DEA/DESS-Doctorat). The former Maîtrise degree corresponds to a master's I level and the former DEA/DESS degree to a master's II. At the time of the implementation of the Bologna Process, French labour markets were more interested in hiring young graduates with a four-year university degree than with a five-year university degree (reflected in the drop in Figure 15.5). With the shift towards the BA-MA-PhD structure, the vacant positions initially aimed at four-year graduates were progressively opened to MA graduates, in addition to the vacant positions traditionally aimed at five-year graduates, which explains the significant increase in MA employability after 2007.

15.4.2   The model of prediction

As we have already mentioned, the model is the keystone of the predictive control approach since it anticipates the behaviour of the process in the future and makes possible a successful tracking of the reference trajectory. The more faithful the model is to the process, the better the model prediction.

The model, based on an econometric model, is assumed to be dynamically representative of the dynamics of employability, and to take into account the uncertain dynamics of the human factor thanks to the use of probabilities. The three outputs of the model are the predicted employability at the three different levels $j = 3, 5, 8$ (bachelor's degree, master's degree, PhD, respectively). For the moment, the more relevant input employed is the number of students enrolled in ES training. The specificities of the model are worth mentioning. First, the evolution of the employability depends on the level. There are, therefore, three dynamics to be identified: one for each of the three degree levels. Secondly, the control inputs act on the model with different time lags (delays). For instance, the effect on employability levels of the number of first-year students in ES university programmes at the current time t will be measurable in three years for bachelor's degrees, in five years for master's degrees, and in eight years for PhDs. In the model, each different delay must be handled in a specific way. Lastly, for a given student cohort, probabilities are employed, so as to take into account the enrollment rate in ES, the completion rate, the retention rate, etc. The model of employability is written in discrete time and can be synthesized by the following non-linear equations:

$$\begin{cases} x(t+1) = f(x(t), u(t-i)), i = 0..7 \\ y_m(t) = h(x(t)). \end{cases} \tag{15.7}$$

The modeled employability y_m estimates the number of students that will complete each degree-level annually by multiplying the number of students enrolled at the beginning of each degree period $(u(t-i))$ by the probability of their pursuits of studies in ES. The probability for a student i to choose to pursue an ES degree is estimated using a multinomial/conditional logit approach (McFadden 1974). At the beginning of each study period, a student can choose between three alternatives: to pursue studies in ES at the upper level, to repeat the same level, or to leave ES studies. This approach assumes that individuals choose their training path, from different alternatives, in order to maximize their utility.

15.4.3   The cost function

The cost function, also called the optimization criterion, is usually a quadratic function of the tracking error. As already mentioned, thanks to the IMC structure (see Figure 15.6), the tracking of the reference trajectory y_{ref} by the process output y_p is equivalent to the tracking of the desired trajectory y_d by the model output y_m. The tracking error is consequently written as

$$e_{tra}(t) = y_d(t) - y_m(t). \tag{15.8}$$

At each sampling time, the process output is measured and compared to the predicted model output, defining the error signal $e(t) = y_p(t) - y_m(t)$ at the current time. The error $e(k)$, $k \in [t+1, t+N_p]$ is assumed to be constant over the prediction horizon and equal to the measurement $e(t)$. This error signal is updated at each measurement. Since the reference y_{ref} is assumed to be known over the whole working horizon, the desired trajectory can be computed:

$$y_d(k) = y_{ref}(k) - e(k), \ k \in [t+1, t+N_p]. \tag{15.9}$$

The cost function can be written in discrete time as

$$J(u) = \sum_{k=t+1}^{t+N_p} e_{tra}(k)^T Q(k) e_{tra}(k) + \Delta u(k-1)^T R \, \Delta u(k-1) \tag{15.10}$$

where Q and R are symmetric definite positive matrices. We also have $\Delta u(k-1) = u(k-1) - u(k-2)$.

Remark: the term $\Delta u(k-1)^T R \, \Delta u(k-1)$ *allows providing a smooth variation of the control variables by penalizing sudden changes.*

15.4.4   The optimization method

The cost function J is to be minimized with respect to a future sequence of control inputs, denoted by $\widetilde{u}$.

The mathematical formulation of the MPC is then given by the following optimization problem:

$$\min_{\widetilde{u}} \; J(u). \tag{15.11}$$

$\widetilde{u} = \{u(t), u(t+1), ..., u(t+N_c), ..., u(t+N_p-1)\}$ is composed of N_c different controls ($N_c < N_p$), with N_c is the control horizon. From $u(t+N_c+1)$ to $u(t+N_p-1)$, the inputs are constant and equal to $u(t+N_c)$. Although the prediction and optimization steps are performed over the prediction horizon, only the value of the input for the current time $u(t)$ is used and applied to the process.

Given its formulation as an optimization problem, MPC is well suited to take into account constraints. In practice, processes are generally subject to constraints on their states, inputs, or outputs, which can easily and explicitly be added to the optimization problem (15.11). It is the most effective way to satisfy all kinds of constraints. Many constrained optimization routines are available in software libraries to solve constrained optimization problems.

15.5   Simulations

All the presented simulations are performed with MATLAB® software. The constrained optimization problem is solved by using the MATLAB® function *fmincon*. This chapter only reports the results of the employability forecast at the level $j = 5$ (i.e. master's degree).

15.5.1   Data and modeling

The internal model uses data from a student tracking survey collected by the Students' Life Observatory (OVE), which describes the transition trajectories of master's degrees three years after graduation, by degree field; university administrative records of the number of intakes and graduates, per year; and complementary data from Varet (2008).

The data collected are represented in Figure 15.6. The red line represents the measured employability. The blue line in the upper graph is the reference employability and the blue line in the lower graph is the observed number of enrolled students. As explained in section 15.4.1, the reference trajectory is calculated using INSEE and E3ME data. The inputs (i.e. number of students enrolled in ES training) are obtained from the OVE data and the administrative data. Thanks to an identification procedure, we obtained a model of employability that simulates the process with a relative error of only 9.33% (see Figure 15.7). The non-linearity of the trajectory of the measured employability reflects the significant increase in the number of employed ES graduates with a master's degree between 2008 and 2009, caused by the Bologna reform of 2004. As explained in section 15.4.1, that reform harmonized the university degrees across Europe into a three-level structure (bachelor-master-PhD), which opened further job perspectives for master's students that earlier had tended to suffer labour competition from four-year graduates.

Over 2004–2012, the three main sectors of activities employing ES master's graduates were education, professional services, and public administration and defence. While ES graduates used to have a relatively high probability of being employed in the sector of mechanical engineering before 2008, with the crisis, their chances of employment have decreased in that sector. Instead, we observe an increase in the number of ES graduates employed in the service sector, largely correlated to a global inflation in consultancy jobs and fixed-term contracts since 2008 (Cutuli and Guetto 2013, Oliver 2012, O'Connor 2013).

15.5.2   Predictive control

The econometric model-based predictive control described in section 15.4 has been implemented. We consider the econometric model identified below. The simulation was performed under the following conditions: $N_p = 5$, $N_c = 4$, $R = [10; 0.1]$ and $Q(k) = Q(1)^k$ with $Q(1) = 2$. The future tracking errors are weighted more and more highly, in order to give importance to the final objective, i.e. the desired employability at the end of the prediction horizon, and this, at each sampling time. The reference employability is still obtained applying the model described in section 15.2.3, using INSEE and E3ME data. Several simulations were carried out, according to different horizons of control and prediction. The prediction horizon $N_p = 5$ seems to be the best compromise between the tracking accuracy, the intrinsic dynamics (the current control will affect the employability of master's degree graduates in at least five years) and the stability of the controlled system.

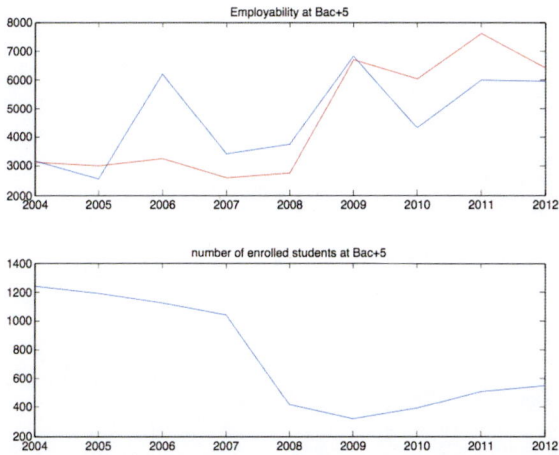

Figure 15.6: Data collected: reference employability (in blue) and measured employability (in red) in ES at master's degree level (upper graph); number of enrolled students (lower graph).

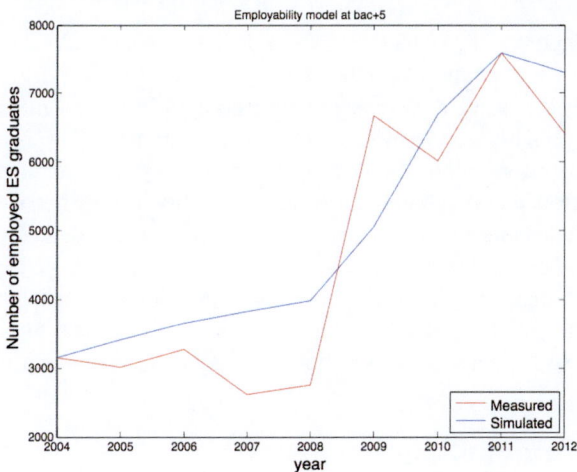

Figure 15.7: Simulated model of the employability in ES at master's degree level.

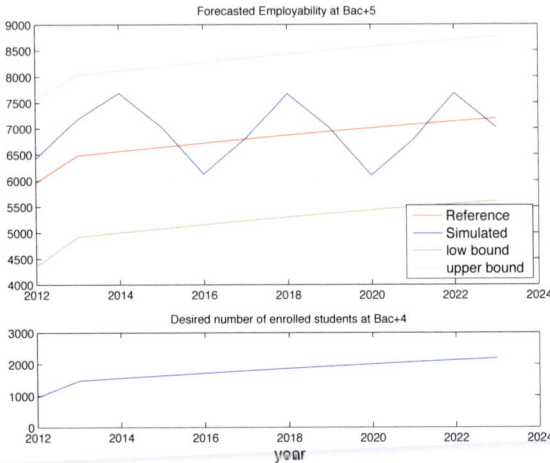

Figure 15.8: Forecast employability in ES at master's degree level: 2012–2023.

From Figure 15.8, we can see that the process output tracks the reference trajectory by remaining within the range of uncertainty.

On the other hand, the number of students enrolled seems more relevant and gives valuable information to policy makers to anticipate universities' future adjustments in human and physical resources. If more students' intakes are needed to reach the desired employability levels, then universities will need to ensure that they have the capacity to welcome and train efficiently this large influx of students.

Overall, these simulations show that the application of such a predictive control strategy to economic issues is feasible and can lead to potentially useful results from a social, educational and economic point of view.

15.6 Conclusions

This paper presents an innovative tool to predict the employability of French graduates in Earth Sciences. Rather than formulating the control objective as a classic computational general equilibrium (CGE) problem, it has been formulated as an optimization problem to take into account the complexity of the system it embeds. For the first time, Model Predictive Control was combined with an econometric model of employability. The MPC enables taking into account disturbances and modeling errors through an internal model control structure, which complements efficiently the error correction model (ECM) implemented in the econometric model used to generate the reference trajectory, at the sectoral and occupational levels.

The potential political implications of the CIPEGE tool are four-fold. First, the disaggregated reference trajectory of the CIPEGE tool can inform about the future labour market's demand by sector of activity, by type of occupation, and by level of degree. Although the detailed results of the model developed to generate the reference trajectory were not presented in this chapter, here are some examples of the descriptive capacity of the CIPEGE tool. At the horizon 2020, the CIPEGE tool anticipates a drop in the demand for higher education skills in the coal industry and in the oil and gas sector. In the case of the basic metal industry, it anticipates instead a smooth increase in the demand for higher education skills. Such a detailed mapping of the demand for skills is strategic for policy makers to anticipate the needs in terms of the provision of education. The second potential political implication of the CIPEGE tool is the capacity of its reference trajectory to capture the impact of different shocks at the energetic and environmental levels on the demand for skills. Third, combining the econometrics approach and the MPC yields a predictive tool where the control inputs are held exogenous to the optimization process, which makes it possible to test for an unlimited range of possible interventions (e.g., intake rates, specialization modules, international experience). The calculated control inputs can then serve as potential action-tools for policy makers. Finally, because this approach is very flexible, it can easily be adapted to other disciplines (e.g., chemistry, medicine) as well as to other countries.

Hence, the statistical capacity (in terms of error control), the economic relevance (in terms of macroeconomic scope and the exogenous nature of the control inputs) and the unlimited potentialities for expanding the application of the CIPEGE tool, makes it an attractive and valuable decision tool for universities and policy makers.

As with any approach of predictive control, the model is the cornerstone of the strategy and needs to be clearly identified from consistent data. At this early stage of the project, the statistical robustness of the results obtained from different simulations is very encouraging. However, given the lack of points of observation at hand, any economic and political interpretation of these early results would be irrelevant and adventurous. Additional data will not only improve the model and the tracking accuracy of the reference employability, but also provide sufficient information to draw pertinent economic conclusions and political recommendations. So far, what comes out of our simulations is the key role that the attractiveness of the ES programs seems to play to ensure sufficient enrollment rates and adequate completion rates. One possible option could be an increase in the investments devoted to the communication of the CIPEGE tool results to the main stakeholders (i.e., Higher Education Ministry, universities, students and industries).

Acknowledgements

The authors would like to thank E. Vergès and Y. Touré, initiators of the project, for having given them the opportunity to work on this challenging project, and M. Fodha for fruitful discussions. This work was supported by the "Investissements d'Avenir" initiative of the French Ministry of Higher Education and Research, through the project LABEX VOLTAIRE (Laboratoire d'Excellence sur les VOLatils, Terre, Atmosphère et Interactions - Ressources et Environnement) (ANR-10-LABX-100-01).

Bibliography

Alessio A., and A. Bemporad, 2009, Nonlinear Model Predictive Control. *Lecture Notes in Control and Information Sciences*, 384.

Arjona Perez, E., C. Garrouste, and K. Kozovska, 2010, Towards a benchmark on the contribution of education and training to employability: A discussion note. *JRC Scientific and Technical Reports*, EUR24147EN.

Boateng, S. K., C. Garrouste, and S. Jouhette, 2011, "Measuring Transition from School to Work in the EU: Role of the data source". Talk presented at the conference *Catch the Train: Skills, Education and Jobs, Brussels, June 2011*, http://crell.jrc.ec.europa.eu/download/Boateng1.pdf.

Camacho E., and C. Bordons, 2007, "Nonlinear model predictive control: An introductory review". *Assessment and Future Directions of Nonlinear Model Predictive Control*, Vol. 358 of Lecture Notes in Control and Information Sciences. Berlin: Springer-Verlag, pp. 1–16.

Cambridge Econometrics, 2011, *Studies on Sustainability Issues: Green Jobs, Trade and Labour*. Cambridge, UK: Cambridge Econometrics.

CEDEFOP, 2008, *Terminology of European Education and Training Policy*. Luxembourg: European Union Publications Office.

Cole H., and R. Rogerson, 1996, "Can the Mortensen–Pissarides Matching Model Match the Business Cycle Facts?" U. of Minnesota Working Paper.

Cutuli G., and R. Guetto, 2013, "Fixed-Term Contracts, Economic Conjuncture, and Training Opportunities: A Comparative Analysis Across European Labour Markets". *European Sociological Review* **29**(3), 616–629.

Davis, S. J., J. Haltiwanger, and S. Schuh, 1996, *Job Creation and Destruction*. Cambridge, Mass.: MIT Press.

De Grip, A., J. van Loo, and J. Sanders, 2004, "The Industry Employability Index: Taking Account of Supply and Demand Characteristics". *International Labour Review* **143**(3), 211–233.

Gazier, B., 1998, *Employability: Concepts and Policies*. Brussels, Belgium: European Commission, European Employment Observatory Research Network.

GHK Consulting, 2010, *Skills for green jobs ‚Äì European synthesis report*. Thessaloniki, Greece: Cedefeop.

Harvey, L., V. Geall, and S. Moon, 1998, *Work Experience: Expanding Opportunities for Undergraduates*. Birmingham, England: Centre for Research into Quality.

Hillage, J., and E. Pollard, 1998, *Employability: Developing a Framework for Policy Analysis*. Department for Education and Employment, Research Report No. RR85.

McFadden, D., 1974, "Conditional logit analysis of qualitative choice behavior". In P. Zarembka (ed.), *Frontiers in Econometrics*. New York: Academic Press, pp. 105–142.

McQuaid, R., and C. Lindsay, 2005, "The Concept of Employability". *Urban Studies* **42**(2), 197–219.

Mortensen, D. T., 1994, "The Cyclical Behavior of Job and Worker Flows". *Journal of Economic Dynamics and Control* **18**, 1121–1142.

Mortensen, D. T., and C. Pissarides, 1994, "Job creation and job destruction in the theory of unemployment". *Review of Economic Studies* **61**(3), 397–415.

Oliver, E. A., 2012, "Living flexibly? How Europe's science researchers manage mobility, fixed-term employment and life outside work". *The International Journal of Human Resource Management* **23**(18), 3856–3871.

O'Connor, J. S., 2013, "Non-Standard Employment and European Union Employment Regulation". *Non-Standard Employment in Europe*, vol. 46.

Petrongolo, B., and C. Pissarides, 2001, "Looking into the black box: A survey of the matching function". *Journal of Economic Literature* **39**(2), 390–431.

Pissarides, C., and D. Mortensen, 1999, "New Developments in Models of Search in the Labour Market". *CEPR Discussion Paper*, No. 2053 (January).

Qin, S. J., and T. A. Badgwell, 2003, "A survey of industrial model predictive control technology". *Control Engineering Practice* **11**, 733–764.

Varet, J., 2008, *Prospective de l'emploi dans le domaine des géosciences à l'horizon 2020.* Orléans, France: BRGM.

Wilson, R. A., 2010, *Skills supply and demand in Europe: Medium-term forecast up to 2020.* Thessaloniki, Greece: Cedefop.

Chapter 16

Language, interaction, information and the modelling of market interactions

François Nemo[1]

16.1 Introduction

From ethology, the science of animal and human behavior (and a branch of biology) to pragmatics, a science which studies the constraints that govern verbal interactions (and a branch of linguistics) to the sociologies of inter-actions, half a century of empirical scrutiny and theoretical studies of inter-actions will be the background of what follows, even if only some aspects of its main outcomes will be considered.

Similarly, from Cournot's first mathematical theory of price formation to general equilibrium and game theory, the study of market interactions within economics also has a long history, with the specificity of not embracing directly an 'interactionist' – and hence neither individualistic nor holistic – stance, as is always the case in the above-mentioned disciplines.

[1]LLL, UMR 7270, Université d'Orléans. Email: francois.nemo@univ-orleans.fr.

Given this double background, our ultimate goal will be to consider and discuss three issues.

The first issue concerns the way the modelling of the complexity of market interactions could be transformed by taking into account some of the parameters which have proved, within interactionist frameworks, to be crucial to a correct understanding of the relation between language, communication and information. We shall in particular discuss the way the performative, attentional and sequential dimensions of interactions may prove useful or necessary when it comes to the description of market interactions.

The second issue concerns what is arguably the most original aspect of the way interactions have been approached within economic questioning – namely by also questioning the "interactions between interactions" – could the other way round transform the way "individual/local"[2] interactions are to be described, notably by forcing the analyst to clarify the way such inter dependencies could be modelled.

Our last issue will be to discuss somehow programmatically the way it could become possible to build models of market interactions which would both fully admit the ultimate autonomy of micro–micro interactions and acknowledge the nonetheless real interdependency between them.

As for the first objective, we shall first recall the reasons which have led to a profound transformation of the study of language and languages after the late 1960s, when linguists started to take seriously into account the enunciative[3] and interactional dimension of language and specifically the fact that there is nothing in languages that was not primarily in an interlocutive and interactive setting and thus that has not been shaped by interactional constraints. We shall then question the heuristic value of what has been learnt in this process and wonder if the discovery of the strongly interactional nature of utterances could prove insightful for the study of market interactions. Specifically, due to the limitation of space here, we shall limit ourselves to the discussion of three of the features and constraints which could arguably be considered as both inherent to all human interactions and important in the study of market interactions, namely their performative, attentional and sequential dimensions.

We shall also ask ourselves whether any global or local representation

---

[2] Interactionist approaches tend to consider individual interactions as the fundamental ground of analytical work.

[3] The concept of *énonciation* describes the fact of something's being said (uttered). It has proven to be the case (see Nemo, 1999) that because of the existence of pragmatic constraints on the fact of saying something, the interpretation of the utterance is deeply altered by its enunciation. It also describes the fact that this dimension of language actually shapes all linguistic realities.

of market interactions may ignore or neglect these dimensions, bearing in mind that when it comes to the role of language in the description of market interactions, it is the case that one must consider:

i. the use of language (including in computerized and algorithmized form) in the very process of market price setting;

ii. the fact that prices themselves are linguistic utterances (and trading utterances);

iii. the use of language (including in mathematical form) in the model itself;

iv. the fact that the interpretation associated to these models is itself language based;

v. the fact that all those layers are actually interdependent, which makes it impossible to make a specific assumption concerning one layer, without making implicit assumptions about other layers.

Our answer to this question will be to claim that the relationship between language and what is commonly called information appears to be such that the notion of information just cannot be defined and modelled outside and independently of language interactions. And also claim that this is an opportunity when it comes to the modelling of market interactions, because it allows and forces the integration of parameters which cannot reasonably be overlooked.

As for clarifying what point iv) concretely means, we can provide a clear-cut example of this interpretational layer in the interpretation of the supply curve in partial equilibrium models, which nowadays is routinely interpreted and taught as the mere aggregation of the suppliers' answers to the question "which quantity would you offer at such or such price?", whereas in Marshall's work it was actually obtained by asking a very different question, namely "what price would be necessary for you to offer such or such quantity?". Meanwhile, it must further be noticed that those interpretations would be either inconsistent or contradicting if the interpretation of the question itself was not actually "which quantity would you offer at such or such price, **if you were sure to sell this quantity?**"[4] and "what price would be necessary for you to offer such or such quantity, **if you were sure to sell this quantity?**". Otherwise, in both cases, it could be predicted that the only

---

[4]The very same presupposition is also present in the interpretation of the Marshallian variant of the question.

consistent answers to the question would and could only be "it depends on how much I can sell at this price". As a consequence, it can be proven to be the case that *the curves of supply and demand cannot in reality co-exist* and be presented together in a single model, and that they must be considered as "incompossible", to use a logical terminology. One would indeed have to presuppose[5] that all participants, when they answer both questions will assume that no price will ever be adopted which doesn't allow supply to be met by the corresponding demand. But in that case, it would both be the case that the existence of a market equilibrium would be **a premise** of the model and not **its conclusion**, and that the mere existence of the demand curve would imply that the participants assumption is ill-founded and based on incorrect and poor information.

16.2   Performativity, price modelling, and market interactions

Starting with performativity, which concerns what is called the direction of fit between the words and the world and which opposes constative utterances[6] (in which the world precedes the words, which must adjust to it[7]) and performative ones (in which words come first and it is the world which has adjust to them[8]), it may be recalled (Nemo, 1995) that while there is no doubt whatsoever that in reality:

- prices are utterances;

---

[5] At all steps of the construction of the partial equilibrium model, the same assumption is made, for instance when it comes to quantify the expected income of an individual supplier, the parameter used (i.e. a "price × quantity" multiplication) makes sense only if it is assumed that the quantity at stake will be sold, otherwise the information provided by the demand curve can only be described as illusory or irrational.

[6] The distinction between constative and performative utterances in pragmatics is a legacy of John Austin's book *How To Do Things with Words* (1962). It opposes utterances which simply describe something which exist prior to the utterance itself, and utterances with a potential to create the reality which they are talking about. A single sentence such as «the session is opened» may thus be used as a constative utterance, for instance when uttered by a journalist who is just reporting the fact that the session is open, and as a performative utterance when uttered by the chairman who is, literally, opening the session by declaring that it is open.

[7] Constative utterances are thus either true or false depending on whether what is said is consistent with a pre-existing world.

[8] Performative utterances are thus said (Austin, *How To Do Things with Words* 1962) to be either *happy* or *unhappy* depending on whether what is said becomes a reality following being said, or not. While constative utterances have as mentioned truth-conditional values and are true or false, performative utterances are described in terms of *conditions of felicity* and are said to be either *felicitous*/(happy) or *unfelicitous*/(unhappy).

- prices are performative utterances and not constative ones. This stringently forbids considering them as true or false in the logical sense of the term;

- prices do not describe a pre-existing and non interactional object that could be called "value" or anything that could be measured;

- a price is actually something "to take or leave", and thus is a reality only insofar as it is accepted by someone;

- prices are therefore not entities which could be concerned or associated with the logical principle of non-contradiction;

the main models used to account for price formation routinely take for granted the opposite assumptions, namely:

- the idea that prices are propositions (in the logical sense of the term);

- the idea that market prices are not only propositions but true propositions;

- the idea that market prices form a system (which can be modelled as a system of equations) in which they have to satisfy the logical principle of non-contradiction, with the extraordinary outcome of allowing calling "general equilibrium" something which is actually nothing else than the logical principle of non-contradiction, and to call "price-taking" that which is nothing else than the negation of performativity.

This negation of performativity has moreover been extensively axiomatized over time by a set of assumptions and prohibitions which make it to some degree unthinkable.

The most radical form of the eradication of the performative dimensions of prices appears to be the possibly involuntary consequence of the formulation of the question which has shaped all price theory, namely Cournot's foundational questioning of the effect of the number of suppliers on the setting of price, and the way it was answered, namely by considering successively a market with one supplier (monopoly), with full performative capacity to set the price, a market with two or a small number suppliers (duopoly or oligopoly), with a relative capacity to influence the price etc., and finally the so-called perfectly competitive market with plenty of small suppliers unable to have any influence on price-fixation, which are considered as price-takers and are thus devoid of any performative price-setting capacity.

It appears, however, (Nemo, 1994; 1995) that Cournot's questioning and line of reasoning is flawed by the fact of being based on the tacit (and uncontrolled) presupposition that the number of suppliers in a market may be considered constant, when it cannot. This presupposition requires, in order to be acceptable, nothing less than to assume the price level to be completely unable to affect or modify the number of suppliers present on the market. It turns out that because the opposite is true in a true (Darwinian) market, the number of suppliers which are able to stay in the market decreases steadily with the price level, such traditionally unthinkable realities as a perfectly competitive monopoly may occur, in which at a (low) price, only one supplier will be present in the market, while at higher prices, more suppliers would be present. It also turns out that performativity is essential for the description of such truly competitive markets, and that performativity cannot be reduced to the desire to fix high prices but must actually be considered as a means (and weapon) to exclude competitors from the market. Cournot's approach and all its legacy in mainstream economics may thus be characterized as leading to ignoring what is possibly the core nature of market competitiveness: the capacity to remain in a market when prices are going down. As a consequence, "price-taking" models (and so-called perfect competition) are associated with a non-Darwinian notion of competition which actually excludes any competition of any kind between competitors.

A second way of ignoring performativity and pragmatics altogether concerns the status of the information/declaration provided by market participants, about which it is either simply assumed that interactants will never lie when asked to provide information about their market intentions, even if lying would in fact be arguably profitable, or simply assumed that they will be forced to "stick to their words", without considering that it is not possible to call *information* what is in fact a performative *commitment* (and thus nothing less than a verbal contract).

An even more problematic aspect of General Equilibrium theory is to actually prohibit any two individuals (or groups)[9] to reach any direct performative agreement outside of a general equilibrium, a prohibition that is the consequence of the model's incapacity to recognize any autonomy and self sufficiency in the local interactions.

A major issue and challenge in the modelling of market interactions is thus to fully admit and integrate the performative dimension of prices as an indisputable and important reality, without being led to consider local interactions as independent one of another, since they are not.

---

[9] For a postulation of the opposite, see the "islands" of Phelps (1969).

16.3   Attentional constraints on interaction

Together with ethologists, psychologists and pragmatists, we may first admit that any human interaction requires attention from the people involved, and that attention at an individual level is a scarce resource which is limited both quantitatively and qualitatively, humans being unable to pay attention to many things at the same time, let alone to everything.

We may further admit it to be true that all humans need attention from other humans, and that this need to get and receive attention from others, and to have their existence and value recognized, is a reality which is at the heart of the human condition[10] (Todorov, 1995) to the point of contradicting and countering selfishness[11].

Once combined, it follows firstly from these two sets of assumptions that since all human life involves an allocation of scarce attentional time, not only between production, consumption, and trade, but also between economic activity and non-economic activity, any postulation of unlimited attention and overestimation of the (attentional) capacity to process information in a given time should be banned in the modelling of interactions. They also imply that it is not possible to describe human behavior in purely individualistic terms, let alone to equate rationality with the maximization of individual interest, once the fact is acknowledged that the satisfaction something represents for someone will be ultimately dependent (and this to a large extent) on its attentional value at a collective/social level in human interactions. So that as long as individuals are sensitive to the way they and their choices are considered, choices cannot be solipsistic.

Miles away from such an attentional understanding of the nature of information and interactional information, it is important to realize that what has been called "perfect information" in economic theory may (and in fact must) be considered as an "absolute informational chaos" once attentional constraints are taken into account. Because of the limited ability of individual information processing due to the limited character of attention[12], because of the fact that being bombarded with irrelevant information will

---

[10] And is, as mentioned by Todorov, virtually insatiable.

[11] More precisely, it can be shown that many if not most of the behaviors associated with so-called selfishness may be proven to ultimately be aimed at social recognition and competition for recognition. The same could be proven also and to the same extent for altruistic behavior.

[12] The idea associated with so-called perfect information being that everyone should be informed of everything is in this respect both quantitatively impossible in purely cognitive terms and irrational since most of this information will be devoid of any interest for the receivers. The scarcity of attentional resources forces rational information to be always goal-oriented, which makes the knowledge of the goals themselves a crucial form of information.

automatically lead to the non-treatment of the information in question, and to the suspension of interaction, as well as because the circulation of information cannot but be limited to *relevant* information and be dependent on the pragmatic constraints on relevance.

As importantly, it may be said that any modelling of interactions which postulates of a population of n individuals that everyone is related to everyone, and that everyone is aware of everything, does not describe either a possible world or an ideal world, but on the contrary, an impossible world and an absolute chaos. "Islands", if they are to be modelled at all, should not be conceived as imperfect fragment or a larger reality but as reasonable aggregates in the first place.

This is, firstly, because it implies that because the mere existence of any offer – be it to sell/supply or to buy/demand – requires an allocation of attention by a hearer, struggling for such attention is no less important in the economy, to use an euphemism, than it is in other areas;

Secondly and most importantly, this is because the organization of mercantile exchanges in markets, trade fairs, shows, etc., is already based on this logic of visibility and minimization of attentional costs.

And finally, this is because efficient communication and information requires attentional coordination, which, when it concerns large groups of individuals, may emerge only in collectively organized forms.

Thus, this implies that each interactant must constantly adjust to his interlocutors and select the relevant interlocutors, and that it is not possible to consider interaction and information as separate issues, the existence and maintenance of interaction being dependent on attention and relevance and the mere existence of information presupposing interaction.[13]

This also means that capacities to win the battle for attention and relevance, such as empathy and mind-reading, are crucial properties of individuals, as are the capacity to know, to guess, to influence, and to manipulate the attention of others and, through doing this, their choices. This has the result that choices, ultimately, form an attentional system, whose interactional and social nature cannot be ignored or minimized: people, for instance, not only buy goods for what they are, but to a considerable extent through the way these goods are seen and valued by others. And in such a context, postulating only psychologically solipsistic agents with the sole capacity of maximizing solipsistic wellness appears thus arguably as disputable as ignoring the interdependency of markets.

---

[13] In a certain sense, one cannot separate information from its interactional attentional dimension any more than it would be possible to separate the genetic code from its DNA materiality. Extra-interactional information simply does not exist.

Attentional constraints thus are a parameter which cannot be ignored in the description of market interactions, be it when they are considered as a whole, because they quantitatively limit the global flow of information, or at the level of each individual, because they considerably constrain his/her economic and non-economic behavior.

16.4   Sequentiality and market interactions

Another fundamental characteristics of any interaction from a language–oriented or social perspective is sequentiality, i.e. not only the fact that everybody's actions are also reactions to previous actions – and also reactions to reactions, etc. – but also the fact that they are ordered chronologically and that such scheduling is indeed crucial to the analysis of market interactions and its outcome.

Considering the apparently trivial fact that a mercantile interaction can begin with either an offer or a demand, both of which may be associated with a price or not, and that this first step either leads to an effective exchange or doesn't, not to mention the possible bargaining processes in between, it is worth wondering whether this sequentiality may be completely ignored in the analysis of what is going on – for instance, by trying to model the meeting of supply and demand as a non-sequential event in which two "forces" would meet exactly as do simultaneous physical forces (Mirowski, 1989) – or whether we should instead take seriously the time gap between supply and demand (or vice versa).

Given that there are numerous reasons to adopt this last view, we shall only be able here to spell out some of them.

What has to be remembered first is the fact that for an individual with no possible knowledge of all current interactions, *the concrete form taken by the reactions to their offers or demands is the primary and more reliable source of information available, and that this reaction is to a large extent a function of what in pragmatics is called common ground, in other words, any form of shared knowledge about previous market exchanges.*

An illustration of the first aspect of the question is that if the performative nature of prices implies accepting the idea that exchange-moves (offers or demands) have to be ratified, it is both the case that the sequential nature of ratification allows it to convey crucial information, and that *describing verbal exchanges that do not result in a commercial exchange is at least as important for the description of market economic interactions as describing those that do.*

An illustration of the second aspect of this question is that while the gap between offer and demand and vice-versa is typically associated with both considering previous and alternative offers, so that for instance the more important the transaction the bigger the time gap between a proposal and its acceptance will typically be, information is time and time is money and this in both temporal directions[14];

As for the markets whose appearance is one of absolute simultaneity (and thus should be considered as counter-examples to the importance of sequentiality), it can be shown that computerized markets, for example, far from making a case for the possibility to theoretically ignore sequentiality, are on the contrary the clearest proof ot the opposite, with huge profits relying on the capacity to have a quicker reaction than other interactants, to the point of replacing humans in the decision process by algorithms, to gain an edge in the sequential competition.

As for the interlocutive and linguistic aspects of economic interactions, it must be emphasized that the actual diversity of the communicative form they may take (e.g. ad, poster, price tag, price list, catalog, price quotation, individual bargaining, auction, auctioneer, organized market, computerized market, etc.) is embedded in the organization of sequentiality, typically by institutionalizing a time frame to regulate the individual interventions. All of which need not be considered as irrelevant or negligible when it comes to accounting for price formation or for the parameterization of interactions.

This is, firstly, because of the fact that these communicative forms not only co-exist and compete one with another, but are actually quite often deeply dependent on each other, so that the idea, for instance, that market theory should consider such and such interactional and linguistic form of market exchanges (for instance those that rule the wheat market in Chicago) as in any sense prototypical or ideal-typical of a market, is empirically falsified by the fact that a market such as the wheat market is only the visible part of an interactional iceberg in which market interactions take the most varied linguistic forms. Similarly, because the very idea of a prototypical or ideal-typical market is associated with the idea that some markets would be a better incarnation/exemplar of markets in general than others, it must be abandoned. Economics may study the competition between the various interactional forms of market interactions, but it cannot axiomatize such

---

[14] To give but one ordinary example, one may consider this excerpt from a *New York Times* article entitled "In a Seller's Market, Every Minute Counts" (Higgins, 2013): "The rules of engagement for buying an apartment in the city have changed. Negotiation, brokers say, is no longer part of the equation. Forget about taking time to mull over your decision. Serious buyers need to be prepared to pounce."

and such an institutionalized form as the best or standard form of the market, for markets have existed for thousands of years under the most varied forms, and, more importantly, are strongly and structurally dependent on each other: the Chicago wheat market cannot be opposed to retail trade, for instance, whose linguistic/informational and interactional organization is radically different, because the two interactional frames are in fact elements of a single continuum.

Secondly, this is because this diversity should not be considered as an obstacle to the description of market interactions in a unified way, since it is both possible, once our three parameters are taken into account, to consider them in a contrasting fashion, and to spell out invariants.

Thirdly, this is because there is nothing odd about the constant coexistence of new and old forms and patterns of interaction, ranging from the most technologically advanced ones such as algorithmized markets or politically motivated ones to immemorial forms such as face to face bargaining, being constantly created and institutionalized.

Finally, as mentioned earlier, it is also the case that because among other things individual interactions have their own time frame[15], they are necessarily related to a memory of previous interactions (and of their outcomes) which is called "common ground" and which is for instance necessary in order to be able to answer such basic questions as "which quantity would you buy at price p?". Pretending to ignore this reality, for example by affecting to ignore that the information provided by the demand curve, and which serves in the model as a premise for a representation of exchanges which claims to describe the interdependency in terms of absolute simultaneity, cannot be maintained[16] without a knowledge of *previous* prices[17], is not an option but a fraud. Meanwhile, taking on this reality, and acknowledging the fact that it is inherent in the informational reality whose modelling is at stake, could allow understanding that individuals may only position themselves in rela-

---

[15] Which to some degree may become institutionalized in collectively accepted forms. The term *market* itself originally described a traditional form of collective coordination and in that sense, instead of presupposing the existence of a market, or instead of presupposing the existence of thousands of markets, it is important to understand that what is labeled *market* is nothing else than a form of collective organization of economic interactions shaped by the scarcity of attentional constraints and the necessity to enhance all the participants' visibility and accessibility. In a certain sense, such and such a market was initially little else than a meeting.

[16] Without the existence of this common ground, the only answers one could receive to a question such as "which quantity would you buy for price p?" would be questions such as "what can you buy for price p?", etc.

[17] It is worth noting that one cannot ask someone to forget this knowledge in his/her answers, with the consequence that no common-ground free information can ever be produced.

tion with a system of reference (and by assuming in their answers that this system of reference is shared by their interlocutors) and that whatever they say can only be interpreted in relation to this assumption; information is complex by nature and embedded in a common ground. Whisking away the premises of a model is not a scientific option, nor is the modelling of these informational premises possible by simply relying on the intuitive interpretation of question/answer pairs of ordinary people without explicating its content.

16.5   Interactional autonomy and interdependencies in the modelling of market exchanges

So far we have shown that when dealing with interactions, it is not possible to merely "postulate" and presuppose their existence without taking into account the constraints and parameters which define them and allow them to exist. We have also shown that because any modelling is build on a linguistic frame[18], it inevitably has its blind spots that can literally render "unthinkable" many realities whose existence is beyond doubt[19]. It is thus time now both to introduce a more positive perspective and to show that widening our understanding of what interactions are, and notably acknowledging both the strongly interactional and interlocutive nature of interactions and its linguistic dimension, may be done without renouncing the modelling of the interdependencies between interactions, which is arguably the most interesting contribution of economics to their analysis.

Decades after the most explicit discussion (Kaldor, 1972) of the contradiction between Equilibrium theory's claim to be a model of a decentralized market and its formal reality as an extremely centralized market

---

[18] The formal dimension of models doesn't alter this reality, both because each of their elements has its own linguistic interpretation and because models are framed by the issue they try to address. Meanwhile, because this issue is a question, it is necessarily linguistic, and dependent on non-explicit assumptions (presuppositions), as we could see with Cournot's assumption that the number of suppliers in a market may be considered constant and independent of the price level, and as can also be illustrated by the partial equilibrium interpretations of the supply curve which were described above. It is indeed indisputable both that formalization cannot dissolve presuppositional issues (first identified by Frege himself) and that scientific utterances, because they are answers to questions, inherit most of their semantic frames from the linguistic framing of the questions.

[19] In that sense, it must be understood that models cannot be considered as mere simplifications or idealizations of reality, for they are indeed semantic filters which determine what is worth considering or not, and in that sense may actually be "wrong" when they ignore dimensions of reality which cannot be overlooked.

in which interdependencies are strongly and, as far as information is concerned, counter-factually overstated, we are still looking for a model that would be able to combine relative autonomy and relative interdependence, and we know from experience that attempting to approach this issue by trying to introduce some autonomy in equilibrium interdependence, as was repeatedly attempted in the last 45 years by somehow trying to disaggregate general equilibrium, and to avoid a Walrasian "commissaire-priseur"[20] are bound to fail if an interaction-based conception of interdependency is not spelled out.

When it comes to the co-existence of interactional autonomy and the interdependence of the interactions, it is indeed important to realize that the modelling of interdependency should require neither a centralized market nor the so-called perfect information (and its pre-condition, the absolute interdependency of all interactions one with another). But once there are acknowledged the facts i) that attention is scarce and thus that the number of interactions is necessarily and strongly limited[21]; ii) that selling or buying is a decision and that as a performative act it must therefore be considered as axiomatically free and unpredictable; iii) information is accessible only through interactions and thus limited and enhanced by the individual capacity to enter into interactions; it would be possible to completely reconsider the way we see and understand market interdependencies and market information, and to realize that in interactions, *autonomy and interdependency come as the two sides of a single coin,* interdependency being "embedded" in autonomy and vice versa[22].

To provide a concrete illustration of this reality, one may consider for instance the modelling of demand in an attentionally constrained market ultimately defined by performative autonomy. Departing from the idea that aggregated demands would meet aggregated supplies producing an equilibrium market price which the individual interactants would have to accept as a fatality they cannot escape, we may consider that at the level of the indi-

---

[20] In mainstream market theory, the fixing of a market price is described as a process involving a auctioneer whose role is to gather information for the determination of the equilibrium price. The term used is the French *commissaire-priseur*, in which *priseur* means litterally "pric-er" (which would be translated as "price-maker" in English).

[21] Even admitting a twitter-like connection between all interactants, in other words even assuming the possibility for a piece of information to be shared by all, it follows from the scarcity of attention that only a limited number of topics (and information) could be widely shared, and that perfect attention to one subject would necessarily mean imperfect or no attention to others. Modelling the allocation of attention is thus a pre-condition of any modelling of exchanges.

[22] Among other things, because performative prices are autonomous decisions which are interactionally constrained.

vidual interactant, demand[23] is basically nothing but the name of either the reaction which an interactant encounters when he/she proposes something, or his/her reaction to a proposition which another interactant has made. Consequently, describing economic encounters and understanding economic interdependencies imposes:

- considering the individual offer/demand exchange as the primary form and precondition of economic decisions in a merchant society;

- describing a decentralized and non-institutionalized market as ultimately based on a set of "individual demand curves" (IDCs from now on[24], representing all the credible responses to his/her offers that an individual may observe or expect, a curve of reaction) forming a set of coexisting but not aggregated[25] demand curves.

Further admitting that these curves have a temporal dimension, which is variable, and that they are the main and often only source of information available to the initiator of the interaction, we may define "perfect information" at the interactant level as the exact knowledge of the form of these curves and "imperfect information" at the interactant level as ignorance or uncertainty of the form of these curves.

Being aware of the fact that the outcomes of the trading process is determined equally by the individual demand curve and the "offerer"[26] constraints and reactions to this curve, it is important to understand that the reactions which an individual demand curve synthesizes are in fact reactions *not only to a specific offer but to all existing alternatives to that offer.*

It follows from this that no matter how autonomous an interaction later is when it comes to economic decisions and actual exchanges, it must be

---

[23] A point which must be clarified is that equality between demand and supply may both be said to be an analytic truth as a consequence of the fact that all effective exchange is simultaneously an offer and a demand, and to be problematic if at a given price not all demand or supply is satisfied.

[24] An IDC here is thus distinct from what is called an individual demand curve in neoclassical theory, the individual being the supplier/offerer and the IDC resulting from the aggregation of each demander's demand from him/her.

[25] In a decentralized market, and in a situation of strictly balanced competition, it may be assumed that individual IDCs for a commodity are, *at each price*, obtained by the *division* of the quantity which is demanded *globally* by the number of suppliers *which are present at that price*.

[26] "Offerer" here could not be replaced by "supplier" because this last word presupposes an exchange which is not certain at that stage and may never occur. Far from any false symmetry between the two sides of the market, it must be stressed that the ultimate decision always belong to the offerer, who even though he/she may be cornered and left with little choice, is the only one with the performative power of making the transfer happen, or of transferring this power to his/her interlocutor(s).

understood that the very form of a demand curve is **the trace of its current state of relatedness between the on-going interactions and all other interactions**, and that consequently "islands" are not truly "islands" because autonomy does not mean independence.

It follows from this that there is no such thing as a purely local interaction, once it is understood that each interaction is indeed connected to the global interactional web through the form of the IDC demand curve. It also follows that because each autonomous decision potentially affects the later form of possibly all other demand curves, be it marginally, we may further realize that the detailed/interactional form of the demand curves allows the offerer to gain direct information and also to make interpretational inferences, providing him/her in both cases with a precise mapping of the current situation, this being a consequence of the difference between "aggregative demand", which provides only access to the global quantity which is demanded at such or such price, and "interactional/detailed demand", which also provides more precise information about "who want what at what price" as we shall now see.

16.5.1   Interactional complexity of IDCs

The fact that what has been labeled here an "individual demand curve" is simultaneously deeply interactional is clearly its most interesting feature. It must be indeed stressed that the IDC is individual only from the perspective of an individual offerer F_1: it aggregates in fact a set of reactions for each distinct individual D_1, D_1, etc., who are in interaction[27] with F_1, and which we shall refer to as DDCs (demander demand curves) to avoid any confusion. Since these DDCs are interpersonal reaction functions, it must further be stressed, as mentioned, that the IDC provides F_1 with much more precise information than the purely quantitative figure that classical demand curves used to provide. For each price, S_1 will indeed know which quantity (s)he could sell if (s)he wanted to, but (s)he will also know individually each interlocutor's price sensitivity so that the information which is available is both much more precise and much more relevant than the aggregate information postulated in classical price theory.

As to this last point, it must be remarked that the problem of knowing which degree of sincerity can be attributed to each DDC is not a parameter which the model has to decide on, for it can vary. But what can be said about

---

[27] It must be stressed indeed that each IDC aggregates a variable number of interactors, and a variable set of interactors, due to the attentional constraint which strictly forbid all offerers to be in interaction with everyone.

it is that there are some interactional rules about it and they are interestingly asymmetrical: a basic fact about a DDC is that saying something such as "I would buy such quantity if it was offered at such price" is an utterance with a commissive value. This means, in other words, that it constitutes a commitment from the individual demander, and should be considered as a pre-contract[28]. This is not only because commitment to the truth of what is said is a general commitment in human conversation (e.g. Grice's Maxim of quality), but because it concerns the credibility and the reputation of the individual demander involved whose credibility is a social and interactional capital whose value must not be underestimated.

It does not, however, follow from this that what is said may be considered as a reliable and indisputable premise when it comes to exchange modelling, because the commissivity of the utterance is in fact strongly asymmetrical: the necessity for the demander involved to stick to his/her word in order to preserve his/her reputation being restricted to the necessity not to demand *less* than stated previously but does not forbid demanding *more* than previously stated. One may indeed pretend for instance that (s)he is not interested in something at such a price, and it may simply be false, the reality being that (s)he would still be interested at that price, as the future may later prove. This is consistent with the fact that bluffing may be an essential and normal part of what will then be called a bargaining or negotiation process, for instance by pretending that one does not want something at a certain price or that selling under a certain price would be inconceivable. To this extent, DDCs may have a manipulatory dimension which must not be ignored or outlawed. But the above-mentioned constraint of sticking to one's word does not apply to changes which are conceived as favorable to the other participant. So that finally accepting something which was initially refused will rather be considered as a defeat than in terms of a lack of reliability. Commitment thus will be restricted to forbidding any backtracking that would be disadvantageous to the receiver of the "comitmental" utterance. Another important dimension of DDCs is that what an individual demander will answer may prove to be, to some extent, dependent on the hypotheses (s)he will make about the global IDC of the supplier. So that on the one hand, it may often be the case that what will be answered will be based on self-conscious beliefs transforming it into a kind of bet that one may perfectly well lose[29].

---

[28] This is if the utterance is addressed to a single offerer. If addressed to various offerers, it becomes a call for a price proposal but remains a commitment, not to someone in particular but to anyone who will react to it.

[29] In that sense, it is the case on the one hand that equilibrium between expressed demand

16.5.2 From IDC as a reaction curve to OFC as a decision curve

But once all this background is acknowledged, the main issue will be to know how and to which extent an IDC will determine or influence what ultimately will take place in the exchange. Modeling this issue starts with the necessity of predicting which reaction the individual offerer will have to his/her IDC, but supposes admitting that this reaction will be a two-step reaction in which the ordering of constraints will be the opposite of the ordering which is postulated by the centralized approach to price fixation.

The information provided by his/her IDC allows each individual offerer to know for each price the quantity that (s)he *could* sell at that price. But since this quantity is not most of the time the quantity that (s)he wants or can provide at that price, and since *(s)he knows* which quantity (s)he could sell at each price if the corresponding demand existed, **our performative offerer is informationally in a position to merge information about demand and information about constraints on supply into a unique decision curve** in which each price is associated with a quantity (which measure the aggregative felicity of a price, i.e. the number of times the performative utterance will become true). As for the modelling of price fixation, the fact that each performative interactant is in terms of information perfectly able to adjust demand and supply at all prices *before* having to choose between them is a feature which radically transform the way price-fixation can be approached and modelled and implies the construction of what we propose to call an offerer felicity-curve (OFC from now on) from the IDC base, prior to the price fixation itself, which comes as the last step and is described in terms of making a choice between equally possible prices, all of whom being so to say "equilibrium"[30] prices.

---

and supply may perfectly coexist with unsatisfied (because unexpressed) demand and effective disequilibrium, and also the case that because bets are speculations, the idea that expressed demand should always be satisfied is not rational.

[30] The use of *equilibrium* here is similar to its use in game theory, where it is interpreted as "possible" and not as "necessary" (see Cadiot and Nemo, 1997), due to the fact that no assumption is made that there would be only one equilibrium price, and to the more positive observation that nothing could prohibit this price from becoming true if ever for any reason, it was chosen.

A felicity curve may be defined, constrained by the alignment/adjustment of possible supply on the one hand and IDC on the other hand. This curve aims at providing a realistic representation of the options in terms of exchange which are available to the offerer involved. It may further be defined by an alignment rule which states that it comprises:

for each price, the quantity that may be exchanged will be either the quantity which is demanded if it is superior or equal to what can be supplied, or in the maximal quantity which may be supplied if it is inferior to what is demanded. [31]

and this for all prices for which this quantity is not null.[32]

In the arguably standard situation in which the potential demand decreases with increasing prices and the potential supply increases with prices, there will further be a price which may be called an inflection price and may be characterized as the threshold between the supply constrained part of the felicity curve associated with lower prices and the demand constrained part of the OFC associated with higher prices, but this price of inflection will play no specific role in what ultimately will determine the final setting of the offerer's price, which will be a matter of choosing between all the *realistic* options represented by the felicity curve. For its role is only to provide a complete set of "price/quantity" pairs which all can be felicitous, even though not all prices will later prove to be equally advantageous for the offerer.

It follows from this that the freedom of the offerer to fix a price is limited to the felicity curve, so that (s)he is not a price-taker but a felicity-curve taker. An individual cannot define what is possible, but (s)he may define what takes place.

16.5.3   From decision-curve OFC to price fixation

The final step of the process, and of its modelling, concerns the choice which our offerer will make between all the equally realistic options which are available to him/her and which are synthesized in the OFC. In this respect,

---

[31] In classical terms, this means that the felicity curve represent the association of a price with the shorter side of the market. But it plays a completely different role, being a decision curve.

[32] This implies that the felicity curve starts at the minimal price for which something could be sold and ends at the maximal price at which something can be sold, thus representing at the level of the individual the possibilities of exchange (and hence his/her "market").

considering this process to be driven by maximization and optimality is a first approximation, which should not mask the fact that what exactly an optimal price stands for might prove to be quite variable, not least because of the variabilility of IDCs in the first place, the variability and dynamics of production capacities and also because maximization is strongly time-framed.

It is important nevertheless to stress that such an optimality model of decentralized performative deeply questions the relevance of transforming the constraints of adjusting supply with demand into a price-setting mechanism, for it assumes that maximization takes place first (being a central assumption for the modelling of supply) and *then* that the adjustment of supply and demand will decide on the price, whose role is thus to ensure this adjustment. Whereas in an economy in which individual performative price fixation is the rule, adjustment between supply and offer will (only) be a key constraint for the determination through the OFC of the exchange options, with maximization taking place only in the comparison between these options.

It must also be stressed that the ultimate right to fix one's own price cannot be denied to an offerer under any circumstances or assumptions, for two complementary reasons, neither of which can be ignored in any attempt to model price fixation. The first is that the term *price-taker* is actually used to denote a *market-price taker* and associated with a Cournotian line of reasoning according to which when offerers become too small they lose all capacity to influence the *market*-price, notably to set it at a higher price than this market-price. But in reality and axiomatically any individual will always maintain the right to propose any price for what he has to offer and in particular to propose a lower price than any existing market-price. So that no model can ever make the opposite claim, let alone transform price-taking into an axiom and overlook the complex assumptions which are implicitly associated with it.

Proving this is easy, for it is possible to describe within our three-step model of performative pricing (IDC, OFC, price fixation) the kind of situations which allegedly are associated with price-taking. The first feature of such a situation would require that in the IDC, demand would be non-existent for any price higher than the price offered by the market (or competitors). Another feature would then be that for lower prices[33], it would require them to be less interesting than the price offered by competitors. So that, ultimately, it may be shown on the one hand that "price-taking" could occur in many circumstances unrelated to the Cournotian assump-

---

[33] The same would be the case for higher prices if the demand were not non-existent.

tions, and more importantly that the interpretation associated with price-taking (namely that the offerer has literally no option but to adopt the "market" price) is ill-founded, since such price-taking can only take place[34] in a situation in which adopting the "market" price is the best option, and thus remains a *choice* and something quite distinct from the Cournotian powerless price-taker whose modelling requires performative autonomous pricing.

The same could be said about the idea that price fixation should be a clearing process, leaving no demand unsatisfied and no supply in stock at the fixed price, whereas, in reality, it is not part of anyone's responsibility to set his/her own price to allow the satisfaction of all demand at a given price, nor acceptable to assume that anyone should be especially concerned about clearance per se and the inflection price in particular.[35] Wondering what could happen in an IDC/OFC model if clearance were to become an exchange issue, for instance if there was frustration or competition among the demanders for an available supply and not only transfer of the unsatisfied demand to someone or to something else, may nevertheless be interesting. In such a case, one issue would indeed be to know to whom the available supply would be attributed and its solution would be widely variable: first come first served, priority to regular customers, contacting only a subset of interactants to avoid frustrating the others, temporal queuing, delayed delivery, ordering or reservation, redirecting demand to someone else, subcontracting, etc. All of which may take place without any necessity of changing the price itself, unless the situation leads to a modification of certain DDCs and hence of the IDC and OFC; as, for instance, if unsatisfied demanders try to avoid being excluded from exchange by accepting higher prices than stated previously, especially if their non-sincere DDC failed, thus changing their DDC (and hence the IDC, the OFC) or finally if they propose to pay

---

[34] Another situation in which something similar to price-taking can be observed is what might be called "mimetic pricing", i.e. a situation in which an offerer will choose as its price the same price as its competitors. The rationality of such a strategy could be explained by admitting the simultaneous existence of an MDC (a mimetic demand curve), i.e. of a set of answers to the question "which quantity would you buy from me if everybody's price was p?" (versus the "Which quantity would you buy from me if my price was p?" for the IDC) and of a favorable comparison of the MDC's outcome with the IDC's outcome. In any case, it must be stressed that mimetic pricing is unrelated with Cournotian price-taking notably because the copied price is a performative one and because, typically, mimetic pricing consists in not proposing a higher price than the copied one and implies that the "market" price-maker will be the proposer of the lower price, whoever he might be.

[35] Under specific but standard assumptions about the shape of the IDC, cost, and maximization, it can be shown that the quantity sold at this inflection price being greater than at lower prices, the choice of a price lower than the inflection price can thus be excluded, excluding hence the existence of unsatisfied demand. Higher prices than the inflection price remain possible and cannot be excluded.

a higher price than others to obtain priority, thus creating a dual price system. But once again, because an important dimension of any interaction is that the interactional settings cannot be unilaterally changed[36], and because the choice of a price is made with perfect knowledge of the OFC, adopting a higher price in case of shortage is ruled out by the fact that the shortage was perfectly known before price fixation, and cannot thus be considered as a possible driver of price change. It may thus be said that price per se is not the adjustment parameter, contrary to IDCs, which may adjust to the situation associated with a chosen price. This has the consequence that the DDC–IDC–OFC–Optimality sequence in itself as a model of price fixation will not be affected by such variations, which are the simple result of the fact that DDCs are to some extent revisable.

In any case, what we hope to have shown is that performativity and its modelling is not a problem but a solution.

16.5.4   Global adjustment to individual price decisions

Admitting that the capacity for an individual to fix a price is neither absolute nor non-existent, since it is absolute within the OFC and non-existent outside of it, what remains to be described is the way the choice of a price within an OFC may affect other interactions and thus what we propose to call the exchange web.

The most immediate consequence of the fact that a price–quantity pair is chosen is to clarify which commitments will have to be fulfilled and which commitments and provisions are dissolved by this choice.

Another consequence is that this clarification will leave many with no other options than to turn to another offerer or product, thus leading to a modification of some, at least, of their DDCs.

A last but crucial consequence is that since all exchange imposes a transfer of what we shall call "validation rights", in other words, of the power for the offerer to become a demander and to produce his/her own DDCs, it is also the case that in a closed economy they will keep on being transferred indefinitely.

As a whole, it is thus the case that each individual exchange will modify the premises of at least some of the subsequent exchanges, through a pattern of propagation which deserves to be modelled and whose minimal form is a web of exchanges, and its maximal form, an exchange system.

---

[36] Which means, in other words, that the rules of the game cannot be changed during the game without recrimination.

16.5.5   From exchange web to exchange system

So far we have assumed that the performative power to make a price true was dependent on and ultimately constrained by a set of DDCs measuring the acceptability of a set of prices for a set of demanders with validation rights. And it was shown that any exchange implied a transfer of such validation rights.

What we shall try to explain now in the short space available here will be that to fully understand the performativity of price, it is also necessary to investigate the often circular pattern of propagation of the validation rights themselves.

The most basic example of interdependencies between DDCs is the division of labour with a single standard of consumption, in other words people consuming A and B but producing A or B. What is important for us in such a case is that the propagation of validation rights is necessarily circular: the buying of A by producers of B is the condition for the buying of B by producers of A, leading to a situation in which, if hoarding is excluded, whatever is spent by producers of A will later on become their income and vice-versa for producers of B. So that in performative terms, the performativity of a price will ultimately be self perpetuating and based on a performativity cycle, allowing for the clarification of the relationship between the micro-economics of performativity and its macro-economic dimension.

With no possibility of addressing this last issue in any detail, it must be stressed that an important dimension of performativity appears to emerge from constraints which are invisible in a strictly individual perspective but which are crucial to the understanding of the overall relationships between performative prices.

A major constraint in that respect, which has been central to the explanation of profits, i.e. to the explanation of the capacity of offerers to receive more validation rights than they have initially transferred, the difference being called a profit, is based on a complex chain of transfers of validation rights whose paradoxical starting point is the impossibility for some validation rights to be used for all that is produced, resulting in the concentration of these validation rights on a part P of what is produced, which in turns allow the offerers of P to choose price–quantity pairs whose total value is superior to what they have spent (and thus transferred), leaving them with a difference that they can use to buy non-P products, with the ultimate constraint of a closure of the system by validation of its starting point, which thus appears to be self-perpetuating. The description of what takes place by reversing this sequence is also possible, as is apparent in the claim attributed to Kalecki by Kaldor (1956, p. 96) according to which "capitalists earn what they spent, workers spend what they earn" and in Keynes's use of the term Widow's cruse[37], and as can be illustrated by introducing in our A+B economy, a new production C, C being for instance a 50-meter long luxury yacht (or the building of a bridge whose exploitation/use will start only years later[38]) in a situation where the producers of A+B cannot afford to buy the luxury yacht. What happens then is that any transfer of validation rights in the production of C will only be usable to buy A or B, forcing, through a rise of prices, the producers of A and B to share the quantities of A and B with the producers of C and allowing the value of these quantities to rise. What is worth understanding in any case is that such eviction effects will translate directly into the existence of additional DDCs associated with the production of C[39], which will contribute to modify the IDC and the OFC of A and B[40] in the direction of an increased demand at all prices and an elevation of the point of inflection.

The conclusion of all this is that behind the IDCs and OFCs, important forces are in action which might not be visible as such but which form a crucial background to account for the performative power of fixing the price at a certain level.

When it comes to the modelling of the performative dimension of prices and market interactions, what can be generalized from the Kalecki-Kaldor-Keynes[41] macroeconomic principles is that given that the performativity of prices is directly constrained by effective validation by validation rights detainers, and that the role of validation rights is to provide access to what is produced (or sold), it can be shown that whatever affects and restricts access to what is produced will similarly affect and restrict the power of validation rights to provide such an access. Which means ultimately that the

---

[37] Keynes (1930, p. 139) argued that profits necessarily emerge with increases in investment and increases in consumption out of profits, stating that "however much of their profits entrepreneurs spend on consumption, the increment of wealth belonging to entrepreneurs remains the same as before. Thus profits, as a source of capital increment for entrepreneurs, are a widow's cruse which remains undepleted however much of them may be devoted to riotous living".

[38] Access eviction is a polymorphous phenomenon, which can be the result among other things of: i) the production of something which will never be sold (and financed through taxation for instance); ii) the production of something which will take years to be amortized; iii) the production of luxury goods; iv) the mere reservation of a part of ordinary production; v) the power to create money.

[39] Notably wages.

[40] We may assume here that if all producers of A (and B) were in a situation of perfect competition, the IDC of each of them would be identical.

[41] For more recent modelling of circuit transfers, see Schmitt (1966; 1996)

role of validation rights in the price fixation process that we have described
in the previous section is itself further embedded on eviction mechanisms
and practices, a reality which can be illustrated by the fact that in the 18th
century (see Thompson, 1971), the population was well and violently aware
of the fact that the power of wages to buy cereals was governed not only
by climatic hazards and objective rarity but also by the concrete capacity
of suppliers to hide cereals or export them away from those who had produced them. At a local level, this reality was indeed indisputable and shows
that saying that prices reflect rarity, saying that prices are performative, and
saying that eviction is crucial to both, are not contradictory claims, once
the fact is acknowledged that rarity may prove to a large extent to be artificial and that prices are not mere constative utterances describing rareness
as a pre-existing reality, but performative utterances whose conditions of
felicity to some extent are eviction mechanisms and practices and the capacity to create alone the rarity which they could seem to describe. The
temporary conclusion of all this is that the study and modelling of the performativity of prices could well allow associating in a single model of price
fixation, the modelling of micro-micro-economic interactions, and the modelling of macro-economic interdependencies, shedding considerable light
on the macroeconomic dimension of micro-economic interactions, and vice-
versa.

16.6   Conclusions

Our starting point was that because attention is scarce, because interactions
consequently are limited, because information can only emerge in interactions and has to be relevant, and because prices are performative utterances
which can become realities only insofar as they are accepted, any model of
market interactions which assumes or axiomatizes the opposite claims cannot but fail to be a model of market interactions, for it will overlook their
inherent complexity and will have to postulate, under the name of perfect
information, an informational chaos.

What we have then started to demonstrate is that accepting all these parameters as realities, far from turning the modelling of market interactions
into an impossible task, provides a reference frame in which its complexity
and variability can be expressed – notably, by admitting that price theory
should start with accounting for the performative capacity of individuals to
set their own prices, and with describing the fact that such a capacity is primarily constrained by the curve of reaction (IDC) of the other interactants,

followed by the merging of this curve with one's own knowledge of the cost and capacity constraints to form a decision curve (OFC) which serves as the informational basis for individual price fixation. Thus showing that the modelling of autonomous agents may come at no cost in terms of the capacity to account for interdependencies.

We have finally shown that the performativity of prices may also reflect or create eviction mechanisms, and more generally that its study and modelling cannot ultimately be separated from the study of validation rights, both in terms of the capacity for the transfer of validation rights, which is inherent to any exchange, the capacity to form circuits and thus to become the condition of their own existence, and in terms of the performative nature of the validation rights themselves, notably of money itself.

Ultimately, however, what is worth considering is not only that the modelling of autonomous performative prices under the constraint of the scarcity of attention and sequentiality is possible and insightful, but that the difference between this modelling and models which postulate the centralized and simultaneous fixation of constative prices under no constraint of attention or relevance, can be shown to be reducible to the way the same fundamental constraints are distinctly ordered, with macroeconomic constraints being upstream of microeconomic ones in the second model and considered as negligible (or "overlookable") in the first one, and with the maximization constraint being a downstream constraint in the second model and an intermediate one (i.e. a premise) in the first one. In this respect, we hope to have shown that making realistic hypotheses about the nature and complexity of interactions could be the only way to understand and model the way they interfere.

Bibliography

Austin, J., 1962, *How To Do Things with Words: The William James Lectures delivered at Harvard University in 1955*, Oxford: Urmson.

Cadiot, P., and F. Nemo, 1997, "Propriétés extrinsèques en sémantique lexicale", *Journal of French Language Studies* **7**, 1–19.

Cournot, A., 1838, *Recherches sur les principes mathématiques de la théorie des richesses*, Paris: Hachette.

De Vroey, M., 1999a, "Equilibrium and disequilibrium in economic theory. A confrontation of the classical, Marshallian and Walras–Hicksian conceptions", *Economics and Philosophy* **15**, 161–185.

De Vroey, M., 1999b, "The Marshallian market and the Walrasian economy. Two incompatible bedfellows", *The Scottish Journal of Political Economy* **46**, 319–338.

Friedman, M., 1949. "The Marshallian demand curve", *The Journal of Political Economy* **57**(6), 463–495.

Higgins, M., 2013. "In a Seller's Market, Every Minute Counts", *New York Times*. May 31, 2013.

Ingrao, B., and I. Giorgio, 1990 *The Invisible Hand. Economic Equilibrium and the History of Science*, Cambridge, Mass.: MIT Press.

Kaldor, N., 1956, "Alternative Theories of Distribution", *The Review of Economic Studies* **23**, 94–100.

Kalecki, N., 1954, "The determinants of profits". In *Selected Essays on the Dynamics of the Capitalist Economy. 1933–1970*, Cambridge University Press, pp. 78–92.

Marshall, A., 1961, *Principles of Economics*, Ninth (Variorum) Edition, Volumes I and II. Text. London: Macmillan.

Mirowski, P., 1989, *More Heat Than Light: Economics as Social Physics, Physics as Nature's Economics*. Cambridge University Press.

Nemo, F., 1994, "Énoncés marchands - Où il est montré que les prix ne sont pas ce que l'on croit", *Revue du MAUSS* **3**, 182–193.

Nemo, F., 1995, "Une alternative à la loi de l'offre et de la demande", *Revue du MAUSS* **6**, 166–176.

Nemo, F., 1999, "The Pragmatics of Signs, The Semantics of Relevance, and The Semantic/Pragmatic Interface", in: *The Semantics–Pragmatics Interface from Different Points of View*, CRiSPI Series, Amsterdam: Elsevier, pp. 343–417.

Phelps, E. S. 1969. "The New Microeconomics in Inflation and Employment Theory," *American Economic Review* bf59, 147–60.

Schmitt, B., 1966, *A New Paradigm for the Determination of Money Prices*, Paris: Presses Universitaires de France.

Schmitt, B., 1996, "A New Paradigm for the Determination of Money Prices". In: G. Deleplace and E. J. Nell (eds.), *Money in Motion*, London: Macmillan and St. Martin's Press.

Todorov, T. 1995, *La vie commune : essai d'anthropologie générale*. Paris: Seuil. Translated in 2001. *Life in Common: An Essay in General Anthropology* University of Nebraska Press.

Tomasello, M. 2008, *Origins of Human Communication*, Cambridge, Mass.: MIT Press.

Chapter 17

Kinetic collision models and interacting agents in complex socio-economic systems

Stéphane Cordier[1]

17.1 Introduction

The emergence of collective phenomena and self-organization in systems composed of a huge number of agents had received increasing interest from various research communities, in economics, sociology, biology, ecology, and robotics. This has been illustrated by several lectures during the ISC conference in June 2013 in Orléans, France, in particular plenary ones by Beresticky "Propagation in inhomogeneous media: From epidemics to contagion of ideas", Fischer "Complex processes in human–robot interaction", Galam "Interacting with a few random liars can jeopardize the democratic balance of a public debate", Goles "Regulatory and segregation networks", and Kirman "Ants and Non-Optimal Self Organization: Lessons for Macroe-

[1]Department of Mathematics and Applications, Mathematical Physics (MAPMO) UMR 7349 University of Orléans and CNRS, 45067 Orléans, France. Email: cordier@math.cnrs.fr.

conomics". Let us mention the recent and complete book by Pareschi and Toscani that describes in detail the mathematical tools and the numerical methods that can be used, such as multi-agent or Monte Carlo methods Pareschi and Toscani (2013).

These so called multiagent systems can be studied using methods which originate in statistical physics. The first attempts made in this direction date back 20 years, when the term 'econophysics' was introduced to describe an interdisciplinary research field Bouchaud and Mézard (2000), Mantegna and Stanley (2000) which aims to solve problems in economics by means of well-established methods of physics. The main idea is that the description of emerging phenomena could be obtained by assuming that the collective behaviour of a group composed of a sufficiently large number of individuals could be treated using the laws of statistical mechanics as happens in a physical system composed of many interacting particles. Let us mention that statistical physics has also been used recently for sociological models, such as, for example, a model of opinion formation Boudin and Salvarani (2009), Boudin, Mercier, and Salvarani (2012) for which the asymptotics is called the "quasi-invariant opinion" limit. Let us refer to the recent book by Galam (2012) for an interesting and personal presentation of the use for sociological issues of approaches originating in physics, that began more than 30 years ago.

These methods allow constructing complex systems composed of autonomous agents who, as a result of their mutual interactions, exhibit a well-defined collective behaviour. The mathematical analysis of the collisional kinetic theory has been a very active field within the last twenty years. Let us refer to the complete survey written by Villani (2002).

In this article, we will briefly present an example of such a method applied to a simplified model of interacting economic agents introduced in Cordier, Pareschi, and Toscani (2005). It starts from microscopic models of simple market economies that describe the interaction between a pair of individuals and the goal is to extract from this microscopic model information about the averaged wealth distribution.

The study of wealth distributions has a long history, going back to the Italian sociologist and economist Pareto, who studied the distribution of income among people of different Western countries and found an inverse power law Pareto (1897). More precisely, if $f(w)$ is the probability density function of agents with wealth w we have

$$F(w) = \int_w^\infty f(w_*) \, dw_* \sim w^{-\mu},$$

which means that the part, $F(w)$, of the population with an income w_* larger than w decays, for large values of w, like some power of w. Pareto mistakenly believed that such power law behaviour would apply to the whole distribution with an universal exponent μ approximately equal to 1.5. Later, Mandelbrot (1960) proposed a weak Pareto law that applies only to high incomes.

The starting point of the present modelling is to describe the so-called microscopic model, i.e. the process of trading between a couple of agents. Then, we describe the evolution of the associated probability distribution function (pdf) which is described by a Boltzmann type collision operator. In other words, the exchange of money during a trade is represented in the model like the change of momentum during a collision between particles. Then, the idea is to perform an asymptotic analysis in the limit of a large number of small exchanges. This leads to a partial differential equations (PDE) for which we are able to find the large time behaviour and to describe the stationary or equilibrium solutions. This equilibrium state can be computed explicitly and is of Pareto type, namely it is characterized by a power-law tail for the richest individuals.

This asymptotic limit, hereafter called the "continuous trading limit", has been inspired by the one used in the context of kinetic theory for granular flows, where the limit procedure is known as the "quasi-elastic" asymptotics Toscani (2000). It is also connected to the so called "grazing collision" limit that permits passing from the Boltzmann equation for Coulombian particles to Fokker–Planck–Landau equations (see Villani 2002).

17.2   A kinetic model of money asset exchanges

We consider now a very simple model of an open market economy involving both asset exchanges between individuals and speculative trading, following Cordier, Pareschi, and Toscani (2005). In this non-stationary economy, the total wealth is not conserved due to a stochastic dynamics which describes the spontaneous growth or decrease of wealth due to investments, e.g. in the stock market. It is important to note that this mechanism corresponds to the effects of an open market economy where the investments cause the total economy to grow (more precisely the rich would get richer and the poor would get poorer). The exchange dynamics between individuals redistributes the wealth between people.

Thus, from a microscopic viewpoint, the binary interaction is described by the following rules

$$w' = (1 - \gamma)w + \gamma w_* + \eta w \qquad (17.1)$$
$$w'_* = (1 - \gamma)w_* + \gamma w + \eta_* w_*$$

where (w, w_*) denote the (positive) money of two arbitrary individuals before the trade and (w', w'_*) the money after the trade. In (17.2) we will not allow agents to have debts, and thus the interaction takes place only if $w' \geq 0$ and $w'_* \geq 0$. In (17.2) the transaction coefficient $\gamma \in [0, 1]$ is a given constant, while η and η_* are random variables with the same distribution (for example normal) with variance σ^2 and zero mean.

Let us describe the three terms on the right hand side. The first term is related to the marginal saving propensity of the agents, the second corresponds to the money transaction, and the last contains the effects of an open economy describing the market returns. Note that since debts are not allowed, the total amount of money in the system is increasing.

This binary interaction model is related to (Bouchaud and Mézard 2000, Levy, Levy, and Solomon 2000). In a closed economical system, it is assumed that the total amount of money is conserved ($\eta, \eta_* \equiv 0$). This conservation law is reminiscent of analogous conservations which take place in kinetic theory. In such a situation, the stationary state is a Dirac measure centred on the average wealth. Thus all agents will end up in the market with exactly the same amount of money.

The kinetic model associated with this simple market economy describes the evolution of the statistical distribution of money by means of these *microscopic* interactions between agents or individuals which exchange money. Each trade can indeed be interpreted as an interaction where a fraction of the money changes hands. We will assume that this wealth after the interaction is non-negative, which corresponds to imposing that no debts are allowed. This rule emphasizes the difference between economic interactions, where not all outcomes are permitted, and the classical interactions between molecules.

Let $f(w, t)$ denote the distribution of money $w \in \mathbb{R}_+$ at time $t \geq 0$. By standard methods of kinetic theory (Villani 2002), the time evolution of f is driven by the following integro-differential equation of Boltzmann type,

$$
\frac{\partial f(w)}{\partial t} =
$$

$$
\int \left(\beta_{('w,'w_*)\to(w,w_*)} J f('w)(f('w_*) - \beta_{(w,w_*)\to(w',w'_*)} f(w) f(w_*) \right) dw_* \, d\eta \, d\eta_*
$$

$$(17.2)$$

where $('w,'w_*)$ are the amounts of pre-trade money that generate the couple (w, w_*) after the interaction. In (17.2), J is the Jacobian of the transformation of (w, w_*) to (w', w'_*), which is a usual technical term, and the kernel β is related to the details of the binary interaction and represents the probability of changing from the pre-trading state to the actual one. As usual in kinetic theory, we write $f(w)$ instead of $f(w, t)$ to simplify the notation. Refer to Cordier, Pareschi, and Toscani (2005), Pareschi and Toscani (2013) for a more detailed presentation.

We shall restrict ourselves here to a transition rate of the form

$$
\beta_{(w,w_*)\to(w',w'_*)} = \mu(\eta)\mu(\eta_*)\Psi(w' \geq 0)\Psi(w'_* \geq 0),
$$

where $\Psi(A)$ is the indicator function of the set A, and $\mu(\cdot)$ is a symmetric probability density with zero mean and variance σ^2. The rate function $\beta_{(w',w'_*)\to(w,w_*)}$ characterizes the effects of the open economy through the distribution of the random variables η and η_* and takes into account the hypothesis that no debts are allowed. Refer to Cordier, Pareschi, and Toscani (2005) for a detailed interpretation of the chosen form for β. The above equation can be included in a more general setting, where the trade rule has a more complex structure, including, for example, risk, taxes, and subsidies (Pareschi and Toscani 2013).

We remark that, for a general probability density $\mu(\cdot)$, the rate function β depends on the wealth variables (w, w_*) through the indicator functions Ψ. This is analogous to what happens in the classical Boltzmann equation Villani (2002), where the rate function depends on the relative velocity. A simplified situation occurs when the random variables take values in the set $(-(1 - \gamma), 1 - \gamma)$. In this case, in fact, both $w' \geq 0$ and $w'_* \geq 0$, and the kernel β does not depend on the wealth variables (w, w_*). In this case, the kinetic equation (17.2) is the analogue of the classical Boltzmann equation for Maxwellian molecules, which presents several mathematical simplifications. In all cases, however, methods borrowed from kinetic theory of a rarefied gas can be used to study the evolution of the function f.

17.3   The continuous trading limit

As explained in the Introduction, we are interested in the repartition of wealth due to a large number of such interactions or, equivalently, the large time behaviour of the equation, i.e. the associated stationary states.

It is proven in Cordier, Pareschi, and Toscani (2005) that the kinetic model (17.2) is well posed. However, it is not possible to explicitly describe the equilibrium or stationary solution of this equation. As is usual in kinetic theory and explained in the Introduction, particular asymptotics of the equation result in simplified models (generally of Fokker–Planck type), for which it is possible to find steady states.

We will consider the situation in which most of the trades correspond to a very small exchange of money ($\gamma \to 0$), rescaling the time scale accordingly ($\tau = \gamma t$) such that both $\gamma \to 0$ and $\sigma \to 0$ in such a way that $\sigma^2/\gamma \to \lambda$.

As proved in Cordier, Pareschi, and Toscani (2005), the scaled density $g(v, \tau) = f(v, t)$ obeys a Fokker–Planck model derived from the Boltzmann equation, introducing a Taylor expansion in the weak formulation, ϕ being a test function with bounded moments

$$
\frac{d}{d\tau} \int_0^\infty g\phi \, dw =
$$
$$
\frac{1}{\gamma} \int_{\mathbb{R}^2} \int_{\mathbb{R}_+^2} \mu(\eta)\mu(\eta_*)g(w)g(w_*)(\phi(w') - \phi(w))dw_* dw d\eta \, d\eta_*. \quad (17.3)
$$

This derivation is similar to the quasi-elastic limit of granular gases of Toscani (2000) and is of major relevance for the study of the asymptotic equilibrium states of the kinetic model. The right-hand side is nothing but the weak form of the Fokker–Planck equation

$$
\frac{\partial g}{\partial \tau} = \frac{\lambda}{2} \frac{\partial^2}{\partial w^2} \left(w^2 g \right) + \frac{\partial}{\partial w} \left((w - m)g \right). \quad (17.4)
$$

The limit Fokker–Planck equation can be rewritten as

$$
\frac{\partial g}{\partial \tau} = \frac{\partial}{\partial w}[(w - m) + \frac{\lambda}{2} w)g + \frac{\lambda}{2} w \frac{\partial}{\partial w}(wg)]. \quad (17.5)
$$

The stationary state of the Fokker–Planck equation can be directly computed and, assuming for simplicity

$$
m = \int_{\mathbb{R}_+} f(w, t) \, dw = 1,
$$

it can be written as

$$g_\infty(w) = \frac{(\mu-1)^\mu}{\Gamma(\mu)} \frac{\exp\left(-\frac{\mu-1}{w}\right)}{w^{1+\mu}} \qquad (17.6)$$

where $\mu = 1 + \frac{2}{\lambda} > 1$. Therefore the stationary distribution exhibits a Pareto power law tail for large w's as observed by Pareto (1897) on real economical data as an "universal" behaviour. Thus, the proposed model agrees with this observed repartition of weath for the richest individuals.

17.4   Conclusions

Note that the numerical simulations of a Boltzmann equation (17.2) to compute its steady state are usually based on Monte Carlo methods, as explained in Pareschi and Toscani (2013). Note that such Monte Carlo methods rely on the underlying microscopic process (17.2). Some numerical test are presented in Cordier, Pareschi, and Toscani (2005) that illustrate numerically the "continuous trading asymptotics".

This simple market economy model, based on binary money exchanges and speculative trading, becomes, at suitably large times, and in the presence of a large number of trades in which agents exchange only a small amount of money, a linear Fokker–Planck type equation, which admits a stationary steady state with Pareto tails. The analogy between the trade rule (17.2) and a one-dimensional molecular dissipative collision suggests in a natural way the continuous trading asymptotic, which is well-understood in kinetic theory as the quasi-elastic asymptotics (Toscani 2000). Furthermore, the formation of overpopulated energy tails for large times in the kinetic model is in accord with the analogous result valid for the Boltzmann equation for a dissipative granular Maxwellian gas.

Let us mention another model where the interactions between individuals is replaced by exchanges through a financial market where a finite number of assets are available (Cordier, Pareschi, and Piatecki 2009). In this case, similar asymptotic approachs yield a log-normal distribution behaviour for large wealth.

This short note presents a very particular case of a complex system: a population of agents that interact by trades. It illustrates how the mathematical framework of kinetic theory of collisios (see Villani 2002) can be useful to understand and predict the formation of equilibria. We refer to Pareschi and Toscani (2013) for a recent book on related topics with numerous other applications.

Bibliography

Bouchaud, J. P., and M. Mézard, 2000, "Wealth condensation in a simple model of economy", *Physica A* **282**, 536–545.

Boudin, L., A. Mercier, and F. Salvarani, 2012, "Conciliatory and contradictory dynamics in opinion formation", *Phys. A* **391**, 5672–5684. doi: 10.1016/j.physa.2012.05.070

Boudin, L., and F. Salvarani, 2009, "A kinetic approach to the study of opinion formation", *Math. Model. Numer. Anal.* **43**(3), 507–522. doi: 10.1051/m2an/2009004

Cordier, S., L. Pareschi, and C. Piatecki, 2009, "Mesoscopic modelling of financial markets", *Journal of Statistical Physics* **134**(1), 161–184.

Cordier, S., L. Pareschi, and G. Toscani, 2005, "On a Kinetic Model for a Simple Market Economy", *Journal of Statistical Physics* **120**(1–2), 253–277.

Galam, S., 2012, *Sociophysics: A Physicist's Modeling of Psycho-political Phenomena*, Series: Understanding Complex Systems, XXIII. Berlin: Springer-Verlag.

Levy, H., M. Levy, and S. Solomon, 2000,*Microscopic Simulations of Financial Markets*, New York: Academic Press.

Mandelbrot, B., 1960, "The Pareto–Lévy law and the distribution of income", *International Economic Review*, **1**, 79–106.

Mantegna, R. N., and H. E. Stanley, 2000, *An Introduction to Econophysics*, Cambridge Univ. Press.

Pareschi, L., and G. Toscani, 2013, *Interacting Multiagent Systems: Kinetic Equations and Monte Carlo Methods*, Oxford University Press.

Pareto, V., 1897, *Cours d'Economie Politique*, Lausanne, Switzerland.

Toscani, G., 2000, "One-dimensional kinetic models of granular flows", *RAIRO Modél Math. Anal. Numér.* **34**, 1277–1292.

Villani, C., 2002, "A review of mathematical topics in collisional kinetic theory", in *Handbook of Mathematical Fluid Dynamics*, S. Friedlander and D. Serre, Eds., Amsterdam: Elsevier.

Chapter 18

The choice of studies as an evolutionary game

Cyrille Piatecki[1]

18.1   Introduction

In economics, many topics should be considered from the complexity point of view.

First of all, while it took approximately 400 years to establish a utility theory from the early works by Mariotte[2], (see Mariotte 1717), 30 years were sufficient to make the building totter on its basis.

Among other sciences, Behavioral Psychology and Neuroscience question the very existence of a utility function derived from preferences and used to make practical decisions. The assumption that human behaviour can generate rational decisions [3] has been denied by new experimental evidence from the recent access to the brain through magnetic resonance imaging (MRI) and molecular biology. Thanks to the MRI, it is now proven that

---

[1]LEO, UMR CNRS 7322 University of Orléans and ALPTIS. Email: cyrille.piatecki@univ-orleans.fr.
[2]Mariotte is the second name to appear in the appendix of Jevons (1871) dedicated to mathematico-economic books, memoirs, and other published writings.

[3]This assumption originated from the great Greek philosophers (essentially Aristotle), followed by Bernoulli, Kant (from the moral point of view), Bentham, Jevons, and Von Neumann, among many others.

decisions can not be taken without reference to emotions and that we are sensitive to chemical substances which alter profoundly our behaviour and decisions (see Damasio 1994, Lehrer 2009).

At the same time, with the emergence of a new type of society where information is huge and plethoric, we have begun to understand that the brain is, in some aspect, a wired electrical network with limited transmission speed[4] and a limited capacity to treat information[5]. Furthermore, since Poincaré, but more effectively since Lorenz, we know that linearity, through which we believe we can understand the most essential structure of our universe, is misleading and that a very small departure from it can produce processes that are truly indistinguishable from randomness (Poincaré 1890, Lorenz 1964, Devaney 2003).

At last, we have begun to understand that strategic interactions, which are mainly described in the setting of standard game theory, may not be intelligible from rationality itself.

Nevertheless, despite all this new evidence, many analysts are satisfied over all with their inherited instrument and are not yet ready to change paradigms. For instance, the core of the modelling of economic markets has not been fundamentally changed in 30 years.

For instance, let us have a look at labour economics. Since the development of the *Human Capital Theory* (Schultz 1961, Becker 1964), economists have concentrated their attention essentially on the main advantage that education confers on individuals: an increase in their income.

The mechanics highlighted under the original human capital theory allowed, in a standard neo-classic framework, proposing an explanation alternative to the theory of compensatory differences that was first advanced by Adam Smith and in which individuals were considered as indistinguishable.

Two research areas have caught the attention of economists. On the

---

[4] Here, we can refer to the limited capacity channel in the Shannon–Hartley theorem on the maximum rate at which information can be transmitted over a communication channel of a specified bandwidth in the presence of noise, see Pierce (1980). But we can also refer to the problem of the acquisition of the expert performance, as in Ericsson et al. (1993) or Shim et al. (2005).

[5] According to Marois and Ivanoff (2005), despite the impressive complexity and processing power of the human brain, it is severely capacity limited. There are many way to approach this fact but an interesting one is linked to the magical number seven± two, see Miller (1956), which has been recently replaced by the number four: see Parker (2012). The problem is to know how many element can be placed in the human memory. We must stress that the working memory is one of the central concepts of current cognitive psychology. As the place of holding the information necessary for ongoing treatment, the limits of working memory constrain a large part of our thinking activity and the evolution of these boundaries with age plays a vital role in intellectual development, (see Barrouillet and Camos 2008).

one hand, following Mincer's works, cf. Mincer (1974), an effort has been made to empirically estimate returns to education, through some earning functions. That approach enables estimating a returns curve for each educational level and making international comparisons. On the other hand, more recently, human capital has been introduced in models of endogenous growth.

Following Lucas (1988), macroeconomists have favoured the stock of human capital over technology to explain growth, and the differences of rates in its accumulation to explain the persistence of international differences in growth rates. In particular, because of the externalities generated by its human capital, for a given level of qualification, a worker is more productive in a country already strongly endowed with human capital, and therefore better paid. This mechanism could explain the strong migratory pressures of the South on the North.

Without questioning the legitimacy of the Beckerian individualistic approach to human capital accumulation, it seems, nevertheless, important to reinterpret the contemporary logic of its accumulation in the context of job competition. In the Beckerian logic, a rise in the level of education appears as a real solution for unemployment, from the individual point of view, because the level of education is positively correlated with the speed of transition towards employment. This transition advantage is also assumed by the filter theory developed by Arrow (1973) and the signal theory developed by Spence (1974).

Other things being equal, assuming a homogeneous population, this transition advantage given by education is expected to motivate all individuals to apply for further education and training. Hence, the ultimate consequence of this dynamic process is the deletion of the number of non-graduates.

In practice, we observe a conflicting logic, which substitutes itself for the Beckerian individualistic approach as soon as such a situation is reached or, at least, as soon as a critical threshold of graduates is exceeded. Indeed, the competitive advantage of a diploma tends to disappear with an increase in the number of its holders. Outside a conflicting approach of human capital accumulation, we can only predict a desertion of the educational sphere: education no longer generates a return on investment which can be evaluated as better than that coming from an investment in stocks.

However, from the empirical point of view, it is rare to observe parents preferring to donate the cost of their education to their children in the form of a financial endowment because it seems to them higher and less risky than that of the studies that their children undertake. Indeed, in spite of its depreciation, human capital can become the key to entry to active life. So, contrary to the predictions of the Beckerian model, the model with the conflicting logic seems to predict a wild race towards the accumulation of education, its non-accumulation appearing as an almost insuperable handicap.

However, if a model which rests on a conflicting logic ends in an extreme distribution – either every player graduates or nobody graduates – it must be rejected as empirically irrelevant because, as far as one can observe, in populations where the individual characteristics are only with difficulty distinguishable, there coexist the two logics of accumulation of human capital, i.e. the maximalist logic of the acquisition of the highest diploma and the minimalist logic of the acquisition of the lowest diploma.

This paper proposes examining the evolution of the graduate population within the framework of the theory of evolutionary games. For every generation, for a given unemployment rate, the percentage of graduates is predetermined by their expected remuneration relative to the previous generation. The graduates, in a context where the diploma gives a priority access to jobs, are all the stronger when there are few qualified people. Consequently, and *a priori*, we should expect that the global educational level does not stop growing, thus excluding the possibility of a mixed balance, that is, a population balance in which both categories of agents coexist. However, when the trained population is numerous, the high training costs decreases the relative advantage conferred by studies, and can incite a disinvestment in education.

The problem hereby underlined is a real problem that all developed countries seem to encounter. For instance, in France, for the year 2000, it has been revealed that one-third of youth found their credentials and consequent opportunities downgraded. Even if the apex of the phenomena was observed between 1986 and 1995 for the graduates at the Baccalauréat (High School exit examination) level, and between 2001 and 2004 at the BAC+2 level (two-year post-secondary or tertiary level degree), due to the overall decrease in employment, even the higher-level graduates were heavily affected. The chances for a new graduate to get a position as an executive fell from 85% to 70%. Only 26% of the youth have a perfectly matched job, in accordance with their formation, in terms of level and speciality. At the same time, we still observe that the higher the level of the diploma, the smoother the transition towards the first job (see Mazari and Recotillet 2013).

In the first section of the present paper, we will consider a static evolutionary game and its dynamics when the matching of the young job-seekers is realized pairwise. The second section considers explicitly the effect of the growth of unemployment on the population dynamics. The third section studies the formation of long-term balances when irrational mutations appear. The fourth (the last) section is devoted to the controllability of the education system[6].

18.2   Pair matching dynamics

Here we study more particularly a game in which the players have two strategies.

① Undertake short studies: strategy (**S**)[7].

② Undertake long studies: strategy (**L**).

At first, we are interested in two players who compete for a job, both of them assigning to it a common present value V. The agent having adopted the **L** strategy will be given priority access to the job compared to the agent having adopted the **S** strategy. This preference for the most qualified refers to the theory of human capital: whether in terms of a filter or in terms of a signal. Because our focus is on the advantage conferred by education, we suppose that the level of the remuneration V is independent of the educational level.

Labour markets are segmented and it is notable that the multiplication of the graduates at every level of study has entailed a toughening of the dualism (for a definition of this dualism, see[8], for example, Taubman and Wachter (1986). Young people that are excluded from the markets, to which the conditions of access correspond exactly to their diplomas, are thus brought to look for an outlet on a market for which they are overqualified. This downgrading search creates an additional bottleneck for their companions who had acquired the diploma required to enter the second market.

We can unwind this reasoning all the way to the markets which require no specific training and which are, in turn, assailed by young graduates. This

---

[6]See the survey published by the CÉREQ at http://www.cereq.fr/gsenew/concours2008/cereq/G98ind/premierepage.htm.

[7]On the sociological motivation for the short studies and its evolution in time see Jaoul-Grammare and Nakhili (2010).

[8]Nevertheless, one can follow Piore (1978). According this theory, the labour market is divided into two sectors: a *primary* which give the right to better paying, promotion, more stable opportunities, and a *secondary*, which contains the poor paying insecure and otherwise unattractive jobs.

logic leads to a more or less important eviction of the non-graduates from the labour market[9]. Naturally, later in their active life, some individuals will manage to return to their desired market, but they will certainly represent an exception to the rule of the strict dualism. So, on an isolated market, we can expect to observe applications emanating from young graduates possessing the minimum level of required diplomas (see Figure 18.1).

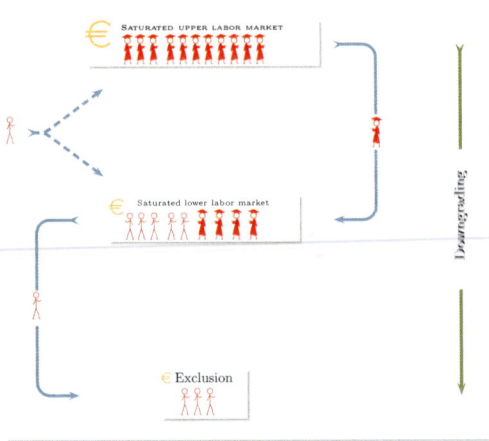

Figure 18.1: The graduates' downgrading process.

A rather reliable indicator of these observations would be established by the ratio between the level of education of the candidates at public service job competitions and the education level required by the examination[10]. We observe that a large share of the candidates have a much higher level of education and qualification than required by the examination. As a consequence, these overqualified candidates systematically evict the least endowed without benefiting from a more generous salary scale. On the other hand, from the analytical point of view, we can imagine that the status of the labour market is such that companies are in a position to monopolize a rent

---

[9]It is not part of our objective to explain why the non-qualified market can not be itself balanced, while we describe a balancing process by downgrading rejection of populations towards the least qualified jobs, and while the existence of involuntary unemployment is a necessary condition for the justification of the model.

[10]It's not a red herring to say that in the French public functionary examination, a great many of the candidates are overqualified (see Kopel (2005)). Obviously, this is not a phenomenon of French exceptionalism (see Green and Zhu (2010)). So, in a first approximation, we can postulate that V is independent of the education level.

or to make sure of the honesty of their employees. This point was developed in particular by Shapiro and Stiglitz (1984). Returning to the model, let us assume that, initially, all players were used to short studies. As, *ex ante*, none of the players is able to determine what will be needed to prevail in the coming conflict, we will adopt Laplace's insufficient reason principle and suppose that they anticipate they each have a 1/2 probability of monopolizing the desired job. The couple of payoffs in a conflict between two individuals having adopted the short strategy is thus: $(0.5V, 0.5V)$.

Let us suppose that after a punctual mutation[11], a fraction $1 - \epsilon$ of the population decides to opt for a new strategy consisting of a cycle of longer studies. These players can be perceived as predators by all those players who have decided not to change strategy. Indeed, we suppose that it is common knowledge that if a long player is accidentally mated to a short player, the first one will obtain the employment and will be paid[12] V, while the second will have to pursue its job search. However, we suppose that, compared to strategy **S** (short studies), the long studies have a fixed additional cost C. The couple of payoffs bound to such a confrontation is then $(V - C, 0)$.

Obviously, a mutant can be mated with another mutant. As previously, in this case, they will obtain the couple of payoffs $(0.5(V - C); 0.5(V - C))$. Presented under normal form, the bi-matrix of the payoffs, associated with this game, can then be written as in Figure 18.2.

In what follows, we will postulate that $0 \leq C \leq V \leq 2C$. This hypothesis is as fundamental as it is natural. First of all, we consider a negative cost. A cost lower than the present value of the income of the promised job is a necessary and sufficient condition to ensure that some individuals will wish to bear it. Lastly, one has to impose that $V \geq 2C$. Indeed, let us suppose the opposite: $V > 2C$. In this case, $(V - C)/C$, which is the *ex post* rate of return of longer studies, and is then lower than unity. Now, if the return was greater than 100%, there would be no student who would not undertake longer studies.

---

[11] The term 'mutation', borrowed from the vocabulary of the biologist, is here excessive. We can imagine that this phenomenon comes upstream from a human capital investment behaviour or, more simply from the invasion of the labour market for the non-qualified by individuals stemming from a qualified labour market which has reached saturation, a hypothesis which is defended in the body of the text.

[12] We use V because all the monetary values here must be understood as present values since the decision of the length of studies is made for the complete life cycle of the players.

	S	L
S	$\left(\dfrac{V}{2}, \dfrac{V}{2}\right)$	$(0, V - C)$
L	$(V - C, 0)$	$\left(\dfrac{V - C}{2}, \dfrac{V - C}{2}\right)$

Figure 18.2: Payoff matrix for the game.

18.2.1   The game as a perfect information game in normal form

Assume first that each player knows his opponent, that is to say, the other player who will apply for the same position as he will. In pure strategies, under the hypothesis $V < 2\,C$, there are two Nash equilibria in this game[13]: $\text{PSNE}_1 = (\mathbf{S}, \mathbf{S})$ and $\text{PSNE}_2 = (\mathbf{L}, \mathbf{L})$. Both equilibria are symmetric, i.e.: both players must choose the same strategy, which in itself is not so astonishing since the players are indistinguishable. To these pure strategy equilibria, one must add a mixed strategy equilibria. One finally has

$$\text{MSNE}_1 = (0, 0), \quad \text{MSNE}_2 = \left(\frac{V - C}{C}, \frac{V - C}{C}\right) \quad \text{et} \quad \text{MSNE}_3 = (1, 1)$$

Figure 18.3 gives the three equilibria as the intersections of the players' reaction correspondences[14].

In front of the embarrassment induced by the multiplicity of Nash equilibria, game theorists have developed a number of procedures aiming at extracting a single equilibrium that will most probably be played by the player[15]. The first seductive approach consists in postulating that the play-

---

[13] PSNE: Pure Strategies Nash Equilibrium; MSNE: Mixed Strategies Nash Equilibrium.

[14] In Figure 18.3, the line player chose a mixed strategy $(p, 1 - p)$ and the column player a strategy $(q, 1 - q)$. Normally, since players must solve two linear programs to find the mixed equilibrium, we would be obliged to use an algorithm of the Lemke–Howson type to find numerically this equilibrium, or any more modern approach, but in the case of two strategies one can easily find an analytical solution: see, for instance, Nisan et al. (2007).

[15] In a finite game, every pure strategy Nash equilibrium is also a mixed strategy Nash equilibrium.

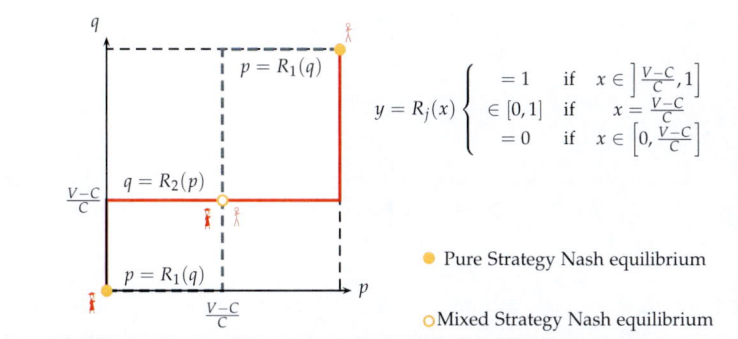

$$y = R_j(x) \begin{cases} = 1 & \text{if} \quad x \in \left]\frac{V-C}{C}, 1\right] \\ \in [0,1] & \text{if} \quad x = \frac{V-C}{C} \\ = 0 & \text{if} \quad x \in \left[0, \frac{V-C}{C}\right] \end{cases}$$

● Pure Strategy Nash equilibrium

○ Mixed Strategy Nash equilibrium

Figure 18.3: The Nash equilibria.

ers are going to notice that one of the two pure equilibria is socially more favourable than the other one, in the sense that the expected payoff is higher there; here, this involves (**S**, **S**), which gives a payoff equal to $V/2$, which is greater than the payoff of $(V - C)/2$ associated to the second pure strategy equilibrium. (**S**, **S**) is called the Pareto dominant equilibrium.

A standard approach for the selection from multiple equilibria was proposed by Harsanyi and Selten (1992). This approach to equilibrium selection is known as *risk dominance*. This approach has the interest of having a double interpretation. A player can realize an experiment in which he tries to guess the strategy adopted by the opponent with whom he will be mated: this is a game theory interpretation. He can also, *ex ante*, try to conceive the characteristics of the distribution of the players between both strategies as members of the population to which they belong: this is a sociological interpretation.

Let us begin by repositioning ourselves in a framework where only pure strategies can be chosen and let us adopt the role of one of the two matched players. If he thinks that his opponent will choose the **S** strategy with probability q and the **L** strategy with probability $1 - q$, his expected utility is

$$\mathbb{EU}_1(s) = \begin{cases} (1-q)\left(\frac{V}{2}\right) & \text{if} \quad s = \mathbf{L} \\ q\left(\frac{V}{2}\right) + \left(\frac{V-C}{2}\right) & \text{if} \quad s = \mathbf{S} \end{cases}$$

The player will prefer to adopt the **S** strategy (the **L** strategy) over the **L** strategy (the **S** strategy) if

$$\mathbb{EU}(\mathbf{S}) < (>) \; \mathbb{EU}(\mathbf{L})$$

Writing $\bar{q}$ for that value of q which makes the player indifferent between both strategies, we find that

$$\bar{q} = \frac{C}{V + C}$$

Now, the player's choice depends on the mental scheme that drives him to postulate a specific value of q. However, it is clear that if $1 - \bar{q} \geq q$, or, more simply, if $\bar{q} \geq 1/2$, which is the present case since $C \leq V$, the opponent will take a greater risk in postulating that the player will choose the **S** strategy over the **L** one. As this is a symmetric game, what is valid for one player is also valid for the other. So one will say with Harsanyi and Selten that the (**L**, **L**) equilibrium *risk dominates* the (**S**, **S**) equilibrium if $\bar{q} > 1 - \bar{q}$, which means that there is a greater likelihood that both players coordinate on the (**L**, **L**) equilibrium than on the (**S**, **S**) one. In this game, according to the *risk dominance concept*, the players should be incited to adopt a coordination over long studies.

18.2.2   The static evolutionary game

The application of traditional game theory to the duration of education raises two problems:

① On the one hand, we cannot consider two isolated players, because the game is played by a very large number of players simultaneously.

② On the other hand, the players have to make decisions without knowing which type of player they will later be mated with: an **S** player or an **L** player. Now, the *ex ante* yield of a strategy depends on the percentage of players who adopt it in the population. In particular, even if there are **S** players in the population, a player can be interested in being an **L** player because the probability of being mated to an **S** player is low and because by choosing to be an **L** player, he does not have to support the additional cost of the extra time of studies.

We begin by studying the pure strategies game. Let us note[16] ϵ, the percentage of players who use the **S** strategy. In such a way, the expected payoff of a **S** player in a mixed population is

$$\mathbb{E}U(\mathbf{S}|\mathbf{L}) = \epsilon U(\mathbf{S}, \mathbf{S}) + (1 - \epsilon)U(\mathbf{S}, \mathbf{L}) = \epsilon\left(\frac{V}{2}\right)$$

---

[16]We must distinguish between p and q, which are variables relating to the personal choice of ϵ that is the percentage in the population of people using the **C** strategy.

where $U(\mathbf{X}, \mathbf{Y})$ is the utility of a $\mathbf{X}$ player matched with a $\mathbf{L}$ player. In the same way, one can write

$$\mathbb{E}U(\mathbf{L}|\mathbf{L}) = \epsilon U(\mathbf{L}, \mathbf{S}) + (1 - \epsilon)U(\mathbf{L}, \mathbf{L})$$

$$= \epsilon(V - C) + (1 - \epsilon)\left(\frac{V - C}{2}\right) = \left(\frac{1 + \epsilon}{2}\right)(V - C)$$

In the evolutionary game terminology, a strategy $\mathbf{X}$ is called *evolutionary stable* if

$$U(\mathbf{X}, \mathbf{Y}) > U(\mathbf{Y}, \mathbf{Y})$$

for every other $\mathbf{Y}$ strategy. This means that $\mathbf{X}$ offers a better fit[17] to the game than every other $\mathbf{Y}$. Maynard-Smith (1974) has shown that $\mathbf{X}$ is evolutionary stable[18] if the two following conditions are satisfied:

① *A Strict Nash Equilibrium Condition*: $U(\mathbf{X}, \mathbf{X}) > U(\mathbf{Y}, \mathbf{X})$.

② *A stability Condition*: if $U(\mathbf{X}, \mathbf{X}) = U(\mathbf{X}, \mathbf{Y})$ then $U(\mathbf{X}, \mathbf{Y}) > U(\mathbf{Y}, \mathbf{Y})$.

It is obvious that the pure strategies Nash equilibria are *ESS*, which here was predictable as far as the game is symmetric. Consequently, by neglecting the fact that they are immersed in a larger population than that of both matched players, and as far as they manage to coordinate on an equilibrium (for example, the risk dominating equilibrium), the players protect themselves against the invasion of mutant players who would have adopted the other strategy. Now, in the same way as in the traditional interpretation, what happens if one allows mixed strategies? If one redefines the utilities in terms of mixed strategies, that is,

$$\mathbb{E}U(\boldsymbol{p}|\boldsymbol{q}) = \epsilon\left(\boldsymbol{p}^\top \boldsymbol{U}\boldsymbol{p}\right) + (1 - \epsilon)\left(\boldsymbol{p}^\top \boldsymbol{U}\boldsymbol{q}\right)$$

where $\boldsymbol{p} = [p, 1-p]^\top$ is the mixed strategy of the first player, $\boldsymbol{q} = [q, 1-q]^\top$ the mixed strategy of the second one and U is defined as earlier. Evaluating in the same way $\mathbb{E}U(\boldsymbol{q}|\boldsymbol{p})$, we can define an *Evolutionary Stable Mixed Strategy (ESMS)* in the same way as for the *ESS*. In that case, the Maynard–Smith theorem gives that $\boldsymbol{p}$ is an *ESMS* if two conditions are verified:

① *A Nash Equilibrium Condition*: $\boldsymbol{p}^\top \boldsymbol{U}\boldsymbol{p} > \boldsymbol{q}^\top \boldsymbol{U}\boldsymbol{p}$.

② *A stability Condition*: if $\boldsymbol{p}^\top \boldsymbol{U}\boldsymbol{p} = \boldsymbol{q}^\top \boldsymbol{U}\boldsymbol{p}$ then $\boldsymbol{p}^\top \boldsymbol{U}\boldsymbol{q} > \boldsymbol{q}^\top \boldsymbol{U}\boldsymbol{q}$.

Now, for exactly the same reason as in the pure strategy case, the three Nash equilibria are *ESMS*.

---

[17]In the sense of greater expected utility.

[18]One speaks of an *Evolutionary Stable Strategy* (ESS).

18.2.3   Population dynamics

The direct *ESS*[19] or *ESMS* approach are clearly insufficient. Indeed, unlike the classic theory of antagonistic games, evolutionary game theory does not postulate any form of rationality on the part of the players. Generally, it considers that they inherit their behaviour. On the labour market, and in an unstable environment, the only way of making a decision more or less justified, is, from one period to the next, when the information is available, to observe the relative fit of a strategy with regard to another one. Therefore, on a macroscopic level, one shall notice an increasing evolution of the population which adopts the strategy **C** if that strategy gives, on average, a satisfactory payoff in the matching with the upper strategy **L**.

If students can observe that the **C** strategy can offer a greater utility than the average utility associated with a population divided into $\epsilon(t)$ percent of their own type and $1 - \epsilon(t)$ percent of the other, then in all likelihood, $\epsilon(t)$ is going to rise. This leads to a Malthusian population dynamics of the **C** players, called *replication dynamics*. This dynamics is described by the following differential equation[20]:

$$\frac{\dot{\epsilon}(t)}{\epsilon(t)} = \mathbb{E}U(\mathbf{C}) - [\epsilon\mathbb{E}U(\mathbf{C}) + (1 - \epsilon(t))\mathbb{E}U(\mathbf{L})]$$

$$= \frac{C}{2}(1 - \epsilon(t))(\epsilon(t) - \bar{\epsilon})$$

with $0 < \bar{\epsilon} = (V - C)/C < 1$. This dynamics[21] models the idea according to which the **C** players' population growth rate is proportional to their relative payoff in a mixed population. Now, the average payoff in a mixed population is defined by the sum of the product of the payoffs with probabilities of the realization of a specific matching. This replication dynamics may be also written as

$$\dot{\epsilon}(t) = \epsilon(t)[1 - \epsilon(t)][\mathbb{E}U(\mathbf{C}) - \mathbb{E}U(\mathbf{L})]$$

which also yields the following interpretation: the increase of the population

---

[19]It is important to distinguish both because the first one is clearly less debated than the second. Numerous economists refuse to consider mixed strategies.

[20]Only a small number of studies have been interested in the justification of the recourse to replication dynamics in the economic environment. Nevertheless, we can quote Weibull (1995), who shows how the dynamics of replication can appear as an approximation to imitation behaviour.

[21]To obtain $\mathbb{E}U(\mathbf{C})$ and $\mathbb{E}U(\mathbf{L})$ one needs only to substitute $\epsilon(t)$ into q cf. *supra*.

of **C** players is proportional to the expected utility differential gap between both strategies, conditioned by the emergence of a mixed matching[22].

Define the function $h(\epsilon(t))$ by

$$h(\epsilon(t)) = \frac{C}{2}\epsilon(t)(1 - \epsilon(t))(\epsilon(t) - \bar{\epsilon})$$

which leads to rewriting the replication dynamics as

$$\dot{\epsilon}(t) = h(\epsilon(t))$$

The function $h(t)$ possesses three fixed points: $\epsilon_0 = 0$, which is the case for the degenerate population composed exclusively of **L** players; $\epsilon_1 = \bar{\epsilon}$, which corresponds to a mixed population; and $\epsilon_2 = 1$, which is the case for the degenerate population composed exclusively of **C** players. These three fixed points are the three *SMES*. As the derivative of $h(\epsilon)$ is $h_\epsilon(\epsilon) = (C/2)[-3\epsilon^2 + 2(1 + \bar{\epsilon})\epsilon - \bar{\epsilon}]$, it is very easy to show that ϵ_0 and ϵ_1 are locally stable equilibria[23] when ϵ_2 is locally unstable, the **C** strategy is no more played. From a formal point of view, we can develop a symmetric line of argument[24]. So there are two attraction manifolds for the locally stable equilibria, see Figure 18.4:

① A first attraction manifold $\mathcal{B}_0 = [0, \bar{\epsilon}[$ for ϵ_0;

② a second attraction manifold $\mathcal{B}_1 =]\bar{\epsilon}, 1]$ for ϵ_1.

Thus, depending on its initial distribution, which is either $\epsilon(0) \in \mathcal{B}_0$, $\epsilon(0) = \epsilon_0$, or $\epsilon(0) = \epsilon_1$, the population converges towards the Pareto-dominant equilibrium (**S**, **S**), remains in internal equilibrium if by chance it was already at this equilibrium, or converges towards the risk dominant equilibrium (**L**, **L**).

We can propose a simple explanation for this result. For an initial population such as $\epsilon(0) < \bar{\epsilon}$, the players are matched with numerous players who, on average, have made long studies rather than short. The cost of longer studies is more than compensated for by the low probability of being mated with another long studies player. The expected relative advantage of a strategy over another determines its adoption.

---

[22]It would be possible to generalize the dynamics by the use of an increasing function of the right side of the replication dynamics, but this does not change the qualitative behaviour of the dynamics.

[23]If we note that $x_j(t) = \epsilon(t) - \epsilon_j$ in a neighbourhood of ϵ_j, we get $\dot{x}_j(t) = h_\epsilon(\epsilon_j)x_j$.

[24]Since $h(0) = 0$ and $h_\epsilon(0) < 0$, as $\epsilon_2 \in]0, 1[$, $h_\epsilon(\epsilon_2)$ is, by necessity, positive.

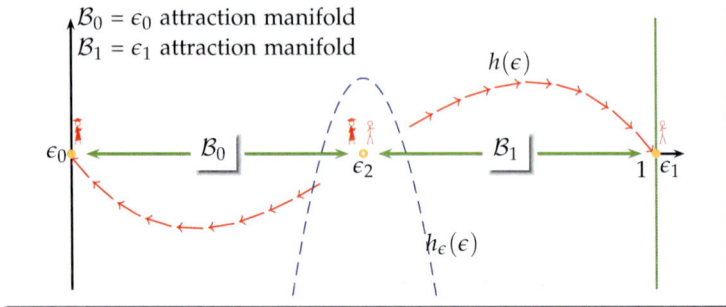

Figure 18.4: The replication dynamics.

Hence, at the final stage of the dynamics, the **S** strategy is no more played. From a formal point of view, one can develop a symmetric line of argument[25] for the case where $\epsilon(0) > \bar{\epsilon}$.

However, this symmetry is only apparent. Indeed, if we suppose that, initially, only the short studies existed and that a spontaneous mutation had moved a small percentage of players towards the **L** strategy, it is not very likely that this percentage tips the initial condition over to the attraction manifold of the other *SMES*, i.e. to the long studies attraction manifold.

While the risk dominance argument is more convincing than the Pareto-dominance argument, in the final equilibrium choice decided by the population of players, it seems that the replication dynamics is clearly in favour of the banishment of long studies from the educational system. Unless it is possible to produce an initial mutation which drives the initial population to the other attraction manifold, i.e., which moves $\epsilon(0)$ over to $\bar{\epsilon}$. A classic argument is then to suggest the appearance of behaviour determined by a rationality still weakened by the number of agents.

18.2.4   Population dynamics with continuous mutations

As pointed out in Orlean (1995), this dynamics leaves aside an important aspect of the evolutionary logic, namely the constant presence of mutational factors. Indeed, two elements can overlap to explain that the moving decision of a number of players is not directly driven by the relative advantage of a strategy. First of all, there is always a certain percentage of the population which has a very restricted access to information: in the replication

[25]When $\epsilon(0) = \bar{\epsilon}$, the population repartition is temporally invariant.

dynamics, the access to information is assumed to be free and without cost for the players. Similarly, a certain percentage of the population is likely to decide to undertake or not to undertake long studies only based on their interest for the studied fields.

Then, the state, for reasons which run from the adequacy of the population for jobs which require an accumulation of human capital always weighting more heavily the prolongment of studies, to the withdrawal of young people from the labour market in a period of under-employment, arranges for the permanent migration of sub-populations having adopted a particular strategy to another one. So, a second dynamic system is built upon the first one, and this new system does not obey the logic of the replication dynamics.

This dynamics translates into the fact that, over a lapse of time of length h, a percentage m of the players go from the **L** strategy to the **S** strategy and a percentage l goes from the **S** strategy to the **L** strategy. In $t + h$, the percentage of the population that has adopted the **S** strategy is then determined by the recursive equation

$$\epsilon(t + h) = \underbrace{[1 - mh\epsilon(t)]}_{\substack{\text{probability not to migrate} \\ \text{for a } \mathbf{C} \text{ player}}} \epsilon(t) + \underbrace{[lh\epsilon(t)]}_{\substack{\text{probability to migrate} \\ \text{for a } \mathbf{L} \text{ player}}} (1 - \epsilon(t))$$

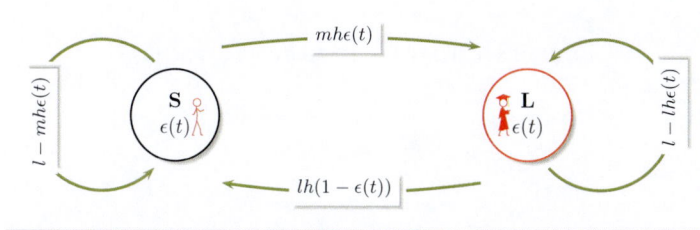

Figure 18.5: The continuous mutations population dynamics.

By making h increase by dt, one finally obtains

$$\dot{\epsilon}(t) = [l - (m + l)\epsilon(t)]\epsilon(t) = k(\epsilon(t))$$

This dynamics has two fixed points $\epsilon_0 = 0$ and $\epsilon_4 = l/(m + l)$. As $k_\epsilon(\epsilon) = l - 2(m + l)\epsilon$, ϵ_0 is locally unstable, unless $l \equiv 0$, and ϵ_4 is locally stable. $\epsilon_1 = 1$ is no longer an equilibrium. It could be under the restriction $m \equiv 0$, in which case $\epsilon_4 = \epsilon_1 = 1$. This has a very simple explanation: if at each period of time some **S** players migrate to become **L** players, it is impossible that the final state of the population be fixed in as an entirely **S** player population. This dynamics, which could be called *a rational extreme dynamics*[26], in opposition to the replication dynamics, insists on the central role of an equilibrium where **S** players and **L** players coexist, see Figure 18.6.

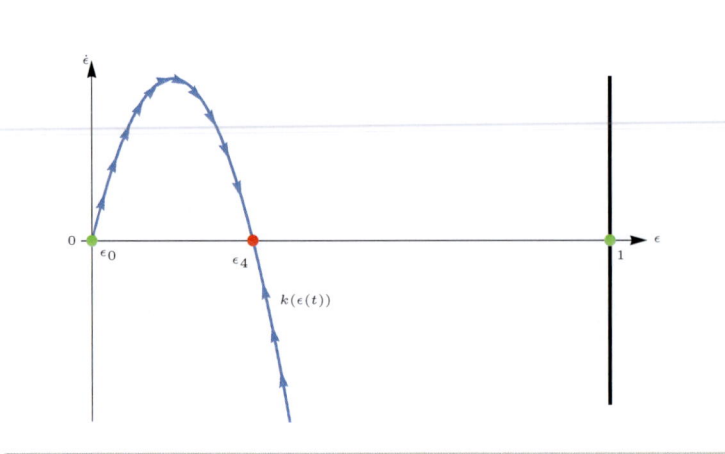

Figure 18.6: The extreme arational dynamics.

Let us consider the general population dynamics corresponding to the juxtaposition of the two types of behaviour: *the strategic behaviour* and the *pre-determined behaviour*. It takes the form

$$\dot{\epsilon}(t) = h(\epsilon(t)) + k(\epsilon(t))$$
$$= \left[\frac{C}{2}\epsilon(t)(1 - \epsilon(t))(\epsilon(t) - \bar{\epsilon})\right] + ([l - (m + l)\epsilon(t)]\epsilon(t))$$
$$= \epsilon(t)\left(\frac{C}{2}\left[(1 + \bar{\epsilon})\epsilon(t) - \bar{\epsilon} + \epsilon^2(t)\right] + [l - (m + l)\epsilon(t)]\right)$$
$$= H(\epsilon(t)|m, l) \tag{dm1}$$

Here the mutations remind one of a *trembling hand process*, see Orlean (1995). They constitute a permanent and autonomous source of variability inside the labour market system.

Then one can also envision a new population dynamics in which α percent of the population share a *quasi-rational* behaviour, and the other $1 - \alpha$ has adopted an *irrational* one. In this case, the dynamics becomes

$$\dot{\epsilon}(t) = \alpha h(\epsilon(t)) + (1 - \alpha)k(\epsilon(t))$$
$$= \alpha\left[\frac{C}{2}\epsilon(t)(1 - \epsilon(t))(\epsilon(t) - \bar{\epsilon})\right] + (1 - \alpha)\left([l - (m + l)\epsilon(t))\epsilon(t)\right]$$
$$= K(\epsilon(t)|m, l) \tag{dm2}$$

It is particularly noticeable that neith dynamics are reducible. Indeed, in (drn1), the irrational dynamics appears as stacked over the replication dynamics, whereas in (dm2) they work in parallel. In this case, for one unit value of α, we get back to the replication dynamics and for $\alpha = 0$, we find the irrational dynamics. However, in spite of these differences, the two dynamics model the same phenomenon with more complexity than either elementary dynamics.

In what follows, we will adopt the subsequent strategy: we will pose $\alpha = 1$, $m = l = 0$ from the replicator dynamics, and study the deformations induced on it from small variations of the parameters.

The (dm1) dynamics

Let us note that the (dm1) dynamics is identical to the replicator dynamics for $m = l = 0$ because $H(\epsilon(t)|0, 0) = h(\epsilon(t))$. As the presence of two new parameters complicates the problem, let us begin by assuming that $l \equiv 0$. In this way, we start with three equilibria: $\epsilon_0 = 0$, ϵ_1 and ϵ_2.

First case: $[l \equiv 0]$

In this case, let us depart from the replication dynamics to study the way a tiny variation alters it. An equilibrium of the (dml) dynamics is characterized by $H(\epsilon(t)|0,0) = 0$. Let us suppose that the value of m increases from $m = 0$.

The equilibria move in the neighbourhood of $\epsilon_i - i = 1, 2, 4 -$ in agreement with the equation

$$H_\epsilon(\epsilon(t)|0,0)d\epsilon + H_m(\epsilon(t)|0,0)dm = 0$$

$$\Longleftrightarrow \quad \left. \frac{d\epsilon(t)}{dm} \right|_{m=0} = -\frac{H_m(\epsilon_i|0,0)}{H_\epsilon(\epsilon_i|0,0)}$$

Now

$$\begin{cases} H_m(\epsilon_i|0,0) = -\epsilon_i \\ H_\epsilon(\epsilon_i|0,0) = \left(\frac{C}{2}\right)\left[-3\epsilon_i^2 + 2(1-\bar{\epsilon})\epsilon_i - \bar{\epsilon}\right] \end{cases}$$

In consequence of this, a small variation of m induces — the results are shown in the (a)-quadrant of Figure 18.7 —:

① an invariance of ϵ_0 since in that case $H_m(\epsilon_i|0,0) = 0$.

② an increase of $\epsilon_2 = \bar{\epsilon}$ since

$$\left. \frac{d\epsilon}{dm} \right|_{m=0} = \frac{\bar{\epsilon}}{(C/2)\bar{\epsilon}(1-\bar{\epsilon})} > 0$$

③ a decrease of $\epsilon_2 = 1$, which is driven inside $[0,1]$ since

$$\left. \frac{d\epsilon}{dm} \right|_{m=0} = \frac{-\bar{\epsilon}}{(C/2)\bar{\epsilon}(1-\bar{\epsilon})} < 0$$

So, a small variation of m in the neighbourhood of $m = 0$ when $l = 0$ eliminates the Pareto-dominant equilibrium. In fact, the vanished equilibrium swerves inward, which gives now two internal equilibria, the new one being stable and at the heart of its attraction manifold.

As long as m remains smaller than $\tilde{m} = (C/2)[(1 - \bar{\epsilon})/2]^2$, nothing changes. Then, at $m = \tilde{m}$, there is a drastic qualitative change[27] because, even if ϵ_0 remains stable, both interior equilibria merge to become a saddle node equilibrium, denoted ϵ_5, which will be reached from every starting point $\epsilon(0) > \epsilon_5$ but not from a starting point $\epsilon(0) < \epsilon_5$; see the (b) quadrant in Figure 18.6. Finally, if $m > \tilde{m}$, ϵ_0 remains an equilibrium. The (d) quadrant in Figure 18.6, called the *equilibrium bifurcation diagram*, summarizes the qualitative properties of the equilibria.

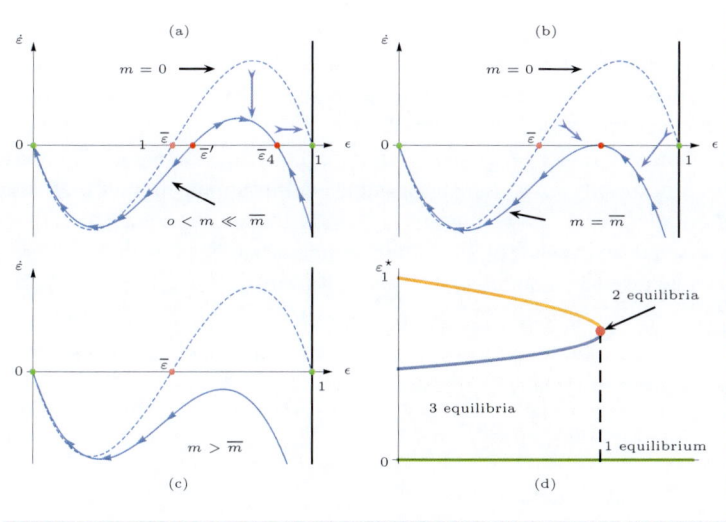

Figure 18.7: The mixed dynamics for $l \equiv 0$.

The main lesson to be learned from the perturbation of the replicator dynamics, with the irrational flight from the **S** players to the **L** players, is that when m increases, the attraction manifold of the all **L**-player equilibrium expands. Furthermore, a mixed stable equilibrium emerges, even if it may vanish with a further increase of m beyond $\tilde{m}$.

In this scenario, the risk dominant equilibrium tends to win over the Pareto equilibrium. One has:

① three equilibria for $m \in [0, \tilde{m}[$,

---

[27] $H(\epsilon|m, 0)$ has $\epsilon_0 = 0$ and $\epsilon_\pm = (2\,C)^{-1}[C(1+\bar{\epsilon})\pm\sqrt{\Delta}]$ with $\Delta = C^2\bar{\epsilon}^2 - 2\,C^2\bar{\epsilon} + C^2 - 8Cm$. One can see that, when $m = \tilde{m}$: $\epsilon_+ = \epsilon_-$.

② two equilibria for $m = \widetilde{m}$,

③ one equilibrium for $m \in]\widetilde{m}, 1]$.

Furthermore, we can observe that it is really the irrational dynamics, a pure mutation dynamics, which perturbs the replication dynamics because the subset of values of m for which one has a stable mixed equilibrium is very small. This tends to confirm that the replication dynamics plays the central role in this scenario.

Second case: $[m \equiv 0]$

Now that the scenario-analysis tools have been set up, we will proceed with a graphical analysis based on Figure 18.7. This time, a small variation of l induces the immediate disappearance of the equilibrium $\epsilon_0 = 0$. In that case, there can no longer be a convergence to a state where nobody chooses to become a **S** player. ϵ_2 remains a stable equilibrium in all the variations of this situation. At first, two internal equilibria coexist, the equilibrium ϵ_5 that has arisen from the shift of the defunct ϵ_0 being stable, while the equilibrium $\bar{\epsilon}'$, born from the $\bar{\epsilon}$ equilibrium, remains unstable. But, when l reaches the limit value $\widetilde{l} = (C/2)(\bar{\epsilon}/2)^2$, there is a fork which combines both internal equilibria into a unique saddle point, which immediately disappears. Only one equilibrium remains. One has:

① three equilibria for $l \in [0, \widetilde{l}[$,

② two equilibria for $l = \widetilde{l}$,

③ one equilibrium for $l \in]\widetilde{l}, 1]$.

Here again, it is the irrational dynamics which perturbs the replication dynamics because when m increases, $\bar{\epsilon}$ and its siblings, the interior equilibria generated by the sliding of $\bar{\epsilon}$ induced by m, change only at the margin in their qualitative behaviour.

Naturally, the most realistic situation corresponds to $m \neq 0$, $l \neq 0$, and $m \neq l$. However, this situation is particularly complex and a more precise approach could be obtained by posing $l = m$. In fact, one can rely on intuition to assume that the values of m and l are not too much apart from each other. First of all, they are mutation rates and, by nature, they are constrained to $[0, 1]$. Then, if those mutations are irrational, i.e. induced by purely subjective elements without any reference to any strategy from mutants, in all likelihood one can suppose that in a large population we will observe very similar percentages of mutants from both populations.

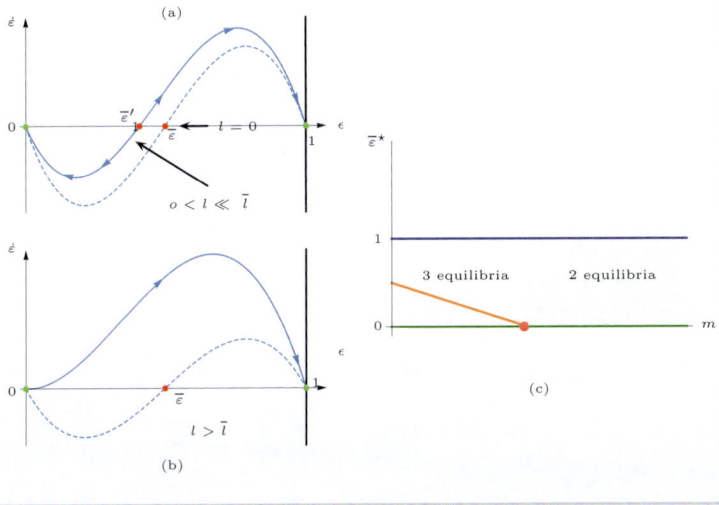

Figure 18.8: The mixed dynamics for $m \equiv 0$.

Third case: $[m = l = k]$

$$\dot{\epsilon}(t) = \frac{C}{2}\epsilon(t)(1 - \epsilon(t))(\epsilon(t) - \bar{\epsilon}) + k[1 - 2\epsilon(t)]$$

Once again, let us study the properties of this case from a purely graph-ical point of view (see Figure 18.8). In contrast with the two polar cases studied above, the focus in this case is concentrated on the internal mixed equilibrium. Indeed, not only does the increase in k drive both extreme equilibria to disappear, but the nature of the internal equilibrium is modified by the variations of k. For $k < \tilde{k} = (C/2)(\bar{\epsilon}^2 - \bar{\epsilon} + 1)$, it is unstable, but stable for $k > \tilde{k}$. In fact, in $\tilde{k}$ there is a bifurcation of the dynamics and the internal equilibrium is the only equilibrium to remain: see Figure 18.9 for the separation hyperbola as a function of $\bar{\epsilon}$ for given C. (One must not forget that $\bar{\epsilon}$ is a function of C and V.)

Consequently, as soon as k exceeds $\tilde{k}$, the replication dynamics is ab-sorbed by the irrational dynamics, which imposes its properties on the repli-cation dynamics equilibrium. In so far as a stable mixed equilibrium seems to better correspond to the observed situation of the labour market than any other equilibrium, the model tells in favour of a significant perturbation of

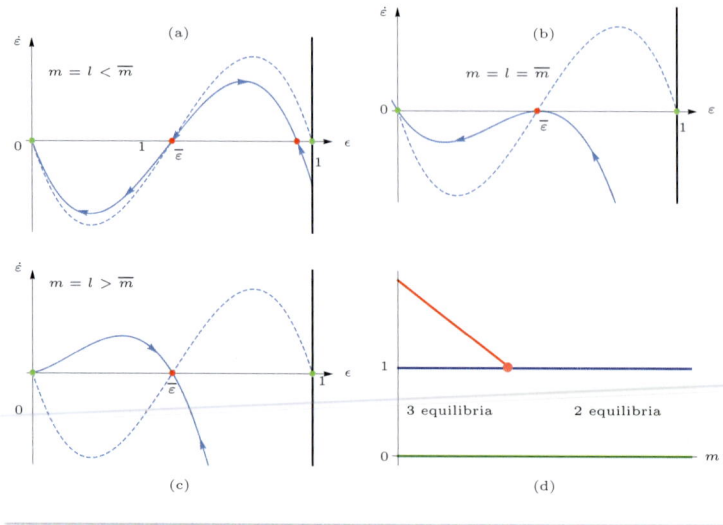

Figure 18.9: The mixed dynamics for $m = l = k$.

the replication dynamics by the irrational dynamics. Contrary to what is asserted in Orlean (1995), with an increase of k, the attraction manifold of the risk dominant equilibrium (but also the attraction manifold of the Pareto-dominant equilibrium), although it might increase for a moment, eventually disappears to make way for a unique stable equilibrium.

This conclusion applies to the general model, i.e. to the model for which $m \neq l$. Indeed, by the method of tiny variations in a neighbourhood of $m = l = 0$, if the measure or the variation of m is equal to that of l, we find the same result as the one from which we had just moved from. It would therefore be useful to explore in detail the case $m \neq l$.

The (dm2) dynamics

We can immediately observe that in the case of this representation of the perturbation of the replication dynamics by the irrational dynamics, only one value of $\alpha = 1$ allows finding anew the Pareto-dominating equilibria and the risk-dominant to exist. The general case tells in favour of an internal equilibria, even for a unique stable internal equilibrium when the irrational dynamics dominates the replication dynamics. Some simulations, which

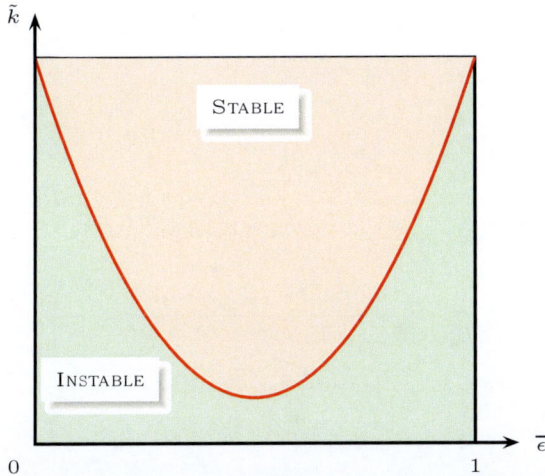

Figure 18.10: The bifurcation frontier for the mixed equilibrium — $l = m$.

are not presented here, prove that the irrational dynamics very quickly dominates the replication dynamics. Consequently, the (dm2) dynamics seems to confirm the results obtained during the study of the (dm1) dynamics (see Figure 18.10).

In fact, this description is only true on the surface, because the dynamics embed a *local emergence phenomenon*. While one should expect that with a decrease of α, the only observed behaviour of the dynamics is the change of sign of its slope, leading the internal equilibrium from instability to stability, we can observe for a very peculiar range of values of α the emergence of an unexpected unstable equilibrium.

Unfortunately, we can not show this emergence other than by specifying the parameter values. We have chosen $\bar{\epsilon} = 0.5$ and $l = m = 0.1$. For this specific range, a simple simulation shows that for the approximate value of their bounds, we have the emergence of an unstable internal equilibrium on $]0, \bar{\epsilon}[$ when $\alpha \in]0.167667, 0.285667[$, see Figure 18.12.

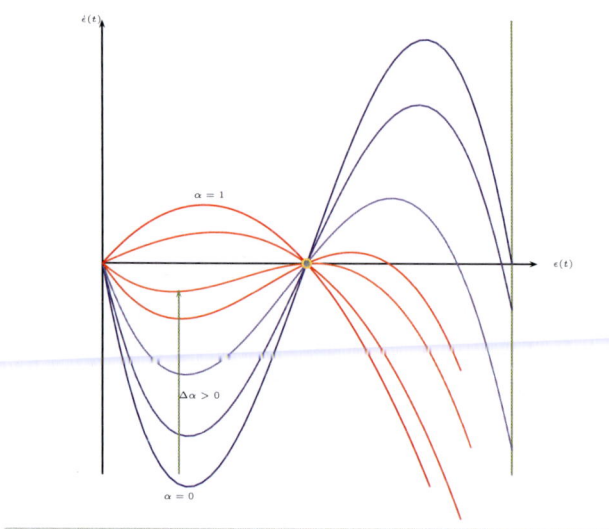

Figure 18.11: the (dm2) dynamics.

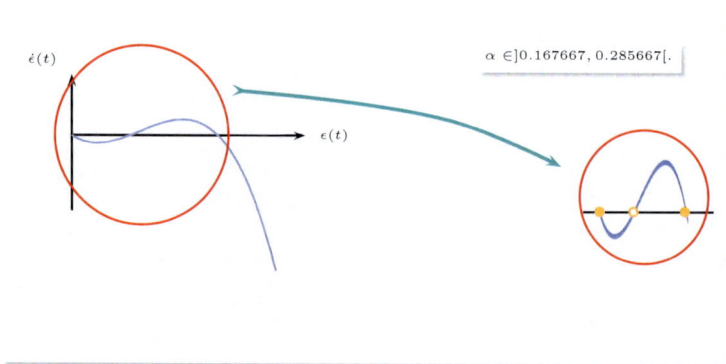

Figure 18.12: Local emergence of an unstable equilibrium.

18.3   Conclusions

In this paper, we have tried to explain the persistence of a mixed equilibrium composed of **S** players and **L** players. This is in no way, at least in the evolutionary game formalism, a trivial affair since one is obliged to perturb the standard evolutionary game by irrational behaviour to generate such an equilibrium. We even show that there is an embedded emergence phenomenon inside the mixed model.

Many extensions to this approach are under study: first of all, a differential wage distinguishing jobs, introduced according to their place in the dual hierarchy of the labour market, can easily be introduced. Then one can also improve the description of the matching by introducing two distinct effects: agents can run several things at once, in that they can not receive one only but several jobs, and / or not be matched in pairs but in N-tuples. One can also introduce rigid agents who never change their mind about to the length of the study. Finally, the information available on the labour market is necessarily dated and it would be interesting to see what is the impact on the evolutionary dynamics of a delayed reaction.

In all cases, this approach reveals the complexity of the specific decision whether to undertake studies even in a stationary world.

Acknowledgements

I would like to thank A. Clark, J.-B. Desquilbet, C. Nieradzik and C. Garrouste for their helpful comments. I would also like to thank two anonymous referees who help to greatly improve the readability of this paper. All remaining errors are my sole responsibility.

Bibliography

Arrow, K. (1973), Higher eduction as a filter, *Journal of Public Economics* **2**, 193–216.

Barrouillet, P., Camos, V. (2008), Le développement de la mémoire de travail, in J. Lautrey, ed., *Psychologie du Développement et de l'éducation*, Presses Universitaires de France, 51–86.

Becker, G. (1964), *Human Capital*, The University of Chicago Press, Chicago

Damasio, A. (1994), *Descartes' Error: Emotion, Reason, and the Human Brain*, Avon Books.

Devaney, R. (2003), *An Introduction to Chaotic Dynamical Systems*, Westview Press.

Ericsson, K., Krampe, R., Tesch-Romer, C. (1993), The role of deliberate practice in the acquisition of expert performance, *Psychological Review* **100**(3), 363–406.

Green, F., Zhu, Y. (2010), Overqualification, job dissatisfaction, and increasing dispersion in the returns to graduate education, *Oxford Economic Papers* **62**(4), 740–763.

Harsanyi, J., Selten, R. (1992), *A General Theory of Equilibrium Selection in Games*, The MIT Press, Cambridge MA.

Jaoul-Grammare, M., Nakhili, N. (2010), Quels facteurs influencent les poursuites d'études dans l'enseignement supérieur ?, *Groupes d'exploitation génération 2004, Net.Doc* **68**.

Jevons, W. (1871), *The Theory of Political Economy*, Kelley and Millman Inc

Kopel, S. (2005), Les surdiplômés de la fonction publique, *Revue Française de Gestion* **3**(156), 17–34.

Lehrer, J. (2009), *How We Decide*, Houghton Mifflin.

Lorenz, E. (1964), The problem of deducing the climate from the governing equations, *Tellus* **16**, 1–11.

Lucas, R. (1988), On the mechanics of economic development, *Journal of Monetary Economics* **22**, 3–42.

Mariotte, E. (1717), Essay de logique contenant les principes de sciences, & la manière de s'en servir pour faire de bons raisonnements, Etienne Michallet.

Marois, R., Ivanoff, J. (2005), Capacity limits of information processing in the brain, *TRENDS in Cognitive Sciences* **9**(6), 296–305.

Maynard-Smith, J. (1974), The theory of games and the evolution of animal conflicts, *Journal of Theoretical Biology* **47**, 209–221.

Mazari, Z., Recotillet, I. (2013), Génération 2004 : des débuts de trajectoire durablement marqués par la crise ?, *Bref du Céreq* (311), 4p.

Miller, G. (1956), The magical number seven, plus or minus two: Some limits on our capacity for processing information, *Psychological Review* **63**(2), 81–97.

Mincer, J. (1974), Schooling experience and earnings, *Columbia University Press*, New York.

Nisan, N., T.Roughgarden, E.Tardos, Vaziran, V. (2007), Algorithmic Game Theory, *Cambridge University Press*, Cambridge MA.

Orlean, A. (1995), De la stabilité évolutionniste à la stabilité stochastique, *Revue Économique* **47**(3), 589–600.

Parker, G. (2012), Acta is a four-letter word, *Acta Psychiatrica Scandinavica* **126**(6), 476–478.

Pierce, J. (1980), *An Introduction to Information Theory: Symbols, Signals & Noise*, Courier Dover Publications.

Piore, M. (1978), Dualism in the labor market : A response to uncertainty and flux. the case of France, *Revue Économique* **1**, 26–48.

Poincaré, H. (1890), Sur le problème des trois corps et les équations de la dynamique, *Acta Mathematica* **13**(1), A3–A270.

Schultz, T. (1961), Investment in human capital, *The American Economic Review* **51**(1), 1–17.

Shapiro, C., Stiglitz, J. (1984), Equilibrium unemployment as a worker discipline device, *The American Economic Review* **74**(3), 433–443.

Shim, J., Chow, J., Carlton, L., Chae W., (2005), The use of anticipatory visual cues for highly skilled tennis players, *Journal of Motor Behavior* **37**, 164–175.

Spence, M. (1974), Job marker signaling, *Quarterly Journal of Economics* **87**, 353–374.

Taubman, P., Wachter, M. (1986), Segmented labor markets, in O. Ashenfelter, R. Layards, eds, *Handbook of Labor Economics*, Vol. II, Elsevier Science Publisher, chapter 21, 1183–1217.

Weibull, J. (1995), *Evolutionary Game Theory*, The MIT Press, London